Mathematics in Computational Science and Engineering

Scrivener Publishing
100 Cummings Center, Suite 541J
Beverly, MA 01915-6106

Publishers at Scrivener
Martin Scrivener (martin@scrivenerpublishing.com)
Phillip Carmical (pcarmical@scrivenerpublishing.com)

Mathematics in Computational Science and Engineering

Edited by
Ramakant Bhardwaj
Jyoti Mishra
Satyendra Narayan
and
Gopalakrishnan Suseendran

Scrivener Publishing

WILEY

This edition first published 2022 by John Wiley & Sons, Inc., 111 River Street, Hoboken, NJ 07030, USA and Scrivener Publishing LLC, 100 Cummings Center, Suite 541J, Beverly, MA 01915, USA

For more information about Scrivener publications please visit www.scrivenerpublishing.com.

Wiley Global Headquarters
111 River Street, Hoboken, NJ 07030, USA

For details of our global editorial offices, customer services, and more information about Wiley products visit us at www.wiley.com.

Library of Congress Cataloging-in-Publication Data

ISBN 978-1-119-77715-1

Cover image: Pixabay.com
Cover design by Russell Richardson

Set in size of 11pt and Minion Pro by Manila Typesetting Company, Makati, Philippines

10 9 8 7 6 5 4 3 2 1

Dedication

Gopalakrishnan Suseendran, Assistant Professor, who is now deceased, as the co-author of this book. He received his PhD in Information Technology-Mathematics from Presidency College, University of Madras, Tamil Nadu, India. He worked as assistant professor in the Department of Information Technology, School of Computing Sciences, Vels Institute of Science, Technology & Advanced Studies (VISTAS). He published more than 75 research papers in various referred journals, authored 11 books and received 6 awards.

Contents

Contents

Preface

Chapter 1

The main aim of Inventory EOQ model is to reduce the Ordering Cost and Holding Cost in the Company. Based on Numerical Example, three proposed models are applied in EOQ. This leads to Brownian Path, which is based on Hausdroff Measure and Levy processes. Hence it is Fractals.

Chapter 2

This chapter gives a good description of ill-posed inverse problems encountered in the field of electrical geophysics. It begins with an overview of the present state of knowledge about electrical resistivity methods for mapping and monitoring in-situ processes that cannot be accessed directly. Based on reciprocity and perturbation analysis, an attempt has been made to introduce generalized multi-dimensional resistivity inversion methods. It may be found highly useful in environmental geophysics and geoengineering discipline to mapping and monitoring in-situ processes where electrical resistivity contrast is encountered.

Chapter 3

In this chapter, theoretical formulations of shadowed sets approximations (SSA) which hinge on ideas of uncertainty balance, average uncertainty and minimum approximation error are presented. Also, decision-theoretic three-way approximation (DTA) models which anchor on principles of minimum distance and least cost are revisited. Subsequently, we give a modified generalized model of decision-theoretic three-way approximation, called system, which does not impose values for and as against the trend in literature where and are chosen to be and respectively. A suitable formula for computing viable threshold from cost-sensitive and minimum distance-based models is derived.

Chapter 4

This chapter depicts a wide survey on Intuitionistic Fuzzy Rough Set theory. Several extensions of intuitionistic fuzzy rough sets and hybridization of intuitionistic fuzzy rough sets with other theories dealing with uncertainties are thoroughly looked over. A detailed discussion on intuitionistic fuzzy rough set theory in various real-world application fields is also presented.

Chapter 5

Air quality of different metropolitan cities of India has worsened over the last decade. Kolkata is among the most polluted urban areas of the country. Particulate matter smaller than 2.5μm ($PM_{2.5}$) is considered as one of the significant parameters for indicating the air quality. Ground based monitoring stations for $PM_{2.5}$ are limited over Kolkata. So, Aerosol optical depth (AOD) obtained by Aqua satellites and Moderate Resolution Imaging Spectroradiometer (MODIS) onboard EOS Terra are used to evaluate the local $PM_{2.5}$ concentration over Kolkata. This work attempts to develop a statistical model to estimate $PM_{2.5}$ concentration using AOD_{MODIS} and meteorological parameters (Temperature, Relative Humidity, Planetary Boundary Layer Height, Total Cloud Cover, Wind speed). The concentration of $PM_{2.5}$ is found to be influenced by various meteorological parameters. It is found that 52% of the variability of the dependent variable $PM_{2.5}$ is explained by the 6 explanatory variables (i.e., AOD_{MODIS}, temperature, relative humidity, average total cloud, planetary boundary layer height and wind speed) whereas only 3.9% of the variability of the dependent variable $PM_{2.5}$ is explained by AOD alone as explanatory variable.

Chapter 6

In this chapter we would see inventory systems and learn to manage it with simulation technique. We would study simulation which is performed manually for better understanding of the topic. Then some merits and demerits of it, Monte Carlo simulation technique and its application in a real life problem would be delved in. We would use Excel software to generate random numbers and perform simulation with it. Inventory sales for a confectionary bakery shop would be predicted for next few days. At the end we would compare and plot a graph so as to see and understand simulation better.

Chapter 7

The chapter gives an idea about change in characteristics of cement mortar while adding some nano admixtures and polymer PVA. Workability of mortar is measured in terms of flow value that is necessary to know how the mortar will behave with these additives. Final strength is dependent on various factors out of which workability is one factor. Also the handling with mortar in the field while mixing, transporting, pouring and compacting the workability plays important role hence various additives are added to achieve proper workability. In this article various nano materials like nano alumina, nano silica, nano zinc oxide, and polyvinyl alcohol has been tried for changing the workability.

Chapter 8

This chapter explores various properties of half-Heusler alloys; LiZnP and LiCdP. All the calculations are carried out based on the density functional theory using pseudopotential plane-wave method as implemented in the Quantum espresso package. The structural and electronic features are well described in result and discussion part along with thermodynamic properties. This chapter exposes the semiconductor and brittle behavior of LiZnP and LiCdP alloys.

Chapter 9

In this chapter, we establish a new common fixed point theorem satisfying the digital topology with Contractive mappings in fuzzy sets. Rather than focussing on mathematical details, we will concentrate on making the concepts as clear as possible. There are several useful technical introductions in fuzzy sets and fuzzy logic with digital space. fuzzy set theory is an analytic framework for handling concepts that are simultaneously categorical and dimensional. starting with a rational for fuzzy sets. in this chapter we provide some basic definitions and results for fuzzy sets. In this chapter we shall discuss two important categories of fuzzy logic with linear applications "digital" and "contractions " with respect to fuzzy digital applications.

Chapter 10

This chapter discusses the bond between mathematics & Artificial intelligence. Mathematics helps to solve the challenging task of hypothetical problems in artificial intelligence using traditional methods and techniques. In the first part the main concern is to demonstrate the mathematical concepts like Linear Algebra in dimensionality reduction of large datasets, Eigen Vectors in ranking of features of dataset, Calculus in optimization task, Statistics for data visualization and so on. The later part discusses the work flow of artificial intelligence and the application areas where Artificial Intelligence is using now a days.

Chapter 11

In this chapter three mathematical models have been used. One is model based on Geometric Progression (G.P.).The second is SIR model and the third one is constructed using differential equations. The model based on G.P. shows how coronavirus is spread using tree chart, assuming that an infected person is capable of infecting two persons who come in contact with the former. It can also be observed that there is a significant difference between the number of covid patients with and without the lockdown.

Chapter 12

This chapter describes about the application of Fuzzy Logic in Mathematical modelling. The chapter is started with explanation about Mathematical modelling with definition and examples. With the help of principles of

Mathematical Models, how real world problems can be solved by it, is described. Fuzzy logic is widely accepted and used term in the light of development for application, tools, techniques as Fuzzy Cognitive Maps, Fuzzy Cluster Means, etc. Here Fuzzy Logic Concept has been studied and tried to explain applications of the concept in various fields as Mathematics, Science, Business, Finance, Controller of Temperature, Home appliances, Aeronautics, Defence, Medical Science and Bioinformatics, Engineering Fields such as Mechanical, Industrial, Production, Electronics, Chemical, Automotives, Signal Processing and Communication, Robotics.

Chapter 13

This chapter explores the different types of recommender techniques with its mathematical foundation and also discusses some of the problems in the prevailing system. It discusses use of sequential patterns of web navigation along with the content information and is based on set and sequence similarity measure (S3M), principle of upper approximation and singular value decomposition for generating recommendations on web data. This chapter makes use of mathematics involved in finding the set and sequence similarity for recommendation to user on CTI news dataset.

Chapter 14

The adhesive anchors are usually installed into un-cracked hardened concrete. The Artificial Intelligence methodology of Neural Network (NN) and Genetic Expression Programming (GEP) are used to develop an explicit equation for estimating and predicting the shear capacity of a single adhesive anchor post. The main objective in this chapter is to provide a mathematical tool to predict shear capacity or shear strength of an anchor without any expansive laboratory testing. The Artificial Intelligence (AI) techniques are well-suited for assessment and prediction purposes.

Chapter 15

Genetic Algorithms is an easy form of learning and improvising design of Algorithm. The foundation of the principle is Darwin's Natural selection, which states survival of the fittest. It offers solutions to various problems and helps to evaluate the value and also permits to combine with each other. This chapter gives confidence to reproduce good solutions, it will steadily produce enhanced solutions. It proves that any range of problem can be supported and solved. Gene is generally made up of a cells which represents individual problems and it represents individual solution of the problem.

Chapter 16

Defects play a key role in the mechanical behavior of reinforced concrete structures during loading. These defects are mainly cracks or small holes. These are usually the results of steel rebar corrosion caused by electrochemical and chemical processes. Defect detections are the prime goal of Non Destructive Testing (NDT) methods. This chapter is mainly focused on the Passive Magnetic Inspection (PMI) method. This is an innovative NDT method, which is highly used to inspect a reinforced concrete sample with three holes, in three different positions and locations of steel reinforcement. Principal Component Analysis (PCA) technique is one of the several signal processing techniques is used herein for corrosion detection.

Chapter 17

This chapter focuses on the control of Plasmodium parasite by studying the stages in the cycle. Backstepping control technique is applied to breakup the life cycle of plasmodium parasites. Lyapunov function is derived for the recursive procedure for the entire system to reduce the energy rate of reproduction among plasmodium parasites. Depending upon the concerned state in the system, 'pseudo controls' are introduced so as to achieve global stability of plasmodium life cycle.

Chapter 18

The purpose of this paper is to introduce a new chaotic oscillator. Although different chaotic systems have been formulated by earlier researchers, only a few chaotic systems exhibit chaotic behaviour. In this work, a new chaotic system with chaotic attractor is introduced for triangular wave nonlinearity. It is worth noting that this striking phenomenon rarely occurs in respect of chaotic systems. The system proposed in this paper has been realized with numerical simulation. The results emanating from the numerical simulation indicate the feasibility of the proposed chaotic system. More over, chaos control, stability, diffusion and synchronization of such a system have been dealt with.

1

Brownian Motion in EOQ

K. Suganthi and G. Jayalalitha*

Department of Mathematics, VELS Institute of Science Technology & Advanced Studies (VISTAS) - Chennai, Tamil Nadu, India

Abstract

EOQ is a fixed factor proposed to assist organizations to reduce the expense of ordering and holding Inventory. It is an estimation exploit inside the zone of Operations, Logistics, and Supply Management. The main goal of the EOQ is the ideal proportion of a thing to be purchased at one time in order to decrease the consolidated yearly costs of the ordering cycle. Shortages are not permitted and Inventory control plays a vital role in reducing the cost of production rates, additional time, subcontracting, inordinate Inventory costs and delay purchase penalties during peak time.

The exact result is achieved by using sensitive analysis, proposed model and numerical examples which are developed in it. In this paper, three techniques are used to resolve the EOQ in Inventory model: (i) Assumptions of the EOQ; (ii) Inventory Control in Instantaneous demand Model under development of the Item; (iii) Classic EOQ in Inventory. These three strategies give Brownian motion as per Trapezoidal Rule which give Brownian Path. This path can be found by external estimate, and it is varied, on the basis of Hausdorff Measure and Levy's processes. Subsequently it is known as Fractals.

Keywords: EOQ, inventory, brownian motion, levy's processes, fractals

**Corresponding author*: g.jayalalithamaths.sbs@velsuniv.ac.in

Ramakant Bhardwaj, Jyoti Mishra, Satyendra Narayan and Gopalakrishnan Suseendran (eds.) Mathematics in Computational Science and Engineering, (1–26) © 2022 Scrivener Publishing LLC

1.1 Introduction

Fractals is a continuous pattern. It is endlessly convoluted examples that are self-comparable across various scales. They are produced by repeating easy procedures again and again in a continuing response loop [20]. Clouds, rivers, mountains, coastlines, seashells, and hurricanes, etc., are a few examples of fractals.The physical system of Brownian movement was first seen in 1827 by botanist Robert Brown while analysing dust grains in water under a magnifying lens. Brownian movement is characterized as the rampant or inconsistent development of molecule in a liquid due to their consistent collision with other quick atoms [10]. This clarification of Brownian movement filled in as persuading proof that molecules exist and was additionally confirmed tentatively by Jean Perrin in 1908. Perrin was granted the Nobel Prize in Physics in 1926 "for his work on the endless structure of matter" [15]. The direction of the power of nuclear bombardment is continually changing, and at various occasions the molecule is hit more on one side than another, prompting the apparently irregular nature of the movement [11].

The many-body connections that yield the Brownian design cannot be tackled by a method each elaborate particle.As a result, just probabilistic models applied to atomic populaces can be utilized to depict it. Two such models of the factual mechanics, because of Einstein and Smoluchowski, are introduced below [12]. The EOQ technique is one of the strategies frequently applied to know the best measure of crude material stock required by an organization to keep up smooth production in an effective expense [6, 18]. This strategy is frequently utilized because it is easy to execute and can give the best answers for organizations. This is proven by utilizing the EOQ technique, which is not only the measure of inventory that is most effective for the organization, but additionally the expenses to be incurred by the organization; its crude material stock is determined by Total Inventory price and the most proper duration to make a repurchase (determined by Re-request Point) [1]. The customary financial request amount EOQ model was intended to take care of the issue experienced by purchase in and sell-out vendors, who decide the stock standard of products toward the start of every period because of the given interest pace of merchandise, in this way limiting the expense per unit time [18]. To grow the uses of the ordinary EOQ model, different sorts of broadened Inventory models have been created by stock administration researchers lately in their investigations on transitory merchandise inventories [7]. Numerical technique is a scientific device intended to solve Numerical issues. The usage of a numerical strategy with a proper combination check in a programming language is known as a numerical calculation. For a subroutine written to register the arrangement of quadratic for a general client, this isn't sufficient [19].

Inventory is the thing of any item or resource utilized in an organization. A Stock Process is the set of strategies that controls and keeps up inventory point [17]. It determines when stock must be replenished, and how large the requirement should be. The primary concern for any manufacturing management is to reduce overall price and thus raise payoff. The Inventory price consists of four expenses – buying expense, ordering price, inventory carrying expense and item out price [3, 16]. It is miles an imperative a part of an employer. Stock conveys an essential desire variable in any regard levels of items assembling, dissemination and pay, not withstanding being a top piece of by and large cutting-edge property of many gathering [2]. The EOQ is the measure of units that a business needs to add to Inventory with each solicitation to diminish the hard and fast tariff of Inventory [9]. As an occasion, ensuring charges, Ordering price. Also, stock out expense. The EOQ is utilized as a quality of standard audit stock framework in which the degree of Inventory is resolved persistently and stable amount is asked at whatever point the Inventory degree appears at a particular substitute point [5]. Mathematical procedures are calculations used to gauge numeric information. They are utilized to offer 'approximate' consequences for the issues that are being handled and their need is felt while it is unimaginable or perceptibly intense to clear up the issue scientifically [8]. From the current writing, it very well may be seen that most investigations of apportioning inventories under the rebate evaluating of the incomplete accumulating framework and different sorts of interest are isolated into two sections: those that emphasize proportioning inventories under different interest and those completed utilizing a solitary, fixed cost without markdown.

Investigations of markdown estimating rarely consider the characteristics of clients when Inventories are not apportioned [4].

1.2 Assumptions in EOQ

1.2.1 Model Formulation

EOQ model is massive for producing or stock control measure.

The assurance of EOQ comprises of subsequent.

1.2.1.1 Assumptions

- The EOQ is chosen for every item in assessment in a business.
- The yearly request is known and constant.
- Consistent and acceptance demand.
- Ordering expense decisively with the number of requests.
- Inventory holding expense is stable.

1.2.1.2 Notations

d: Annual Demand
q: The Optimum order amount.
t: Optimum time duration.
C_1: Unit production price.
H: Storage price
I: Stock Carrying Cost

1.2.1.3 Inventory Ordering Cost

Ordering cost associated in getting an item into the Inventory, these expenses are caused each time a request is made.

$$Number\ of\ Order = \frac{d}{q} \tag{1.1}$$

For each request with a fixed value paying little regard to the scope of stock. The Yearly requesting worth might be determined by increasing the wide arrangement of requests by that steady cost. This is conveyed as follows

$$Annual\ Ordering\ Cost = \frac{d}{q} \times T \tag{1.2}$$

1.2.1.4 Inventory Holding Cost

Holding expenses are the extra price associated with putting away and keeping a bit of Inventory throughout the span of a year. Holding prices are determined by using EOQ method that organizations make to choose an ideal opportunity to arrange new Inventory.

$$H = I\,C_1 \tag{1.3}$$

Assuming that request is steady, the stock amount can be accepted to debilitate at a predictable rate after some time. Exactly when the stock shows up at zero, the order is put and Inventory is renewed as appeared.In that limit, the holding cost of the Inventory is constrained by discovering the proportion of the stock product at whatever point and the holding cost per unit. It will be itemized as follows:

$$Annual\ Holding\ Cost = \frac{d}{2} \times H \tag{1.4}$$

1.2.1.5 Inventory Total Cost in EOQ

Total expenses are a financial measure that sums all expenses paid to create an item, buy and expenditure to acquire a bit of equipment including the underlying money expense as well as the open door cost of the decisions.

The following is shown,

$$Annual\ Total\ Cost = \frac{d}{q} \times T + \frac{d}{2} \times H \tag{1.5}$$

$$EOQ = \frac{dTC}{dQ} = \sqrt{\frac{2dT}{H}} \tag{1.6}$$

1.2.2 Example

For example, the association faces a yearly interest of 2,000 units. This costs the association 2,000 for each request set and 38 for every unit of the medicine.

This faces a passing on expense of 15% of the unit yield. What is the proportion of money-related solicitation?

The components can be planned as below.

Such as Variable a Value is D = 2000, T = 2,000, C = 265,

H=38, I = 15.5%

We attain the ideal solutions of every case tested in Table 1.1. Table 1.1 is displayed in Figure 1.1. This blueprint shows the Order size is settled through Distribution bend that shows changes in the Q, $\frac{D}{Q}$, $\frac{D}{Q}\times T$, $\frac{D}{2}\times H$, and T_C Parameters comparing to Order size Q for the model with changed objective work.

The blue curve shows the adjustment of all together amount price possibility upon Quantity at the confined Company limit. The Nature of the

Table 1.1 Effect of Q at various price.

Parameters	q	$\frac{d}{q}$	$\frac{d}{q}\times T$	$\frac{d}{2}\times H$	T_C
1	112	17.86	35,720	38,000	73,720
2	125	16	32,000	32,000	64,000
3	144	13.8	27,600	40,000	67,600
4	162	12.35	24,700	30,000	54,700
5	201	9.95	19,900	44,000	63,900
6	228	8.77	17,542	28,000	45,562
7	260	7.69	15,380	48,000	63,380
8	314	6.37	12,740	25,000	37,740
9	270	7.41	14,820	50,000	64,820
10	317	6.31	12,620	23,000	35,620
11	402	4.98	9,960	52,000	61,960
12	517	3.87	7,740	20,000	27,740
13	895	2.23	4,460	55,000	59,460

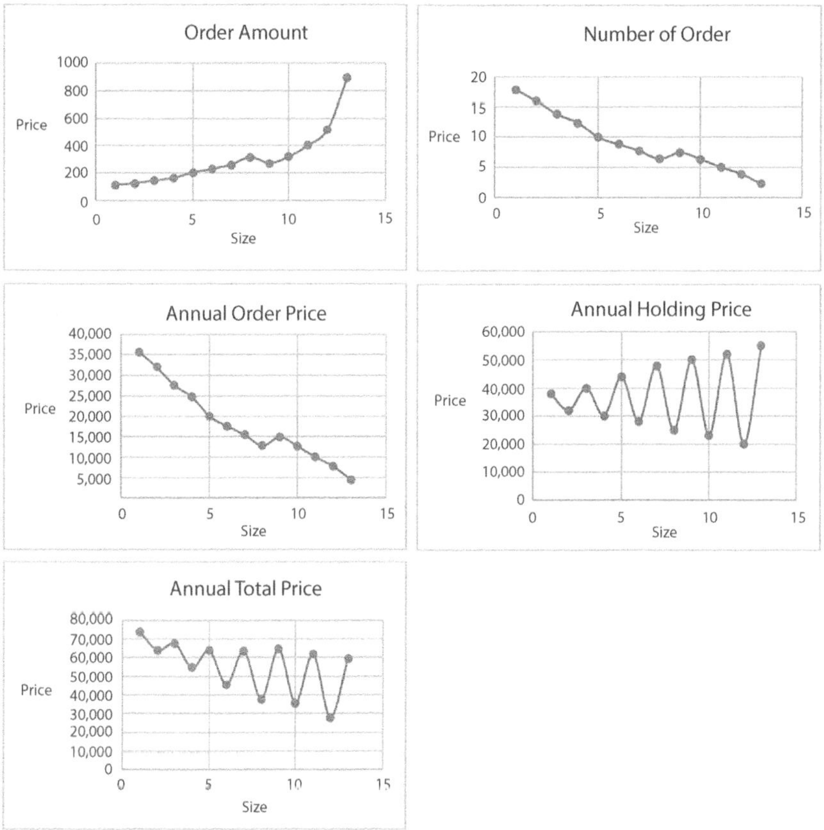

Figure 1.1 Optimal result of the order quantity in EOQ.

given Inventory usage, while moreover considering the exchange time span and fixed nature of the provider to the degree the transport framework.

Numerical Examples

Parameters: Let assume that D = 2000, T = 2,000, C_1 = 260, H =$650, I = 15%. The optimal solution is Q= 112, $\frac{D}{Q} = 17.86$, $\frac{D}{Q} \times T = 35{,}792$, $\frac{D}{2} \times H = 38{,}000$, $T_C = 73{,}720$.

1.2.3 Inventory Control Commodities in Instantaneous Demand Method Under Development of the Stock

An EOQ Inventory technique for falling apart items with Instantaneous Demand and Continuous Replenishment. Further, it shows that limited target cost work, infer the ideal arrangement and Set-up cost esteem is ordinarily thought to be autonomous of the sum requested for the delivered. Parameters. The market request can likewise increment with the selling of the item over the long run when the units don't lose because of disintegration. This model is indistinguishable from that steady set up value, say Fixed set up expense is related buy or making objects in each time span.

The numerical created model for resulting documentation and assumptions.

1.2.3.1 Assumptions

The accompanying assumptions are considered to build up this model.

- The request cost for the thing is Inventory organized.
- Shortages are allowed.
- Instantaneous request and stable Replenishment.
- Stock decayed during the arranging skyline are repairable.
- Holding cost, Set-up cost, Shortage cost and unit cost stay consistent over the long run.
- The dispersion of an opportunity to fall apart follows a four Parameters.
- Replenishment is quick.

1.2.3.2 Notations

This section begins with a listing of the Notations used.

S = Highest Stock stage.
f (d) = Probability density characteristic of Demand.
D = Deterioration.
Q = Optimum production order amount.
C_S = Shortage price.
C_H = Holding Price per unit per unit of duration held in Stock.
Q^* = Back ordering is permitted.
I = units per year
C = Unit price for producing or purchasing every unit,
TEC_1 (Q) = Optimum Inventory achieve local minimum.
TEC_1 (I)= Expected price.
TEC (Q^*) = Conditional for Ordering.

1.2.3.3 *Model Formulation*

This model is the same as fixed setup cost for buying any units to renew Stock at start of period, say K, is related buy or making things in a given timeframe or cost of assembling. Leave I alone the Inventory stage toward the start of the level infers that a request size (Q-I) thing can be set to pass on the available Inventory up to Q. Hence, the anticipated cost transforms into,

$$TEC_1(Q) = K + C(Q-I)f(d)dD + C_s \int_{D-0}^{Q}(Q-D)f(d)dD$$
$$+ C_s \int_{D-Q}^{Q}(Q-D)f(d)dD = K + TEC_1(Q) \tag{1.7}$$

The optimal value of Q says Q* that minimizes TEC_1 (Q)) is given by

$$F(Q*-1) \leq \frac{C_s - C}{C_H + C_s} \leq F(Q*)F \tag{1.8}$$

where

$$F(Q) = \int_0^Q f(d)dD$$

Since K is constant, minimum value of TEC' (Q) ought to additionally accept via the same condition as given in equation

$$\sum_{D=0}^{Q-1} f(d) \leq \frac{C_s}{C_H + C_s} \leq \sum_{D=0}^{Q} f(d) \tag{1.9}$$

Since K is constant, minimum value of

And consequently Q* can even decrease TEC (Q).

Let us present two new control factors S and s, where S represents the extreme Stock stage and s signifies the reordered stage that is while the Stock degree tumbles to s, a request is situated to bring the Stock of Inventory items up to S.

Thus, value of S= Q* and the price of s is determined by the relationship

$$TEC(s) = TEC(S) = K + TEC(S);\ s < S \tag{1.10}$$

As I the fundamental Inventory prior to starting the period, at that point to decide the request size to bring the available Inventory of articles as much as Q*, the ensuing three occasions might be examined.

$$\text{(i) } I < s, \ (ii)\ s \le I \le S, \ and\ (iii)\ I > S$$

Case 1: If we start the length with I unit of Inventory and do not now buy or produce more prominent, at that point TEC (I) is the foreseen cost. In any case, on the off chance that we expect to purchase extra (Q-I) units in the event that you need to convey Inventory stage as much as Q*, at that point TEC' (Q*) will include the set-up expense furthermore. Subsequently, for all I<s, the condition for requesting is

$$\underset{Q>I}{Min} \ \{TEC(Q*)\} = TEC'(S) < \text{TEC1 (I)}), \tag{1.11}$$

That is, while Inventory stage arrives at S=Q*, request for Q-I units of Inventory might be put.

Case 2: For this situation, if I<Q, the request size is controlled by the condition

$$\text{TEC}_1 \text{ (I)} \le \underset{Q>I}{Min} TEC(Q) = TEC(S) \tag{1.12}$$

This implies that no ordering substantially less costly than ordering. Thus Q*=I.

Case 3: If Q>I, at that point foreseen cost for a request up to Q could be extra than generally speaking foreseen cost if no structure is found, that is

$$\text{TEC}_1 \text{ (Q)} \ge \text{TEC}_1 \text{ (I)} \tag{1.13}$$

Consequently, it is better not to put request for acquirement of things and afterward Q*=I.

1.2.3.4 Numerical Examples

In framework to exhibit the proposed model, a Numerical model is settled with the accompanying Parameters value I=10, C_H=0.53, Cs=Rs 5, C=2.5 and K=25. At that point get ideal arrangements are Q*= 45, TEC_1 (I)=202.79, TEC_1 (Q)=168.5, TEC (Q*) =193.5. Figured outcomes are demonstrated in Table 1.2.

Table 1.2 Optimal instantaneous demand solution of the order policy.

Parameters	Q*	TEC_1 (I)	TEC_1 (Q)	(TEC (Q *)
I=10, $C_H = 0.53$ $C_S = Rs\ 5,$ $C = 2.5 and$ $K1 = 25$	45	202.79	168.5	193.5
I=10, $C_H = 0.54$ $C_S = Rs\ 7,$ $C = 3.5 and$ $K2 = 26.5$	46	283.77	233.77	260.27
I=10, $C_H = 0.51$ $C_S = Rs 8.5,$ $C = 4.5 and$ $K3 = 27.5$	44	344.51	291.22	318.72
I=10, $C_H = 0.52$ $C_S = Rs\ 9.5,$ $C = 5.5 and$ $K4 = 28.5$	40	385.01	340.16	368.66
I=10, $C_H = 0.52$ $C_S = Rs\ 10.5,$ $C = 6.5 and$ $K5 = 29.5$	36	425.51	387.41	416.91

1.2.3.5 Sensitivity Analysis

Sensitivity analysis is performed in this section with respect to crucial parameter. We change the value parameter by Q*, TEC_1 (I), TEC_1 (Q), TEC (Q*) individually keeping different boundaries at their unique qualities and noticed its impact on ideal approach. This demonstrates that as the Ordered quantity increases (Q*), Expected cost Increases TEC(I), Optimum Stock achieve local minimum TEC(Q) decreases, Condition for Ordering TEC (Q*) is increments. Thus, the Sensitivity of the ideal outcome to little change in the Parameter esteem is inspected. Legitimate Inventory limit the Ordering cost of the multiplying. Proposed model be thing by taking

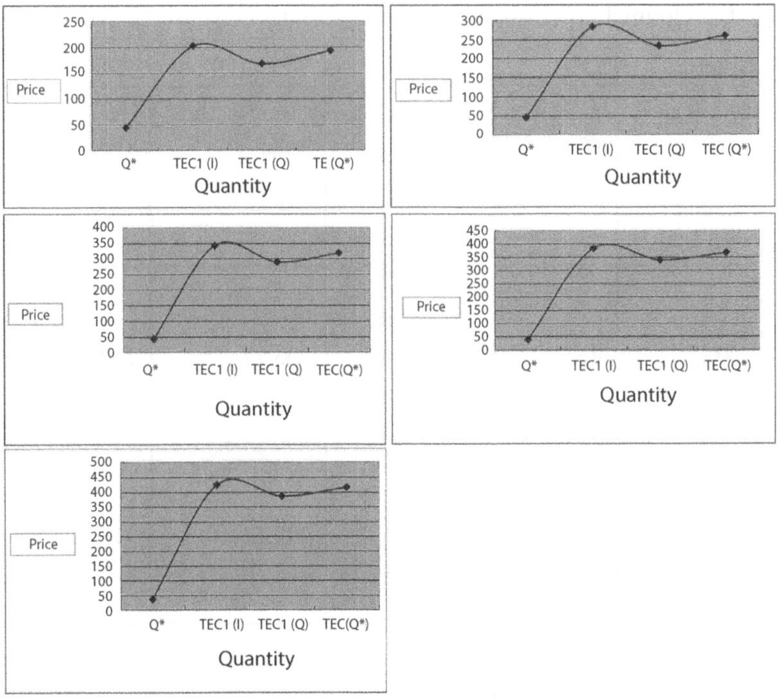

Figure 1.2 Graphical representation of Inventory Instantaneous demand in Brownian movement.

various suppositions cost of Inventory value Parameters. The outcomes are presented in Table 1.2. Table 1.2 is displayed in Figure 1.2.

1.2.4 Classic EOQ Method in Inventory

EOQ model intent to resolve ideal number of units to arrange, so that administration can minimize the total cost associated with the purchase expense, transportation price and storage of a product. In other words, the classic EOQ is the amount of inventory to be requested per time for limiting yearly stock cost. EOQ which is profoundly act as a gadget for Inventory Control.

1.2.4.1 Assumptions

The proposed model is established by the following presumptions.

- The Demand cost for the years is known and resupplied momentarily.

- Ordering cost straight forwardly.
- Inventory when an order shows up.
- The management ordering cost per unit time in dollars.
- Cost of ordering is stable.
- Lead time for the Inventory cycle.
- The Lead time, that is the time between the putting of the request and the receiving of the order is known.
- There is no restraint on order size.
- An order is a request for something to be provided.
- Ordering costs which may be caused an acquiring extra Inventories. The more regularly arranges are put and less the amounts bought on each request.
- There is no quantity concession.
- To survey the hidden suspicions of the EOQ model for the improved apprehension of current Inventory Management.
- Shortages are not permitted.

1.2.4.2 Notations

The accompanying documentation is utilized to build up the model.

d = Total number of units produced.
k_1 = Set up cost related to the arrangement of orders.
L = additionally appear some of the region
Q = Order quantity.
I_c = The Stock processes for this pattern is $T_0 = \frac{Y}{d}$ time units.
$Y*$ = Order the Quantity in every day.
N = the number of highest integers.
H = Holding cost per unit every day.
S = No Shortage is allowed.
R = Reorder point.
$L_E d$ = The reorder point accordingly happens when the Inventory level drops.
A = Sum of the initial and end ordinates.
B = Sum of the final Ordinates as Trapezoidal rule.
C = Item Cost

1.2.4.3 Mathematical Model

The mathematical method confesses the Inventories position and it is expressed as

Thing is also diminished at the ordinary demand amount d.
The ordering period for the models is $T_0^* = \frac{Y*}{d}$.
Put that the Normal Inventory stage is $\frac{Y}{2}$,
The total price per unit time (TCU) is along these lines figured out as
TCU(y) = Set up cost per unit time + Holding Cost per unit time

$$= \frac{Setup\ Cost + Holding\ Cost\ per\ cycle\ t_0}{t_0}$$

$$= \frac{K + H\left(\frac{Y}{2}\right)T_0}{t_0}$$

$$= \frac{K}{t_0} + H\left(\frac{Y}{2}\right)\text{since}\left(T_0 = \frac{Y}{d}\right)$$

$$= \frac{K}{\left(\frac{y}{D}\right)} + H\left(\frac{Y}{2}\right) \tag{1.14}$$

The most helpful assessment putting in a request sum y is controlled with method of reduce TCU(y) concerning y. Consider y is fundamental circumstance for finding the ideal assessment of y.

Here Y assumed as continuous,

$$\frac{dTCU(Y)}{dY} = -\frac{Kd}{Y^2} + \frac{H}{2} = 0 \tag{1.15}$$

The terms are additionally sufficient because of the reality TCU(y) is Convex.

The result of the situation yields the EOQ, y*as

$$Y* = \sqrt{\frac{2Kd}{H}}$$

Subsequently the most ideal Inventory strategy for the propounded model is

$$Y* = \sqrt{\frac{2Kd}{H}} \tag{1.16}$$

Units every $T_0^* = \frac{Y^*}{d}$ time.

A new order needs no longer be acquired in the meanwhile it is ordered. Rather than of high-quality Lead time L, may also additionally appear some of the region and the receipt of an order as Reorder element inside the exemplary EOQ models. In this situation the reorder aspect shows up even as the Inventory degree drops to LD units.

Reorder point inside the conventional EOQ version assumes that the lead time L is an awful lot much less than the cycle period t_0^*, which may not be the case in extensively well known. Lead time that is the quantity of time among placing an order and accepting the stock.

Effective lead time is defined as

$$L_E = L - NT_0^* \tag{1.17}$$

where n is the highest integer not exceeding $\frac{L}{T_0}$

The range of integer cycle consists of in L is

$$\text{N} = \left(\text{Maximum integer} \leq \frac{L}{T_0^*} \right) \tag{1.18}$$

Each the Inventory situation acts as if the interval amongst setting an order and getting another is L_e.

The reorder factor as a result takes area while the Inventory degree drops to $L_e D$.

1.3 Methodology

This research became applied quantitative research design to measured data due to Numerical and to get appropriate and specific statistics the frame of the researchers. This model examines how Inventory model can assist in minimising the total cost of Inventory model. The Trapezoidal guideline works by approximating the region under the graph of the capacity f(x) as

Trapezoid and computing its area. It is ruled to locate the estimation of a positive fundamental utilizing numerical technique.

$$\text{The Trapezoidal Rule is } \int_q^p f(X)dX, \tag{1.19}$$

Let f(x) be continuous in the line [p,q] and DC be the curve Y = f(X) and DC, CB is terminal Ordinates.

Let OA = p and OB = q, then, AB = OB−OA = q−p

Divide AB into n same segment $A, A_1, A_1A_{2........}A_{n-1}B$

So that each segment = $\dfrac{q-p}{\mathrm{H}} = \mathrm{H}$.

Portray the ordinate among

$$A, A_1, A_1A_{2........}A_{N-1}B \tag{1.20}$$

and put then be referred as $Y_1, Y_{2.........}Y_{N,}Y_{N+1}$*respectively,*

Then

$$\int_{x_0}^{x_N} f(X)dx = \frac{H}{2}(A+2B) \text{ approximately} \tag{1.21}$$

where, A = Sum of the first and last ordinates

$$= Y + Y_{N+1} \tag{1.22}$$

B = Sum of the remaining Ordinates as Trapezoidal rule

$$= Y_2 + Y_3 + Y_4 + ... + Y_N \tag{1.23}$$

1.3.1 Brownian Motion

This segment provides information for the simple solution of a Brownian movement B, along with some common variations in terminology which we use for some purposes [14]. The essential definition of B, as a random continuous function with a particular family of finite-dimensional distributions, is motivated in the appearance of this technique as a limit in distribution of rescaled random walk paths. Let (Ω, F, P) be a probability range [13].

A stochastic technique (B (T, ω), T ≥0, ω ∈Ω) is a Brownian movement If,

(i) For stable each T, the random variable $B_T = B(T)$ has Gaussian distribution.
(ii) The procedure B has stationary independent increments.
(iii) For each fixed ω∈Ω, the path T→ B (T, ω) is continuous.

The which means of second point is that if, $0 \leq T_1 \leq T_2 \leq T_3 \leq \ldots\ldots \leq T_n$ then

$$B_{T_1}, B_T - B_{T_1} \ldots\ldots\ldots B_{T_n} - B_{T_{n-1}} \tag{1.24}$$

are independent, and the distribution of $B_{t_i} - B_{t_{i-1}}$ depends only on $T_i - T_{i-1}$. According to (i), this distribution is normal with mean 0 and variance $T_i - T_{i-1}$. The real B is continuous to demonstrate that B has continuous paths as in (iii). Due to the convolution properties of ordinary distributions, the joint distribution of $x_{T_1} \ldots\ldots\ldots x_{T_n}$ are predictable for any $T_1 \leq \ldots\ldots \leq T_n$.

1.4 Results

The Numerical Example is shown in Table 1.3.

To demonstrate the solution manner introduced above, consider an Inventory object with the subsequent associated parameters ($Y*$, T_0, N, L_E, TCU(Y), $L_E\, d$) tabulated in Table 1.4 to Table 1.8.

It is presented in Table 1.9. Table 1.9 is presented in Figure 1.3 by using MATLAB. This model analyses how Inventory model can help in minimising the Total cost of Inventory.

The Reorder is a diffusion of enterprise. It is a price-saving technique that can assist prevent Inventory outs overlooked possibilities in business and a probable interruption in the operational technique.

The whole life pattern of a request from the maximum amount element of offer to select and decide to transportation to client conveyance.

The time it takes a Supplier to convey products after a request is set alongside the time period for a business reordering needs.

The Order Quantity is the best measure of an item to buy at a given time. It is a significant count since holding a lot of Stock is costly.

Reorder element is an approach to choose to decide when to arrange. It does not address how s extraordinary arrangement to arrange when a request is made.

Table 1.3 Optimal results of the Inventory in Various Parameters

Parameters		$Y*$	T_0^*	N	L_E	TCU(Y)	$L_E d$
1	$k_1 = \$105$ H=$.06 d=29 L=29	318.59	10.99	2.64	-0.02	19.12	-0.6
2	$k_2 = \$52$ H=$.04 d=20 L=29	228.04	7.9	3.67	0.007	9.12	0.20
3	$k_3 = \$98$ H=$.02 d=41 L=30	633.88	15.46	1.88	-0.06	12.81	-2.46
4	$k_5 = \$104$ H=$.03 d=22 L=29	372.38	16.93	1.71	0.05	11.73	1.1

Table 1.4 Optimum results of the ordering processes.

T_0^*				
x	0	1	2	3
f(x)	10.99	7.9	15.46	16.93

Table 1.5 Optimum results of the number of integer cycles.

N				
x	0	1	2	3
f(x)	2.60	3.66	1.34	1.9

Table 1.6 Optimum results of the effective lead time.

L_E				
x	0	1	2	3
f(x)	-0.03	-0.012	0.04	-0.039

Table 1.7 Optimal solution of the TCU(Y).

TCU(y)				
x	0	1	2	3
f(x)	17.32	12.25	49.19	69.57

Table 1.8 Optimum results of the reorder in $L_e D$.

$L_e D$				
x	0	1	2	3
f(x)	-0.9	-0.36	1.6	-0.78

Table 1.9 The optimal results of the inventory in trapezoidal rule.

Parameters	T_0^*	N	L_E	TCU(y)	$L_e d$
x	0	1	2	3	4
$\int_0^3 f(x)dx$	37.32	7.73	-0.038	37.36	-2.01

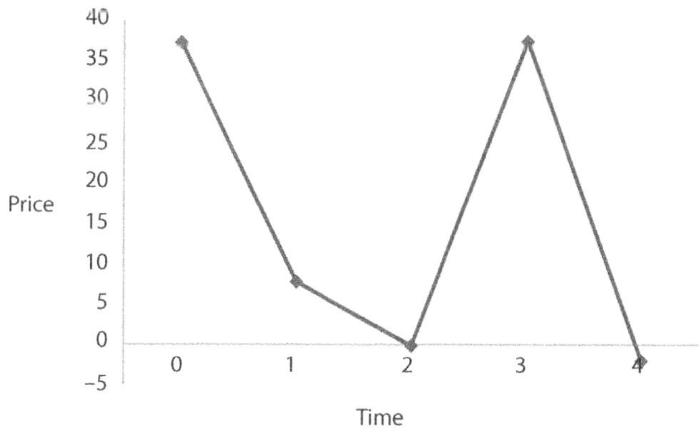

Figure 1.3 Trapezoidal rule in brownian movement.

It became accepted that there might be no time along requesting and buying of materials. The ascertaining Reorder level includes the figuring of utilization cost every day. Consider an association that works with a provider. The organization stores a few items renew by the providers to fulfil its Customers need.

Figure 1.3. Depicts the applicable of Inventories of Parameters $Y*$, T_0, N, L_E, TCU(Y), L_e d calculation and the subsequent optimization of Inventories with the aim of minimizing the proposed goal.

1.4.1 Numerical Examples

This Numerical Example to illustrate the above Mathematical model.

Parameters 1: k_1 = \$105, H = \$.06, d = 29 units per day L = 29 days. Optimal solutions Y* = 346.41, $T_0^* = \frac{Y*}{d}$ = 10.99 days, N = 2.64, $L_{E=-0.02}$ days, $L_e D$ = −0.6, The everyday Inventory price related with the Expected Inventory scheme is TCU(y) = 19.12.

Parameters 2: k_1 = \$52, H = \$.04, d = 20 units per day L = 29 days. Optimal solutions Y* = 228.04, $T_0^* = \frac{Y*}{d}$ = 7.9 days, N= 3.67, $L_{E=0.007}$ days, $L_e D$ = 0.203., The everyday Inventory price related with the Expected Inventory scheme is TCU(y) = 9.12.

Parameters 3: k_1 = \$98, H = \$.02, d = 41 units per day L = 29 days. Optimal solutions Y* = 633.88, $T_0^* = \frac{Y*}{d}$ = 15.46 days, N = 1.88, $L_{E=-0.06}$ days, $L_e D$ = −2.46, The everyday Inventory price related with the Expected Inventory scheme is TCU(y) = 12.81.

Parameters 4: k_1 = \$104, H = \$.03, d=22 units per day L = 29 days. Optimal solutions Y* = 372.38, $T_0^* = \frac{Y*}{d}$ = 16.93 days, N = 1.71, $L_{E=0.05}$ days, $L_e D$ = 1.1, The everyday Inventory price related with the Expected Inventory scheme is TCU(Y) = 11.73.

1.4.2 Sensitivity Analysis

The EOQ model of Inventory Management takes over the day-by-day utilization of Inventories, the optimum estimation of the orders amount y is controlled by minimizing TCU(Y).

The study effects of change in the value of the system are displayed from Table 1.4 to Table 1.9. This is

Important Inventory Parameters ($Y*$, T_0, N, L_E, TCU(Y), $L_E d$) are classified.

By Parameters 1: The set-up cost increases, T_0^* diminishes, n decreases, L_E decreases and $L_E d$ decrease, then The Everyday Inventory expense joined with the propounded Inventory scheme is TCU(y) diminishes.

By Parameters 2: The set-up cost increases, T_0^* decreases, N decreases, L_E Decreases and $L_E d$ decrease, then The Everyday Inventory expense joined with the propounded Inventory plot is TCU(Y) diminishes.

By Parameters 3: The set-up expense increases, T_0^* decreases, N diminishes, L_E Decreases and $L_E d$ decrease, The Everyday Inventory

Cost combined with the propounded Inventory scheme is TCU(Y) diminishes.

By Parameters 4: The set-up cost expands, T_0^* diminishes, N diminishes, L_E Decreases and L_Ed decreases, The Everyday Inventory price joined with the propounded Inventory scheme is TCU(Y) Increases.

This Y* is not Trapezoidal Rule and hence it is not formed a Brownian movement, but remaining Parameters T_0, N, L_E, L_Ed, TCU(Y) *is a* Trapezoidal Rule. It is framed a Brownian Movement.

At last implementing Sensitivity assessment on the decision factors through changing the Inventory parameters ($Y*$, T_0, N, L_E, L_Ed, TCU(Y). In sensitivity investigation make fluctuate to the factors built into that Inventory to provide the Brownian movement. This method is generally sensitivity to change around the EOQ.

1.4.3 Brownian Path in Hausdorff Dimension

Brownian direction is a-Holder a. s. for all a $<\frac{1}{2}$. A Brownian movement is sort of really now not $\frac{1}{2}$ -Holder. However, there does represent a $t = t(w)$ such that

$$|B(T+H)-B(T)| \leq C(W)H^{\frac{1}{2}} \tag{1.25}$$

For every h most definitely. The decided of such t offer degree zero. This is the moderate improvement that is locally possible. Having confirmed that Brownian direction are extremely popular [19]. Make see why they are "peculiar". One rationalization is that the strategies of Brownian movement don't get any time of monotonicity. Indeed, if [a,b] is an time of monotonicity, at that point splitting it into an equivalent sub-period $[a_i, a_{i+1}]$ every expansion $B(a_i) - B(a_{i+1})$ should have a similar indication. This has probability 2.2^{-n} and taking n→ ∞ indicates that the probability that $[a, b]$ is a time of monotonicity ought to be [13]. Let a countable association prove that there may be no gap of monotonicity with rational that Brownian movement solution, however each monotone. Period could have a monotone logical alternate period that Brownian movement is b -Holder for any $b<\frac{1}{2}$ a.S. This will infer us to conclude a higher certain on the Hausdorff measurement of the graph [12]. Reproduce that the graph G_f of a feature f is the fix of factors (t, f (t)) as t degrees over the area off. it will permit us to infer a top positive at the Hausdorff length of the graph [12]. Review outlines G_f of an event f is the fix of things (t, f (t)) as t levels outer the area of f.

1.4.4 The Hausdorff Measure

The Hausdorff degree of a self-affine set ought be either 0 or ∞. Assume the number set d has non-uniform parallel fibers and let γ = dim (K (d), then $H^{\gamma}(K(d)) = \infty$.

In addition, K (d) is not generally s-bounded for H^{γ}. Since the intersections of *K* (*d*) with the d square shapes of the original are interprets of each other and t. The Hausdorff measure H^{γ} is interpretation invariant, its constraint to *K* (*d*) should distribute all these square shapes a comparative measure. Note the assessment with the self-similar sets which have effective and definite Hausdorff estimate in their measurements, increasingly exact data can be communicated utilizing measure capacities.

$$dim(GB) \leq dimM(GB) \leq dim(GB) \leq \frac{3}{2}. \tag{1.26}$$

1.4.5 Levy Processes

An alternative generalization of Brownian movement gives the stable technique presented by Levy

A stochastic system $X = \{XT\}_{T \in \mathbb{R}_+}$ is a Levy procedure beginning at zero if it satisfies,

(i) Increments X(T + H) – X(T) are stationary.
(ii) *X has* Independent additions.
(iii) X has stationary increments that is, for all $0 \leq S < T$, coincides with the law the law X_{T-S}.
(iv) X is stochastically continuous, that is,

$$lim_{S \to T} P(|X_T - X_S| > \in) = 0, \forall \in > 0, T \in {}_{\mathbb{R}+}.$$

Levy manner able to choose a completely unique change whose paths are right non-stop and with left limits. This by using the Brownian motion. Obviously, condition (iii) and (iv) strongly restrict the possible regulation of the technique X and its circle of relatives of finite dimensions distributions. A Levy Process X is determined through using the regulation of X_T, however this regulation cannot be arbitrary. It must be infinitely Processes with stationary independent increments called Levy processes.

Levy processes and Hausdorff measure pertinent for all the three techniques are,

(i). presumptions of the EOQ; (ii) Inventory Control in Instantaneous demand Method under development of the stock; (iii) Exemplary EOQ in Inventory. It is depending on the outer measure. This is obtained from Brownian path. Hence it is known as Fractals.

1.5 Discussion

Without limitations the study is vast. The observation made a major contribution for organization technology and operations management to utilize the Inventory control technique attention on the possibility of optimising the company Inventory to manage method technique in an organization. Thereby creating awareness of the opportunity of optimising the company's Inventory cost for better profitability. By so doing there is realistic signification of the EOQ.

This approach represents a variation from the tradition in EOQ model, where the inventory level change over time an average Inventory can be calculated on basis of physical space limitations which is related to tangible goods, thus requiring the definition of a reorder parameter. EOQ selection duties taking into account: i.e., delivery costs generally based on purchased rate that do not depend on the scale of the party ii. Prices for the maintenance of stocks are proportional to their size, which were before described as assumptions and boundaries in existing models. This model fulfils all necessities at least expense.

1.5.1 Future Research

The Research extended to deal with Inventory models to manage EOQ model. In this case, there is a connected assistance for stock expense which is represented in dollars as per undertaking data per time unit. That is the expense brought about for having the foreseen assignment or data prepared to be utilized in a feature of the administration. As clarified before, basically "storing" these foreseen tasks or data permits to keep forever their ideal level should have been prepared for use without waiting for "reorder" level, as there is no compelling reason to build up a boundary to decide the second where administration undertakings and data ought to be foreseen again in the administration cycle, ensuring the proper assistance arrangement to customer, improving quality, customization, speed or cost and to keep up adequate supply of crude materials and control investment and give proposal for future work.

1.6 Conclusions

EOQ Inventory model engages to keep a track on the nonappearance of substance due to imprudence and theft. There is a more prominent possibility of carelessness and stealing if inventory has not been done in the right way. Company Inventory model will be helpful if those concepts are implemented and it shows an indication of what stage of sales to count on. EOQ models are utilized to choose the ideal inventory policy when the demand is deterministic and the top-quality ordering or manufacturing amount are prompted by using Parameters of prices. These models are used to the Inventory part size that cut-off points Inventory extraordinary cost. Mathematical assessment and generation have shown that it uses the resources and even more gainfully achieves the most extraordinary advantages and can improve shopper reliability. In association's Inventory the executive circumstance will be more apex. The main goal is to limit selling cost and process duration cost out of all business benefit and these were represented by using mathematical models. In this three-methodologies Brownian movement are established by Trapezoidal Rule. These results are showed in Brownian Path and found by the outside measure. Subsequently it is Fractals.

References

1. Makoena Sebatjane, Olufemi,Adetunji, Economic order quantity model for growing items with imperfect quality, Volume 6, 100088 (2019).
2. Mohamed Hassan Dhodi, The Effect of Information Technology on Inventory Management for the Manufacturing Companies in Mogadishu, *European Journal of Logistics, Purchasing and Supply Chain Management*, Vol.6 No.3, pp. 20-29, June (2018).
3. Zohreh Momeni and Amir Azizi, Current Order and Inventory Models in Manufacturing Environments, Feb 02, Paper Id.: IJMPERDFEB2018129 (2018).
4. Rakesh Kumar, Economic Order Quantity (EOQ) Model, *Global Journal of Finance and Economic Management*. ISSN 2249-3158 Volume 5, Number 1 pp. 1-5 (2016).

5. Naser Ghasemi and Behrouz Afshar Nadjafi, EOQ Models with Varying Holding Cost, Hindawi Publishing Corporation *Journal of Industrial Mathematics* Volume Article ID 743921, 7 pages (2013).
6. Ganesh Prasad Shukla and Prashant Kumar Jangde, Determination of Economic Order Quantity and Reorder Point Inventory Control Model for Xyz Retail Enterprises, ISSN (Print): 2393-8374, (Online): 2394-0697, Volume 4, Issue 12 (2017).
7. Jose L. Gonzalez and Daniel González, Analysis of an Economic Order Quantity and Reorder Point Inventory Control Model for Company XYZ, March 10, (2010).
8. Iyengar S.R.K., and Jain R.K.,*Numerical Methods*, ISBN (13) 978-81-224-2707-3, 2009.
9. Rudi Abdullah, Samsul Bahari Bahar, Asrianti Dja'wa, La Ode Dedi Abdullah Inventory Control Analysis Using Economic Order Quantity Method volume 436 2019.
10. A guide to Brownian motion and related stochastic processes Jim Pitman and Marc Yor, [v1] Tue, 27 Feb 01:36:17 UTC (111 KB), 2018.
11. Jörn Dunkel, Peter Hänggi, Relativistic Brownian motion, *Physics Reports* 471 (2009) 1–73.
12. Toshio Yanagida, Masahiro Ueda, Tsutomu Murata, Seiji Esaki a, Yoshiharu Ishii, Brownian motion, fluctuation and life, *BioSystems* 88 (2007) 228–242.
13. Luczkaa J., Talknerb P., Hanggi P., Diffusion of Brownian particles governed by fluctuating friction. *Physica A* 278 (2000) 18–31.
14. Julien Berestycki,ÉricBrunet, Simon. Harris, Piotr Miłos, Branching Brownian motion with absorption and the all-time minimum of branching Brownian motion with drift. *Journal* of *Functional Analysis* 273 (2017) 2107-2143.
15. Huerta-Cuellar G., Jiménez-López E., Campos-Cantón.,Pisarchik, A.N., An approach to generate deterministic Brownian motion, *Commun Nonlinear SciNumerSimular* 19(2014) 2740-2746.
16. Camella Burja, Vasile Burja, Analysis Model for Inventory Management, *Annals of the University of Petrosani, Economics,* 10(1), 43-50, (2010).
17. Aref Gholami and Abolfaz Mirzazadeh, An inventory model with controllable lead time and ordering cost, log-normal-distributed demand, and gamma-distributed available capacity. *Cogent Business & Management* 5: 1469182(2018).
18. Mehmood Khan, Mohamad Y. Jaber, Maurice Bonney, An economic order quantity (EOQ) for items with imperfect quality and inspection. *Int. J. Production Economics* 133(2011)113-118.
19. Christopher J. Bishop and Yuval Peres, *Fractals in Probability and Analysis,* 2017.
20. Mandelbrot B. B. *The Fractals Geometry of Nature,* Freeman, San Francisco, CA, 1982.

2

Ill-Posed Resistivity Inverse Problems and its Application to Geoengineering Solutions

Satyendra Narayan

Department of Applied Computing, Faculty of Applied Science and Technology, Sheridan Institute of Technology and Advanced Learning, Oakville, Ontario, Canada

Abstract

The most important physical properties to study ill-posed inverse problems in physical sciences are electrical conductivity, magnetic permeability, density, wave-velocity, elasticity parameters/modulus, and dielectric permittivity. This paper attempts electrical conductivity of the earth materials and describes some innovative approaches which have been used to solve ill-posed resistivity inverse problems encountered in mapping and monitoring geo-environmental problems.

The paper begins with an overview of the present state of knowledge about electrical resistivity methods for mapping and monitoring *in situ* processes that cannot be accessed directly. The current study indicates that a generalized mathematical approach has not been developed to investigate the sensitivity of resistivity measurements to changes in resistivity at depth. Therefore, the paper also presents a generalized mathematical formulation for sensitivity analysis and describes sensitivity of resistivity measurements. Reciprocity and perturbation analysis form the basis for the mathematical formulation, which has been extended further towards introducing multi-dimensional resistivity inversion useful for mapping and monitoring *in-situ* processes. A generalized multi-dimensional mathematical technique is described herein for computing numerical response over the one-dimensional (1-D), two-dimensional (2-D) and three-dimensional (3-D) resistivity models excited by a three-dimensional (3-D) point source. These problems also described as 1-D/3-D, 2-D/3-D and 3-D/3-D inverse problems in the scientific literature.

Email: narayan.satyendra@gmail.com

Ramakant Bhardwaj, Jyoti Mishra, Satyendra Narayan and Gopalakrishnan Suseendran (eds.)
Mathematics in Computational Science and Engineering, (27–54) © 2022 Scrivener Publishing LLC

Keywords: Resistivity inversion, resistivity inverse problems, sensitivity of resistivity measurements, multi dimensional resistivity inversion, *in-situ* resistivity monitoring, design strategy for resistivity monitoring, geo-environmental resistivity problems, geo-engineering resistivity problems, linearization of nonlinear resistivity problems

2.1 Introduction

The keywords ill-posed inverse problems have been reported in the scientific literatures since the beginning of the 20th century. In physics, it is reported as an inverse problems of quantum scattering theory. In geophysics, it is reported as an "ill-posed inverse problem" of electrical resistivity mapping, seismic mapping, and gravitational-potential field mapping. It has also appeared in astrophysics and other areas of science and engineering. With recent advances in mathematical computing and powerful computers over the past decades, application of inverse and ill-posed problems, its theories and the associated mathematical methods has been extended to almost every field of science and engineering. In forward numerical modelling of physical sciences, researchers attempt to formulate appropriate functions. These functions are used to describe different physical processes involving propagation of seismic waves, sound waves, electromagnetic waves, and heat waves, etc.

To understand ill-posed inverse problems thoroughly in the field of science and engineering, it is important to understand the meaning of ill-posed and well-posed problems, and the concept that may be applied to solve the ill-posed problems. In mathematical notation, a well-posed problem has a system of partial differential equations that can be solved uniquely, and it has a unique solution and depends continuously on the input data. In other words, well-posed mathematical models of the physical processes (e.g., Dirichlet problem for Laplace's equation, heat flow equation with given the initial conditions, etc.) in science and engineering have three well-defined properties: a solution exists for the model, solution of the model is unique, and the model solution's is continuous with the changes in initial conditions (also called parameters or input data). These three properties are also described as existence, uniqueness and stability. It is important to note that the ill-posed problems do not meet all these three well-defined properties. A problem that is not well-posed is known as ill-posed. Several first-order differential equations and inverse problems are ill-posed. If the physical problem or mathematical model is not well-posed, it is required to be reformulated for numerical computation. Typically, ill-posed models or problems require additional assumptions (e.g., smoothness of solution) and the process is usually called as regularization [1]. In geophysical literature, Tikhonov regulation is one of the highly used regularizations for the ill-posed problems.

Continuous mathematical models are often discretized to obtain numerical solutions. These solutions may be continuous with respect to the initial parameters. Furthermore, when these problems are solved with a finite precision, it may suffer from a numerical instability. Even though these problems are well-posed, they may be ill-conditioned. Here, the meaning of ill-conditioned refers to a small error in the initial data resulting in a larger error in the solution. In mathematical literature, an ill-conditioned problem is defined by a large condition number, which is a measure of sensitivity of the model. This gives indication quantitatively how much error is in the output from an error in the input. A physical model is called well-conditioned if it has a low condition number. If the condition number is high, it is called ill-conditioned.

An inverse problem in the field of science and engineering is a process of calculating physical model parameters from a set of real or synthetic observations (in other words, computing the input parameters from the output data/results). Examples are computing images in X-ray computed tomography, source reconstruction, calculating density distributions of the Earth material from the measured gravity potential field, etc. It is known as an inverse problem because it starts with the results of the physical model and computes the physical model parameters (called input to the model). In other words, this can be viewed as the inverse of a forward problem, which starts with the causes and then calculates the effects.

Linear or non-linear Inverse problems are very important mathematical problems in the field of science and engineering. This is due to the fact that these problems give us information about the parameters that cannot be accessed or observed directly. These problems have a wide range of applications in system identification in the field of science and engineering including natural language processing, machine learning, nondestructive testing, and many other domains. This paper is focused on in-depth analysis of ill-posed inverse problems that are usually common in electrical geophysics.

2.2 Fundamentals of Ill-Posed Inverse Problems

These problems attempt to describe a system of coefficient matrix from observed data, which is used to estimate physical model parameters of the forward numerical models. This coefficient matrix usually represents important properties of the media or of the model that is under study. These properties in the field of geophysics are density, electrical conductivity, heat conductivity, magnetic susceptibility, etc., of the Earth materials.

In the process of solving such problems, it is possible to delineate many other details such as the structural intrusions, defects, source of contaminations, location, shape, etc. There is a large number of scientific articles and research publications that have dealt with the ill-posed inverse problems directly or indirectly. Since the theory associated with the inverse problems is relatively new, a shortage of textbooks has been felt in this area. This is due to the fact that the new theories, concepts and approaches have been continuously evolving. Kabanikhin [2] has given a very good and detailed overview of the ill-posed inverse problems. In this review paper, definition of ill-posed inverse problems, its types, and several examples of ill-posed inverse problems are described well.

2.3 Brief Historical Development of Resistivity Inversion

According to the literature review in the field of electrical geophysics, interpretation of electrical resistivity data using electrical resistivity inverse methods are commonly done for layered models and geological structures (e.g., groundwater exploration and mapping & monitoring of groundwater). However, in the field of mineral exploration, geothermal exploration, mapping and monitoring of *in-situ* processes, the layered geologic models are inadequate. With the advent of large computers, two-dimensional (2-D) numerical electrical modelling techniques for surface-to-surface electrode and other electrode configurations are used extensively to interpret electrical data. Integral equation method has a limitation because it allows inhomogeneties only inside the homogeneous sounding host media. Three-dimensional (3-D) numerical modelling methods using finite difference and finite element methods are reported in the geophysical literature. These methods are useful to compute electrical model response over a given 3-D geologic structure. A complete overview about the forward and inverse modelling in electrical geophysics may be found in Narayan [3].

Most of the forward modelling methods have attempted to address some aspect of the design of field experiments. These methods are not very useful for the interpretation of electrical field data on two counts. First, it is based on a trial-and-error mode and second, it is time intensive. Furthermore, they do not yield additional information. Therefore, it is important to introduce a new and effective electrical inversion method to interpret electrical field data in terms of 2-D or 3-D geological models for a variety of electrode configurations. This article emphasizes to use a generalized inverse theory for multi-dimensional structures and an attempt has been made to develop a practical way of inverting the resistivity data for mapping and monitoring two-dimensional (2-D) geologic features using a pair of surface and subsurface electrodes [3]. A brief historical development of resistivity inversion (one-dimensional (1-D) resistivity inversion, two-dimensional (2-D) resistivity inversion, and three-dimensional (3-D) resistivity inversion) used in environmental, engineering and hydrological fields has been reported extensively [4–12].

Electrical impedance tomography (EIT) methods are also known as electrical resistivity imaging methods - another version of resistivity inversion. These methods have proven to work nicely in most of the geophysical settings. EIT methods/imaging methods have been gaining momentum rapidly in recent years. This is due to the fact that they are easy to use and they are non-invasive testing tools. EIT methods or electrical resistivity methods are based on a low-frequency electrical current or DC current (unidirectional flow of electric charge) to probe a medium of the system, and measure its resistance or impedance of electrical current flow. EIT methods are highly sensitive to changes in electrical resistivity. In EIT methods, a known amount of electrical current is injected in the medium of the model and resulting electrical potential field is measured around the boundary points of the medium of the model. From these data, it is possible to perform electrical resistivity inversion of these potential field data to determine the electrical conductivity or resistivity inside the medium of the model that is being probed by the currents. In this way, the internal resistivity or electrical conductivity distribution is reconstructed using electrical measurement from the boundary of the medium of the model under investigation. Such electrical resistivity or conductivity distribution gives valuable information about the interior of the medium. Basically, all EIT methods deal with solving forward problem and inverse problem iteratively. EIT methods are used for various shallow depth archaeological prospecting, geothermal resource prospecting, geo-environmental monitoring, hydrogeological invitations, and geotechnical investigations etc. A good account of EIT methods and electrical resistivity imaging tools may be found in many geophysical literatures [13–20]. In brief, all these inverse problems are solved iteratively until a best possible solution is obtained.

2.4 Overview of Inversion Schemes

In the preceding sections, iterative approaches of electrical inversion methods are described. These approaches consist of a function called "objective function". It starts an initial parameter guess, updated at the end of every run iteratively and then applied to the "objective function" in a least-squares manner. This process is continued until a predefined level of acceptable error or a fixed of number of runs is obtained. Narayan *et al.* [9] has explained various versions of objective function effectively. Electrical inversion methods may be further differentiated based on the use of data variances in the objective function. For all practical purposes, electrical field data are not uniformly affected by the model properties and therefore, the use of data variances is rarely justified.

Hohmann and Raiche [21] have given a good description and classification of model properties in terms of relevance, non-relevance, or irrelevance. The importance and relevance of model properties can also be described theoretically using eigenvalues and its size. The "singular value decomposition" (SVD) of a matrix has been found useful in the least-squares electrical inversion methods. Tripp *et al.* [22] have given an excellent overview of this technique in electrical resistivity inversion. It is important to note that the SVD technique needs some special efforts that are computationally demanding and time consuming.

Another computationally attractive technique is described in the electrical resistivity inversion, which uses a "damping factor". Different criteria are described in the research literature for the selection of "damping factor". Narayan *et al.* [9] have given a good account of the various electrical resistivity inversion schemes and methods. The main benefit to use a damping in inversion methods is that it produces a stable solution. Constable *et al.* [23] have introduced an Occam's inversion, which is receiving attention in the electrical geophysics. Without knowing magnitudes of the eigenvalue, it is difficult to quantify an appropriate amount of damping. Oristaglio and Worthington [24] have described a method and criterion for selecting the amount of damping. It is important to note that the selection of the amount of damping is highly debatable. A different approach of selecting relatively high damping in the initial part and gradually low in the later part of the inversion is introduced by Eaton [25]. It may only be done based on a genuine experience.

2.5 Theoretical Basis for Multi-Dimensional Resistivity Inversion Technqiues

There are two important steps for the inversion of electrical data as shown below:

- A fast numerical modelling scheme, which computes electrical data theoretically for a given numerical/physical model (i.e., a fast "forward modelling scheme"), and
- An efficient numerical approach to compute first derivatives of the data (also called "coefficient matrix").

Unfortunately, the second requirement is not easily ready for 2-D or 3-D inverse geophysical and geological models. Furthermore, there is no claim for an optimum inverse technique or approach for the interpretation of 2-D and/or 3-D geophysical data. The purpose of this article is to introduce an effective solution for this problem and to propose multi-dimensional electrical resistivity inversion techniques.

Based on "reciprocity theorem (1961)" and "perturbation analysis", a new method for the multi-dimensional resistivity inversion is introduced herein. This approach for inverse formulation is entirely different from resistivity inversion methods published in the geophysical literature. The main benefits of the proposed method are listed here:

- The first derivatives of the data with respect to the model properties are computed efficiently.
- It provides a mathematical picture of sensitivity analysis. It is expressed as an amount of power that is lost in the anomalous zone.

The governing equations (also known as "Poisson" equation) for potential field distribution due to a 3-D direct current (DC) point source $I_s(x, y, z)$ are given by [26]

$$\Delta \varphi (x, y, z) = -\rho (x, y, z) \, J (x,y,z) \tag{2.1}$$

and

$$\Delta \cdot J (x, y, z) = I_s (x, y, z), \tag{2.2}$$

In the above equations, $\varphi(x, y, z)$ represents electrical potential field distribution, $I_s(x, y, z)$ is a 3-D direct current point source, $J(x, y, z)$ represent current density, and $\rho(x, y, z)$ describes three-dimensional distribution of electrical resistivity of the medium around the point source. The fundamental equations (2.1) and (2.2), representing potential field distribution (φ) and the electrical current density (J) due a point source (I), may be rewritten in different notations as shown below:

$$DV = \beta, \tag{2.3}$$

where

$$\mathrm{D} = \begin{bmatrix} \rho & - \\ \_\cdot & 0 \end{bmatrix}, \quad \mathrm{V} = [\,\mathrm{J}\ \phi\,], \quad \text{and } \beta = \begin{bmatrix} \frac{0}{I} \end{bmatrix}. \tag{2.4}$$

A perturbation analysis is performed here to study or investigate sensitivity of the potential field. Use a vertical cross-section of the 3-D model and changing electrical resistivity at depth by a small amount (Figure 2.1), equation (2.3) may be rewritten as shown in equation (2.5).

$$(\mathrm{D} + \delta\mathrm{D}).(\mathrm{V} + \delta\mathrm{V}) = \beta, \tag{2.5}$$

where

$$\delta\mathrm{D} = \begin{bmatrix} \Delta\rho & 0 \\ 0 & 0 \end{bmatrix}, \quad \text{and} \quad \delta\mathrm{V} = \begin{bmatrix} \delta\mathrm{J} \\ \delta\phi \end{bmatrix}. \tag{2.6}$$

After expanding equation (2.5) completely and upon neglecting second-order terms here, it gives

$$\mathrm{D}\cdot\delta\mathrm{V} = -\delta\mathrm{D}\cdot\mathrm{V}. \tag{2.7}$$

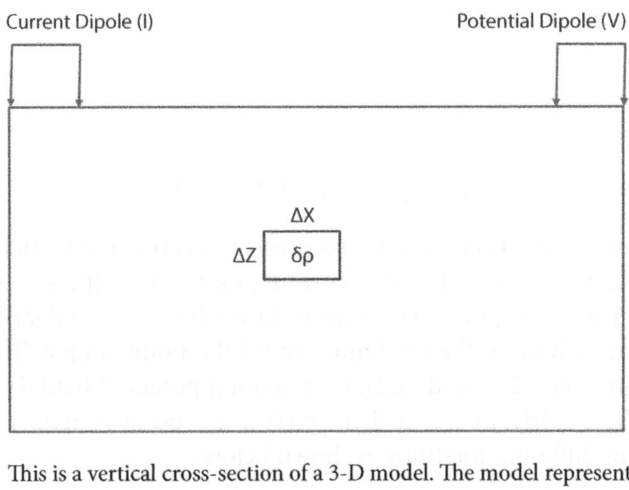

Figure 2.1 This is a vertical cross-section of a 3-D model. The model represents an anomalous zone. Electrical resistivity of the anomalous zone is perturbed. Reciprocity may be achieved by interchanging the current and potential dipoles.

Equations (2.3) and (2.7) are compared here. Upon comparison of these two equations, it is observed that small changes in electrical potential field distribution is related to small changes in electrical resistivity of the medium. Also, $(-\delta D \cdot V)$ in equation (2.7) behaves as current source for changes in potential fields.

Now, let us consider the "generalized Green's identity" of Lanczos [27], which is an integral over volume τ

$$\int_\tau \left[u^*(D\,V) - V^T\,(\bar{D}\,u)^* \right] d\tau = F(u^*, V). \tag{2.8}$$

If the adjoint problem **D** and its boundary conditions are chosen properly, then F($\mathbf{u}^*$,**V**) will vanish and the Green's identity will reduce to the bilinear identity

$$\int_\tau \left[u^*(D\,V) - V^T(\bar{D}\,u)^* \right] d\tau = 0. \tag{2.9}$$

The adjoint system is

$$D^T u^* = \gamma^*. \tag{2.10}$$

Here, ~ denotes Hermitian and * denotes complex conjugate. If

$$u^* = \begin{bmatrix} u_1 \\ u_2 \\ u_3 \end{bmatrix}, \tag{2.11}$$

then we can write

$$u^*\,(DV) = u_1\left(\rho\,J_x + \frac{\partial\phi}{\partial x}\right) + u_2\left(\rho\,J_y + \frac{\partial\phi}{\partial y}\right) + u_3\left(\rho\,J_z + \frac{\partial\phi}{\partial z}\right)$$
$$+ u_4\left(\frac{\partial J_x}{\partial x} + \frac{\partial J_y}{\partial y} + \frac{\partial J_z}{\partial z}\right). \tag{2.12}$$

In order to get expressions for $\mathbf{V}^T\,(\mathbf{Du})^*$, we need to write complete differentials

$$u_1 \rho J_x - u_1 \rho J_x = 0, \quad so\ V^T(D^T u^*) = u_1 \rho J_x, \tag{2.13}$$

$$u_1 \frac{\partial \phi}{\partial x} + \phi \frac{\partial u_1}{\partial x} = \frac{\partial}{\partial x}(\phi u_1), \quad so\ V^T(D^T u^*) = -\phi \frac{\partial u_1}{\partial x}, \tag{2.14}$$

$$u_4 \frac{\partial J_x}{\partial x} + J_x \frac{\partial u_4}{\partial x} = \frac{\partial}{\partial x}(u_4 J_x), \quad so\ V^T(D^T u^*) = -J_x \frac{\partial u_4}{\partial x}. \tag{2.15}$$

Similar expressions for $\mathbf{V}^T(\mathbf{D}^T\mathbf{u}^*)$ can be written for the y and z variables. Thus, the expression for $\mathbf{V}^T(\mathbf{D}^T\mathbf{u}^*)$ will be

$$\begin{aligned} V^T(D^T u^*) &= u_1 \rho J_x + u_2 \rho J_y + u_3 \rho J_z - \phi \frac{\partial u_1}{\partial x} - \phi \frac{\partial u_2}{\partial y} - \phi \frac{\partial u_3}{\partial z} \\ &\quad - J_x \frac{\partial u_4}{\partial x} - J_y \frac{\partial u_4}{\partial u_y} - J_z \frac{\partial u_4}{\partial z}. \end{aligned} \tag{2.16}$$

$$V^T(D^T u^*) = \left[J_x\ J_y\ J_z\ \varphi\right] \begin{bmatrix} \rho u_1 - \frac{\partial u_4}{\partial x} \\ \rho u_2 - \frac{\partial u_4}{\partial y} \\ \rho u_3 - \frac{\partial u_4}{\partial z} \\ -\frac{\partial u_1}{\partial x} - \frac{\partial u_2}{\partial y} - \frac{\partial u_3}{\partial z} \end{bmatrix} \tag{2.17}$$

$$or \quad = V^T \begin{bmatrix} \rho & -\Delta \\ -\Delta\bullet & 0 \end{bmatrix} \begin{bmatrix} u_1 \\ u_2 \\ u_3 \\ u_4 \end{bmatrix}. \tag{2.18}$$

$$\textit{If we consider} \quad \begin{bmatrix} u_1 \\ u_2 \\ u_3 \\ u_4 \end{bmatrix} - \begin{bmatrix} J' \\ \hline \phi' \end{bmatrix}, \tag{2.19}$$

then the adjoint system of equation $\mathbf{D}^T\mathbf{u}^* = \boldsymbol{\gamma}^*$ will be written as

$$\rho J' + \Delta \varphi' = 0, \tag{2.20}$$

and

$$\Delta \bullet J' = - I_{s'}. \tag{2.21}$$

The above analysis will also yield the expression for $F(\mathbf{u}^*,\mathbf{V})$, which is

$$F(u^*,V) = \int_\tau \left[\frac{\partial}{\partial x}(\phi\, u_1) + \frac{\partial}{\partial y}(\phi\, u_2) + \frac{\partial}{\partial z}(\phi\, u_3) + \frac{\partial}{\partial x}(u_4\, J_x) + \frac{\partial}{\partial y}(u_4\, J_y) + \frac{\partial}{\partial z}(u_4\, J_z) \right] d\tau. \tag{2.22}$$

Rewriting the boundary terms

$$F(u^*,V) = \int_\tau \left[\frac{\partial}{\partial x}(\phi\, u_1 + u_4\, J_x) + \frac{\partial}{\partial y}(\phi\, u_2 + u_4\, J_y) + \frac{\partial}{\partial z}(\phi\, u_3 + u_4\, J_z) \right] d\tau. \tag{2.23}$$

Substituting u_1, u_2, u_3, and u_4 values yields

$$F(u^*, V) = \int_\tau \left[\frac{\partial}{\partial x}(\phi\, J_{x'} - \varphi'\, J_x) + \frac{\partial}{\partial y}(\phi\, J_{y'} - \phi'\, J_y) + \frac{\partial}{\partial z}(\varphi\, J_{z'} - \phi'\, J_z) \right] d\tau. \tag{2.24}$$

Rewriting the boundary term again, it gives

$$F(u^*,V) = \int_\tau \Delta \bullet (\varphi J' - \varphi' J)\, d\tau. \tag{2.25}$$

Changing to the surface integral, it becomes

$$F(u^*, V) = \int_s (\phi J' - \phi' J) \bullet ds. \tag{2.26}$$

As the extended Green's theorem is valid regardless of the field substituted for **u** and **V**, we replace **V** by δ**V** in the bilinear identity. If the sides and bottom surfaces are chosen far enough from the anomalous region, then δρ and δ**J** will be zero. At the top surface, **J** and **J**′ are parallel to the surface, and **J·ds** = 0. Therefore, the boundary term vanishes and equation (2.10) becomes

$$\int_\tau \left[u^* \cdot (D\, \delta V) - \delta V^T \cdot (\bar{D}\, u)^*\right] d\tau = 0. \tag{2.27}$$

Substituting equation (2.7) and $(u)^* = \gamma^*$ in the above expression, one obtains

$$-\int_\tau \delta V^T . \gamma^*\, d\tau = \int_\tau u^* (\delta D \cdot V)\, d\tau. \tag{2.28}$$

This expression can be rewritten using expressions for δV^T, γ^*, δD, V, and u^*

$$-\int_\tau \begin{bmatrix} \delta J & \delta\phi \end{bmatrix} \cdot \begin{bmatrix} 0 \\ -I_{s'} \end{bmatrix} \cdot d\tau = \int_\tau \begin{bmatrix} J' & \phi' \end{bmatrix} \cdot \begin{bmatrix} -\delta\rho J \\ 0 \end{bmatrix} \tag{2.29}$$

or

$$\int_\tau \delta\phi \bullet I_{s'}\, d\tau = -\int_\tau \delta\rho J \bullet J'\, d\tau \tag{2.30}$$

Considering a unit point source $I_s' = \delta(x - x_0)\, \delta(y - y_0)\, \delta(z)$ at the observation point, the above equation becomes

$$\delta\phi = -\int_\tau \delta\rho \, \mathrm{J} \cdot \mathrm{J}' \, d\tau. \tag{2.31}$$

Using equation (2.31), small changes in the potential field distribution due to small changes in electrical resistivity of a 3-D model may be expressed in a different format. Assuming the 3-D model is discretized into small individual blocks, equation (2.31) may be rewritten further in algebraic notation representing power loss or dissipated in the blocks. For a given electrode geometry (Figure 2.1), equation (2.31) becomes

$$\delta\varphi = -\sum_{i=1}^{N} \delta\rho_i \int_0^{dx}\int_0^{dy}\int_0^{dz} (J_x J_{x'} + J_y J_{y'} + J_z J_{z'})_i \, dx\, dy\, dz. \tag{2.32}$$

The current densities for each block can be computing using a forward resistivity modelling scheme, so that the equation (2.32) reduces to

$$\delta\phi = \sum_{i=1}^{N} \delta\rho_i \,(J_x J_{x'} + J_y J_{y'} + J_z J_{z'})_i \, \Delta x\, \Delta y\, \Delta z. \tag{2.33}$$

The above relationship may further be simplified for 2-D and 1-D models. The current density in Y-coordinate direction of 2-D geologic model is negligible and the average current density in the strike direction is zero, in which case, equation (2.33) becomes

$$\delta\phi = \sum_{i=1}^{N} \delta\rho_i \,(J_x J_{x'} + J_z J_{z'})_i \, \Delta x\, \Delta z \int_{-\infty}^{\infty} dy \tag{2.34}$$

for 2-D geologic models, or

$$\delta\phi = \sum_{i=1}^{N} \delta\rho_i (J_z J_{z'})_i \Delta z \int_{-\infty}^{\infty} dx \int_{-\infty}^{\infty} dy + \sum_{i=1}^{N} \int_{-\infty}^{\infty}\int_{-\infty}^{\infty} \delta\rho_i \,(J_x J_{x'} + J_y J_{y'})_i \, dx\, dy \tag{2.35}$$

for 1-D geologic models.

2.6 Mathematical Concept for Application to Geoengineering Problems

Resistivity inversion methods have been implemented successfully for a variety of applications. However, the method has not been tested fully in various possible applications, such as for monitoring *in-situ* processes for improved oil recovery (IOR), environmental and geotechnical aspects of landfills and similar retainment structures. This may be because field surveys conducted until recently were done manually. Manual execution involves direct human activity to set up current and potential electrodes, electrode connections, and to take measurements of the induced potential field arising from current injection into the ground; this tends to make long-term investigations uneconomical or impractical. Another reason may be that field data are sometimes difficult to interpret in terms of a geologic model, owing to a lack of an appropriate interpretive tool (inversion model), poor resolution, poor quality data, or poor data coverage. The advent of the personal computer has led to dramatically increased efficiency in data collection. It is now possible to measure and interpret field data with a far better resolution and coverage than could be obtained with manual data collection, particularly if a fixed-electrode strategy is used. This in turn enhances the possibility of obtaining unambiguous geological interpretations of the field data because incomplete or varying locations for data sets over a time interval can be difficult to interpret. Mathematical tool discussed herein believes that the possible applications of direct-current resistivity methods are now limited mainly by our lack of imagination or opportunity, and it is likely that many more applications will be attempted in the future.

Whenever a sufficient resistivity change over a region or at a front is generated as a result of a dynamic process such as groundwater contamination or IOR processes, the induced electrical-field response to that process can be modeled with an appropriate mathematical tool, and an optimum monitoring strategy determined. This monitoring capability can be achieved with currently available technology at relatively low expense, and it may be highly complementary to other monitoring methods (e.g., seismic response, geochemistry changes, surface displacement data, and pressure-volume-temperature (PVT) data in the case of IOR projects).

It is important to note that electrical resistivity methods highly used for pore fluid delineations. Particularly if fluid electrical conductivity is changing but the saturation is not. Seismic properties are effectively insensitive to fluctuations in fluid ionic properties occurring as a result of IOR processes. However, pore pressure changes without resistivity changes clearly affect wave velocities and attenuation, but have less effect on electrical properties. Seismic response differences over time ("4-D" seismic) have been successfully used to map IOR processes. Pore-fluid type, salinity, temperature, and saturation, as well as lithology and to a lesser degree stress and pressure, all influence resistivity. There are rock parameters that affect either electrical response, seismic response, or both.

In this article, applications of direct-current resistivity methods for monitoring *in-situ* processes are investigated and emphasized based on a solid mathematical basis. Attempt is made herein to explain the mathematical concept that can be used for monitoring *in-situ* processes (e.g., processes associated with geotechnical problems, processes of geo-environmental problems and processes of IOR projects involving water flooding and steam flooding).

Referring to a mathematical equation (2.31), J'(x, y, z) is a current density in the region of resistivity change after applying reciprocity between the receiver and transmitter. **J**'(x, y, z) may also be viewed mathematically as a Green's function. The symbol τ in the above equation is the volume of an anomalous region where resistivity has been perturbed. Using equation (2.31), measurement sensitivity of the surface potential field to electrical resistivity changes can be expressed by the inner dot product of current densities in the anomalous zone to be monitored. This is mathematically a fundamental principle and concept that form the basis to introduce new techniques and strategies for resistivity measurements and tools for data interpretation. The physical insight derived from this analysis is that surface measurement sensitivity of the potential field is proportionally related to the amount of power dissipated (current density) around the zone of interest. This theoretical concept used in the derivation of equation (2.31) is commonly known as an adjoint solution or an adjoint state technique in the geophysical literature. This technique involves the transformation of the differential equation (in the case of resistivity, it describes the potential field due to a direct current point source) to yield a Green's function. The Green's function approach is quite common for implementing inversion of geophysical data. After deriving the theoretical formulation, an attempt has been made to interpret it physically. The physical insight derived from the relation has been used to guide sensitivity analyses as well as to introduce a new technique for resistivity measurements [3].

Current density is a function of several parameters such as depth of zone to be monitored, electrode spacing, electrode orientation, resistivity of the sounding medium, and the amount of current injected. To delineate and investigate a deeper anomalous zone, larger currents and sampling electrode separations are required. It is necessary to maximize current dissipated in and around the anomaly at depth to cause perturbations in the electrical-field large enough to detect reliably at the surface [3]. If only a surface configuration is used, it is difficult to have enough current flowing through the plume or anomalous zone of interest even at a moderate depth. However, if the current electrodes are placed at the right depth, a sufficient current flux through the anomalous region can be achieved to cause substantial variations in surface potential field measurements over time intervals.

Narayan [3] has showed that detectability of resistivity changes at depth could be maximized in surface measurements by increasing the amount of power dissipation in the zone of changing resistivity; this is achieved by placing the transmitting dipole adjacent to it or across it. This clearly is a basic aspect of practical recommendations for any subsurface-to-surface resistivity method. The transmitting dipole location, its spacing, and its proximity with respect to the zone control this power dissipation, therefore a parametric study of these factors has been conducted numerically, and these results provide a basis for a new method of resistivity measurements (Figure 2.2). Based on the theory and method of resistivity measurements discussed herein, a hypothetical case example of monitoring fluid progression in IOR projects is discussed in Narayan and Dusseault [28].

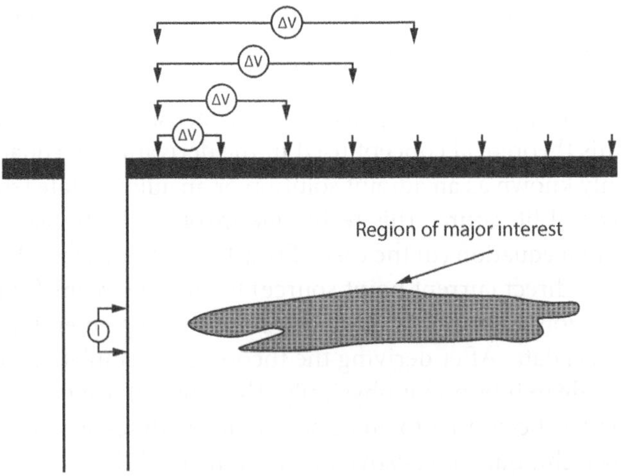

Figure 2.2 Schematic diagram of the new method for sampling and measuring potential field at the surface. Current dipole is fixed at depth closer to the anomalous zone of interest.

Before continuing further, another advantage of a fixed-electrode strategy must be noted. Methods such as "4-D" seismic are inherently expensive because the excitation source typically involves large surface equipment (vibroseis units). Therefore, surveys tend to be at long time intervals, months or years. In resistivity monitoring, excitation is easily achieved using an economical power source, therefore repeated surveys can be taken within a day or a week, examined to see if sufficient changes have occurred, and if so, analyzed in greater detail. Furthermore, we note that arrays of electrodes at the surface are undoubtedly cheaper than geophones or accelerometer arrays. If appropriate well designs are used, such as fiber-glass casings, electrode arrays can be directly combined with seismic receivers, as well as with temperature and pressure monitoring sensors. These advantages and their inherent flexibility imply that resistivity methods can be extremely economic to operate once the array is in place.

2.7 Mathematical Quantification of Resistivity Resolution and Detection

A good description of resolution, or resolvability, may be explained in terms of a minimum separation of two geological features or anomalies to be delineated [28]. Detectability is usually defined in terms of the size of the anomalous zone or electrical contrast of the anomalous zone that can produce a measurable potential field response [28]. On-site geological noise level must be considered in defining detectability. To delineate and quantify changes within the anomalous zone of interest, changes in geophysical parameters (e.g., electrical conductivity or resistivity) must occur over a distance. A detailed description of "resolvability" and "detectability" may be found in the geophysical literature. Greaves *et al.* [29] have discussed about the geometry and spatial distribution of current dipoles and potential dipoles giving a better resolution and detectability of the anomalous zone to be monitored.

Resolution may be further improved if the transmitting dipole or receiving dipole is placed into a borehole. Furthermore, it reduces the range of possible solutions and makes the mathematical analysis and interpretation more tractable and concrete. As indicated above in Figure 2.2, Δ-monitoring by inverting the differenced vector of potential field measurements may further enhance detectability if there are no systematic and non-random sources of noise. Narayan [3] has given a through description of theoretical sensitivity analysis and study for the resolvability and detectability of resistivity anomalies. Accordingly, a large amount of the current must flow through the anomalous region of interest. In this way, potential field measurements at the surface are detected effectively. It is important to focus on maximizing current flow through the anomalous zone. Current flow depends on several factors (e.g., spacing of current dipole, location of current dipole, dimension of anomalous region, electrical resistivity contrasts, etc.). It may be evaluated mathematically via theoretical model simulations and using other available data.

Figure 2.3 illustrates that the changes in potential field response as a function of electrical resistivity contrast [3, 28]. This also verifies equation 2.31 numerically. Also, the Figure shows that electrical potential field response varies exponentially with increasing resistivity contrast. A linear relationship has been found for small resistivity changes. The following important observations from this numerically computed result are summarized below:

- The magnitude of the electrical potential field response increases linearly for small resistivity contrasts of the anomalous zone. From the mathematical formulation for sensitivity analysis, this result is expected here.

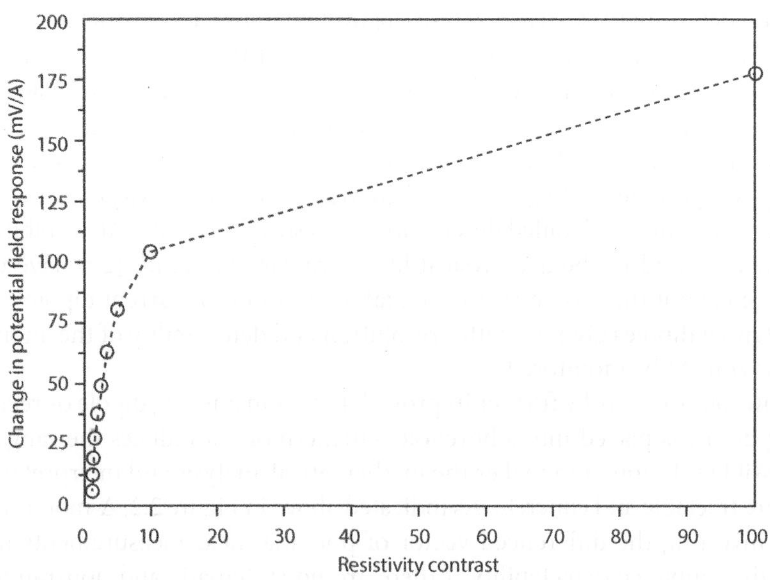

Figure 2.3 Computation of changes in the potential field response with increasing resistivity contrast.

- An important observation here is that a linear relationship exists up to a change in electrical resistivity by a factor of four. This may be found extremely helpful in implementing a linear resistivity inversion in imaging *in-situ* processes using a difference of observed potential field data from time A to time B.
- A linearized inversion will be valid only in situations where the resistivity perturbation in the target region is not more than a factor of four during two consecutive resistivity measurements.
- This linearity for small electrical resistivity changes may also be found useful as a basis to introduce the concept of adaptive resistivity inversion.
- Adaptive resistivity inversion should permit interpretation of difference potential field responses in terms of changes in resistivity at depth.

Practically, it is important and advised to maximize current dissipation in and around the region that is to be monitored so that electrical field perturbations are large enough to detect at the surface. It is usually difficult to achieve enough current flowing through an anomalous region if only surface electrodes are used. However, by placing current dipole electrodes at the right depth with an appropriate current dipole spacing, sufficient current flux through the target zone can be achieved that in turn causes substantial variations in surface potential field measurements over the monitoring periods. In addition to increasing current flux, detectability may be enhanced in surface measurements by placing the current dipole adjacent to or across the anomalous zone that is to be monitored.

2.8 Scheme of Resistivity Data Presentation

When monitoring *in-situ* processes, a number of essential questions arise. What is the minimum signal amplitude that can be detected? How should we quantify detectability? What is the minimum signal repeatability? These are vital questions from the monitoring point of view and must be dealt with quantitatively. We attempt to answer some of these questions using numerical model responses and by defining measurability and detectability consistently.

The issue of detectability of a signal associated with an anomalous zone is related to the depth of exploration. For a given target, this is usually defined as a maximum depth that may be detected with a given electrode configuration. Ward [30] noted that the depth of exploration is a function of several parameters such as sensitivity to inhomogeneity & bedrock topography, lateral effects, general topography, dip, etc., and signal-to-noise ratio. As these issues were not studied systematically for a wide range of electrical resistivity models, he did not define different electrode configurations in terms of detectability. Oldenburg [31] also noted that the without studying a wide range of models, there is no analytical basis for considering one electrode configuration over another for resolving model parameters. However, the approach herein is somewhat different, using a mathematical basis to evaluate placement of a current dipole at depth in the proximity of an anomalous zone to be monitored, which increases the amount of current flux in the zone, leading to a better detectability. Other issues affecting detectability must be studied using a forward model on a case-by-case basis to optimize the array characteristics.

Narayan [3] defined the term "measurability" in terms of percent difference of the measured signal with respect to the background. This gives us an idea as to the signal levels that must be achieved on top of the background signal in order for an anomaly to be measured. Available commercial equipment, for example, is supposedly capable of measuring signal variations with an accuracy of 0.1% of background. The measurability is therefore 0.1% and is equipment-dependent. As an indicator of target detectability, this measure is misleading on two counts. First, it is unlikely that the measurability applies over the entire range of the instrument's response. For example, an instrument may measure ±1 mV on top of a 10 V background signal but cannot measure ±1 μV on top of 10 mV background. Narayan [3] quantified the term "detectability" on the basis of magnitude of the signal observed with respect to the background and measurability of the signal together. Second, cultural and to some extent instrumental noise is usually specified in terms of an absolute voltage (e.g., ±10 mV). Therefore "detectability" has both an absolute and a percentage expression, either of which may be required in a given situation. As a result, we have presented detectability in both forms, percentage and absolute. If we know the geologic noise characteristics and magnitude, we may be able to put a threshold limit, defining the minimum signal to be detected with respect to the anomaly over the zone to be monitored. This concept has been used to develop a design chart of detectability for a hypothetical landfill model [3].

2.9 Design Strategy for Monitoring Processes of IOR Projects, Geo-Engineering, and Geo-Environmental Problems

The IOR processes and integrity of Civil Engineering landfills, mine tailings retention ponds, brine ponds, saline water invasion cases, and other impoundment and waste storage structures must be monitored actively. Integrity cannot be predicted mathematically without uncertainty, nor can integrity be guaranteed unequivocally through the use of existing technology. In addition to these structures, there are a number of other situations where the migration of materials in the subsurface may be of interest. For example, the development of saline plumes underneath large waste embankments, such as in the potash mining industry in Saskatchewan; the migration of thermal plumes associated with hot structures in the ground; the migration of salt contaminated groundwater underneath Ministry of Transportation stockpiles; and so on. In all of these cases, providing that there ls a sufficient electrical resistivity contrast generated in a sufficiently large volume in the ground, the progression of the resistivity change through the porous or fractured medium can in principle be monitored using one of a variety of electrical techniques. One of these techniques, and the one with the most chance of successful and accurate resolution of the location and migration of materials of different resistivity in the pores, is the method of electrical impedance tomography using a fixed-electrode strategy and low frequency current excitation.

Among the complications associated with electrical techniques are the shallow annual variations because of saturation and hydraulic head changes due to rainfall, changes in the ground temperature as a result of seasonal effect, etc. Even if the physical issues of seasonal and other changes have been successfully addressed, and providing that there is a sufficient resistivity contrast, a major issue remains the successful mathematical modeling of the data to give quantitative information on the maximizing the sensitivity of an array to changes requires successful Installation of an appropriate electrode array and precise mathematical modeling.

The electrode array, which is proposed here for the three-dimensional Impedance tomography is a three-dimensional system of conducting points that can be treated as point sources in the ground. These points are Installed in various locations to maximize their usefulness. All the electrodes are linked together to a central switching system, where changes in resistivity between electrode pairs can be measured with time, or more

Interestingly, changes in induced potential field gradient can be measured (Figure 2.4). In this method, an electrical field is induced by passing a strong current through a selected pair of electrodes (a dipole source), and measuring the induced voltage between all possible pairs of monitoring electrodes (sampling dipoles).

The mathematical model required to analyze this voltage data is one where resistivity changes or the potential gradient field changes can be analyzed to give the spatial location of plumes or bodies in the soil substructure that have different electrical properties than surrounding bodies, and are changing with time. The theory required to analyze a field of electrical potential gradients (voltage gradients) is now well developed and discussed herein in [3].

Figure 2.5 illustrates another similar strategy for monitoring EOR projects that requires hot water injection and stream injection or CO2 injection. Narayan [3] has demonstrated that the best approach is to take data at time one (T1) to create an initial base case; then, data at time 2 (T2) are taken and subtracted from those at T1 to give a difference in resistivities or potential gradients. This differenced data set is then analyzed to determine the changes that must have occurred between the two times to give rise to the changes in the observed data. This Is often referred to as the Inversion process in geophysics. Once the problem has been solved, the changes are added to the status at T1 to give the new status of the body that has a different electrical signature. The mathematical procedure requites considerable computer analysis. How it can be done, this is essentially known at this moment. Although there are refinements that can be made in the electrode

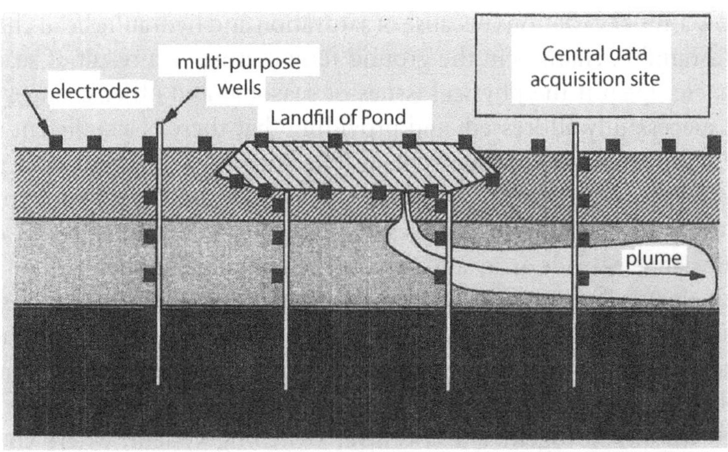

Figure 2.4 Electrode array for landfill monitoring.

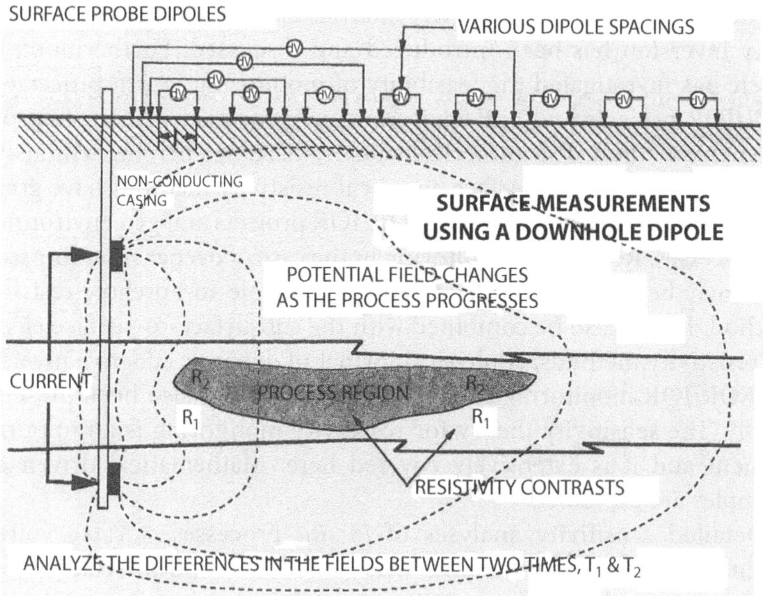

Figure 2.5 Electrode array for monitoring EOR/IOR processes using subsurface current dipole.

array design and in the mathematical formulation of the problem. In order to give more precise and more rapid answers, the major elements are well understood and are in a sufficiently mature state.

The mathematical problem is solved using what is called a forward optimization scheme. The differences between predicted values (differences) and observed values are minimized according to some rule. One of the most common techniques Is the minimization of the sum of the squares of the deviations, known as the least squares technique (L2-norm). This is by no means the only technique possible, there are techniques that minimize the absolute difference of the measurements, or other suitable functions. As well, there are a number of ways to accelerate the computational efficiency of such error minimization algorithms. The electrical monitoring technique should probably be used in all cases where there is a significant chance of a plume of different electrical resistivity being generated by a process.

2.10 Final Remarks and Conclusions

Attempts are made herein to explain ill-posed inverse problems and resistivity inversion in the field of electrical geophysics. A detailed mathematical basis for sensitivity analysis and multi-dimensional resistivity inversion has been introduced and discussed. Furthermore, this article has investigated the feasibility of monitoring *in-situ* processes of EOR/IOR projects and geo-environmental processes with an improved resistivity method. The numerical studies, discussed herein, demonstrate that the subsurface-to-surface electrical resistivity methods have greater promise for monitoring shallow EOR/IOR projects and geo environmental processes. Resistivity monitoring in the case of deeper reservoir structures may be further improved using a borehole-to-borehole resistivity method. It may also be combined with the subsurface-to-surface electrical resistivity methods. A physical format of dynamic adaptive inversion for EOR/IOR monitoring and *in-situ* processes has also been presented briefly. The sensitivity theory for resistivity monitoring is an important element and it is extensively covered here. Mathematical driven case examples are explained.

Detailed sensitivity analyses of *in-situ* processes, surface voltage/potential field measurements with fixed electrodes strategy, new approaches to electrical resistivity anomaly enhancement, and mathematically driven theoretical approach for the interpretation of time differenced field resistivity data or potential field data in terms of resistivity changes indicate that the dynamic adaptive inversion method is feasible. Several numerical modelling studies have clearly demonstrated that the electrical resistivity method is well suited for monitoring *in-situ* processes provided current dipole/excitation source is optimally installed with respect to the zone to be monitored. Sampling potential field with different electrode spacing is an economical and a better choice for EOR/IOR and geo-environmental *in-situ* processes [32]. Additionally, several subsurface-sampling electrodes may further help in accurately delineating propagation of fluid front and identification of swept anomalous zones.

The physics of steam-front movement in IOR projects as a result of an injection process is different and complex, compared to the movement of a hot-water front. On the other hand, hot-water fronts may show large-scale fingering complexity. IOR projects involving steam injection create higher resistivity contrast, compared to IOR projects involving hot-water injection. The numerical studies [32] for the reservoir model involving a simple scenario of fluid-front propagation suggest that the monitoring of hot-water and steam fronts based on the measurement of differenced response is easier to understand only for a simpler resistivity distribution.

References

1. TIKHONOV, A., and ARSENIN, V., *Solution of Ill-Posed Problems*; V.H. Winston and Sons, Washington, 1978.
2. KABANIKHIN, S.I., Definitions and Examples of Inverse and ill-Posed Problems; *J. Inverse Ill-Posed* Problems 16, 317–357 DOI 10.1515 / JIIP.2008.069, 2008.
3. NARAYAN, S., Subsurface-to-surface Resistivity Method for Monitoring *In Situ* Processes Using a Combination of Surface and Subsurface Electrodes; PhD Thesis, Department of Earth Sciences, University of Waterloo, Waterloo, Canada, 1994.
4. BARTEL, C., and SEAVEY, W., Instrumentation for *In-situ* Coal Gasification; III Electrical Techniques for Remote Monitoring; *In Situ*, Vol. 2, pp. 247-269, 1978.
5. WILT, M.J., and TSANG, C.F., Monitoring of Surface Contaminants with Borehole/surface Resistivity Measurements; *Proc. Surface and Borehole Geophysical Methods in Ground Water Investigation, Fort Worth, TX, February 12 - 14, 1985.*
6. BEVC, D., and MORRISON, H.F., Borehole-to-Surface Electrical Resistivity Monitoring of a Salt Water Injection Experiment; *Geophysics*, Vol. 56, pp. 769-777, 1991.
7. NARAYAN, S., Vertical Resolution for Two-dimensional Dipole-Dipole Resistivity Inversion; *Proceedings of 62 International Annual Meeting of Society of Exploration Geophysicists, New Orleans, USA*, EM2.3, pp. 431-434, 1992.
8. DAILY, W., RAMIREZ, A., LABRECQUE, D., and NITAO, J., Electrical Resistivity Tomography of Vadose Water Movement; *Water Resource Research*, Vol. 28, pp. 1429-1442, 1992.
9. NARAYAN, S., DUSSEAULT, M.B., and NOBES, D.C., Inversion Techniques Applied to Resistivity Inverse Problems; *Inverse Problems* 10 (1994) 669-686. Printed in the UK.
10. MADSEN, L.M., FIANDACA, G., and AUKEN, E., 3-D Time-domain Spectral Inversion of Resistivity and Full-decay Induced Polarization Data—Full Solution of Poisson's Equation and Modelling of the Current Waveform; *Geophysical Journal International*, Volume 223, Issue 3, Pages 2101–2116, 2020. https://doi.org/10.1093/gji/ggaa443.
11. PANG, Y., NIE, L., LIU, B., LIU, Z., WANG N., Multiscale Resistivity Inversion Based on Convolutional Wavelet Transform; *Geophysical Journal International*, Volume 223, Issue 1, Pages 132–143, 2020. https://doi.org/10.1093/gji/ggaa302.
12. DEVI, A., ISRAIL, M., SINGH, A., GUPTA, P.K., YOGESHWAR, P., TEZKAN, B., Imaging of Groundwater Contamination Using 3D Joint Inversion of Electrical Resistivity Tomography and Radio Magnetotelluric

Data: A case Study from Northern India; *Near Surface Geophysics*, Volume 18, Issue 3, Pages 261-274, 2020. https://doi.org/10.1002/nsg.12092.

13. DAHLIN T., The Development of DC Resistivity Imaging Techniques; *Computers & Geosciences* Volume 27, 1019–1029, 2001.
14. AL-SAADI, O. S., SCHMIDT, V., BECKEN, M. and FRITSCH, T., Very-high-resolution Electrical Resistivity Imaging of Buried Foundations of a Roman Villa Near Nonnweiler, Germany; *Archaeological Prospection*, Volume 25, Issue 3, 2018. https://onlinelibrary.wiley.com/doi/abs/10.1002/arp.1703.
15. AHMAD, S.U., Analytical and Iterative Regularization Methods for Nonlinear Ill-Posed Inverse Problems: Applications to Diffuse Optical and Electrical Impedance Tomography; In Partial Fulfillment of the Requirements for the Degree Doctor of Philosophy Mathematics, Clemson University, South Carolina, USA, 2019.
16. CARRIER A., FISCHANGER F., GANCE J., COCCHIARARO G., MORELLI G., and LUPI M., Deep Electrical Resistivity Tomography for the Prospection of Low- to- Medium-Enthalpy Geothermal Resources; *Geophysical Journal International,* 219(3):2056-2072, 2019.
17. SCHMIDT-HATTENBERGER, C., BERGMANN, P., BÖSING, D., LABITZKE, T., MÖLLER, M., SCHRÖDER, S., WAGNER, F., SCHÜTT, H., Electrical Resistivity Tomography (ERT) for Monitoring of CO2 Migration - from Tool Development to Reservoir Surveillance at the Ketzin pilot Site, *Energy Procedia* 37, 4268 – 4275, 2013.
18. LIU B., GUO Q., LI S., LIU B., REN Y, PANG, Y., GUO XU, LIU L., and JIANG P., Deep Learning Inversion of Electrical Resistivity Data; *IEEE Transactions on Geoscience and Remote Sensing*, 2020.
19. NAVEEN KUMAR T., RAMA RAO P., and NAGANJANEYULU K., Electrical Resistivity Imaging (ERI) Using Multi-electrodes for Studying Subsurface Formations in Cauvery Plains; *Advances in Applied Science Research*, 6(5): 47-53, 2015.
20. MAJA BRIŠKI M., STROJ A., KOSOVI ́I., and BOROVI S., Characterization of Aquifers in Metamorphic Rocks by Combined Use of Electrical Resistivity Tomography and Monitoring of Spring Hydrodynamics; *Geosciences*, 10, 137, 2020. doi:10.3390/geosciences10040137.
21. HOHMANN, G.W., and RAICHE, A.P., Inversion of Controlled-Source Electromagnetic Data, in Nabighian, M.N., Ed., *Electromagnetic Methods in Geophysics*, 1: Society of Exploration Geophysicist, Investigation in Geophysics, no. 3, p. 469-503, 1988.
22. TRIPP, A.C., HOHMANN, G.W., and SWIFT, C.M., Two dimensional resistivity inversion; *Geophysics*, v. 49, p. 708 1717, 1984.
23. CONSTABLE, S.C., PARKER, R.L., and CONSTABLE, C.G., Occam's Inversion: A Practical Algorithm for Generating Smooth Models from Electromagnetic Sounding Data; *Geophysics*, v. 52, p. 289-300, 1987.

24. ORISTAGLIO, M.L., AND WORTHINGTON, M.H., Inversion of Surface and Borehole Electromagnetic Data for Two-Dimensional Electrical Conductivity Models; *Geophysical Prospecting*, v. 28, p. 633-657, 1980.
25. EATON, P.A., Three-Dimensional Electromagnetic Inversion: Ph.D. thesis, University of Utah, 1987.
26. TELFORD, W. M., GELDART, L. P., and SHERIFF, R.E., *Applied Geophysics*, Cambridge University Press, 1976, Revised Edition, 1990.
27. LANCZOS, C., 1961, Linear Differential Operators: Van Nostrand Company, Ltd., 1961.
28. NARAYAN, S., and DUSSEAULT, M.B., Sensitivity Studies of Resistivity Monitoring for Shallow Enhanced Recovery Processes - A Numerical Case History; *Journal of Canadian Petroleum Technology*, February 2000, Volume 39, No. 2, 2000.
29. GREAVES, R.J., BEYDOUN, W.B., and SPIES, B.R., New Dimensions in Geophysics for Reservoir Monitoring; SPE Formation Evaluation, pp. 141-150, June 1991.
30. WARD, S.H., Resistivity and induced polarization methods; *Geotechnical and Environmental Geophysics* - I, S.H. Ward (ed.), Society of Exploration Geophysicists, Tulsa, Oklahoma, 97-99, 1990.
31. OLDENBURG, D.W., The Interpretation of Direct-Current Resistivity Measurements; *Geophysics* 43, No. 3, 610-25, 1978.
32. NARAYAN, S., and DUSSEAULT, M.B., Subsurface-to-surface Resistivity Methods for Monitoring Steam floods—A Theoretical Study; *Proceedings of 8th International Conf. of the Association for Computer Methods and Advances in Geomechanics, Morgantown, WV, USA*, Vol. 2, pp. 2113-2116, May.

3

Shadowed Set and Decision-Theoretic Three-Way Approximation of Fuzzy Sets

M. A. Ibrahim*, T. O. William-West and D. Singh

Department of Mathematics, Faculty of Physical Sciences, Ahmadu Bello University, Zaria, Nigeria

Abstract

Shadowed set and decision-theoretic approximations are three-way decision-making models that rely on a pair (α, β) of determined thresholds from the interval [0,1] to categorize a fuzzy set F into three regions referred to as *positive, boundary* and *negative*. These models approximate the membership grades of elements in F to a value in $\{0, m, 1\}, m \in (0, 1)$ to describe F with the three regions. In this chapter, theoretical formulations of shadowed sets approximations (SSA) which hinge on ideas of uncertainty balance, average uncertainty and minimum approximation error are presented. Also, decision-theoretic three-way approximation (DTA) models which anchor on principles of minimum distance and least cost are revisited. Subsequently, we give a modified generalized model of decision-theoretic three-way approximation, called $\{n, m, p\}$ system, which does not impose values for n and p as against the trend in literature where n and p are chosen to be 0 and 1 respectively. A suitable formula for computing viable threshold (α, β) from cost-sensitive and minimum distance-based models is derived.

***Keywords*:** Fuzzy set, shadowed set, three-way decision, three-way approximation, decision-theoretic approximation, decision cost, uncertainty balance

3.1 Introduction

Decision-making is essential in human daily life. People often use the idea of thinking in threes: *accept*, *reject* and *non-commitment*, in the

**Corresponding author*: Adekubash1@gmail.com

Ramakant Bhardwaj, Jyoti Mishra, Satyendra Narayan and Gopalakrishnan Suseendran (eds.) Mathematics in Computational Science and Engineering, (55–90) © 2022 Scrivener Publishing LLC

process of making decisions with uncertain phenomenon where an alternative may be *good enough* to be accepted, *bad enough* to be rejected or cannot be expressly determined leading to a non-commitment decision. One of the main problems of decision-making lies in how to determine precise criteria for acceptance, rejection and non-commitment decisions. In order to address this problem, a theory of three-way decision has been proposed by observing the phenomenon of *three-way thinking* prevalent in medicine [16, 24], psychology [12], data mining [16], cognitive science [23], etc.

Given a fuzzy set, F, drawn from a finite nonempty set X, an element x in F is described with the aid of a membership function $\mu F{:}X \rightarrow [0,1]$, which depicts the degree to which it belongs to F by a quantity $\mu_F(x) \in [0,1]$. A three-way decision on $x \in F$ exploits a set of decisive criteria, $\{\alpha, \beta\}$, $0 < \alpha \leq \beta < 1$, to make three types of classification: positive, negative or boundary instance. An element x with membership grade $\mu_F(x) > \beta$, $\mu_F(x) < \alpha$ and $\alpha \leq \mu_F(x) \leq \beta$ is accepted as a positive instance of F, rejected as negative instance of F and classified into the boundary region respectively. Along this line, the original membership grade, $\mu_F(x)$, of the element x is approximated to 1, 0 or 0.5 respectively.

A key aspect of three-way approximation of fuzzy set (TWA) is the determination of optimum value of α and β. The optimality of the partition criteria, (α, β), depends on the decision objective. Usually, in three-way approximation of F, a number of decision objectives: uncertainty balance [8, 19, 20], minimum approximation error [4, 27], minimum distance [25], least cost [4, 25, 27], average uncertainty [7] and mean entropy [5], can be pursued to determine (α, β).

In cluster analysis, patterns *typical, atypical* and patterns with *borderline* can be interpreted as positive, negative and boundary region as in TWA. For details, see [13, 14, 17, 18, 29, etc.]. The regions can also be interpreted in email spam filtering as *legitimate* email, *spam* email and *further-examinable* email [32]. Also, in government decision analysis as *execute* policy, *don't-execute policy* and *defer decision* [33]. Other interpretations and applications of TWA can be found in speech data and medical image segmentation [1], satellite image segmentation [14], recommender systems [7], etc.

Generally, TWA is studied via two main models; shadowed sets [19] and decision-theoretic approximations [4]. The distinction between the models is that while shadowed sets depend on search algorithm to obtain the threshold (α, β), decision-theoretic depends on a *camera-ready* formula of α and β in terms of a subjectively defined decision costs. A common attribute in the models is the use of a mapping on a universe X into a three-element set {0, 0.5, 1} for the approximation of a given fuzzy set.

Yao *et al.* [25] suggested a general $\{n,m,p\}$, for three-way approximation mapping where $n \neq 0$, $m \neq 0.5$ and $p \neq 1$ to handle both normal and sub-normal fuzzy sets.

To address the aforesaid problem on minimization of approximation error, which may also yield reduction of total approximation cost, we present ways of determining appropriate values of n, m and p. This chapter is motivated by the lack of rare comparative studies between the aforesaid three-way decision models and the need to determine appropriate values of n, m and p in three-way approximations.

The aim of this chapter is to carry out investigatory studies on shadowed set and decision-theoretic approximations. The contribution of this chapter is as follows:

1) provide methodological exposition of the framework of shadowed sets,
2) underline the comparative performance of shadowed sets and decision-theoretic three-way approximations,
3) introduce formal foundation of $\{n, m, p\}$ approximation system,
4) provide a generalized formula for calculating (α, β) based on cost sensitive and minimum distance based principles.

3.2 Preliminaries on Three-Way Approximation of Fuzzy Sets

3.2.1 Shadowed Set Approximation

A set $F = \{(x, \mu_F(x)) : x \in X\}$ drawn from a nonempty finite universe X is called a fuzzy set [26] if it allows an element x admission in F as a member as a matter of degree $\mu_F(x)$ in the closed unit interval [0,1].

Given a fuzzy set F drawn from X, a shadowed set S is characterized by a three-valued mapping $S_\alpha : X \to \{0, [0, 1], 1\}$ defined by

$$S_\alpha(\mu_F(x)) = \begin{cases} 1, & \mu_F(x) > 1-\alpha \\ [0,1], & \alpha \le \mu_F(x) \le 1-\alpha \\ 0, & \mu_F(x) < \alpha \end{cases}$$

[19]

An element x is put in the reduced area, $Red(S) = \{x \in F: \mu_F(x) < \alpha\}$, if its membership grade in F is less than α. An element is put in the shadow area, $Shd(S) = \{x \in F: \alpha \leq \mu_F(x) \leq 1 - \alpha\}$, if its membership grade is between the symmetric thresholds α and $1-\alpha$. Subsequently, elements are put in the elevated area, (S) $Cor(S) = \{x \in F: \mu_F(x) > 1 - \alpha\}$, have membership grades greater than $1- \alpha$. Based on the mapping $S_\alpha(\mu_F(x))$, an element in a shadowed set can only assume one of three values 0, [0, 1] or 1 according to the region in which it is placed. Elements in *Cor*(*S*), *Shd*(*S*) and *Red*(*S*) are assigned membership grades 1, [0, 1] and 0, respectively.

When working with shadowed sets, it is computationally efficient to select an appropriate value in [0, 1] as the membership grade of elements in *Shd*(*S*). Cattaneo and Ciucci [2] suggested the intermediate value, 0.5 which captures the maximum value of uncertainty in the unit interval, as the membership grade of elements in *Shd(S)*.

Induction of a shadowed set from a fuzzy set requires determination of the threshold $\alpha \in (0,0.5]$ and can be pursued from many principles, some of which are discussed in this chapter.

3.2.2 Decision-Theoretic Three-Way Approximation

Let F be a fuzzy set drawn from a finite universe X and $K = \{\alpha, \beta\}, 0 \leq \alpha \leq \beta \leq 1$, be a set of conditions or criteria. A three-way decision aim at classifying the elements in F into positive, $POS(F) = \{x \in X: \mu_F(x) > \beta\}$, negative, $NEG(F) = \{x \in X: \mu_F(x) < \alpha\}$ and boundary, $BND(F) = \{x \in X: \alpha \leq \mu_F(x) \leq \beta\}$, at the instance of F based on the elements fulfilment of K. Deng and Yao [4] considered the unit decision costs λt of classification and the approximation error incurred when an element's membership grade is changed from $\mu_F(x)$ into $t \in \{0, 0.5, 1\}$ and proposed three decision rules (please see [22] for details on three-way decision) for minimizing the total cost, C, arising from transforming F into a three-region set $T = POS(F) \cup BND(F) \cup NEG(F)$.

The decision to classify an element into *POS*(*F*) is governed by the rule

i) If $\lambda_1(1 - \mu_F(x)) \leq \lambda_{0.5\downarrow}(\mu_F(x) - 0.5)$ and $\lambda_1(1-\mu_F(x)) \leq \lambda_0(\mu_F(x)-0)$, then the membership grade of x, $T(x)$, in T equals 1. That is, $T(x) = 1$,
ii) If $\lambda_1(1 - \mu_F(x)) \leq \lambda_{0.5\uparrow}(0.5 - \mu_F(x))$ and $\lambda 1(1-\mu_F(x)) \leq \lambda 0(\mu_F(x)-0)$, then the membership grade of x, $T(x)$, in T equals 1. That is, $T(x) = 1$.

This decision comes with the cost $C(POS(F)) = \lambda_1(1-\mu_F(x))$.

Here λ_1, λ_0 and $\lambda_{0.5\uparrow}$ denote the cost of classifying an element x into $POS(F)$, $NEG(F)$ and $BND(F)$, respectively. The value $0.5\uparrow$ indicates that $\mu_F(x) < 0.5$ is elevated to 0.5. Further, $\lambda_{0.5\uparrow}$ is the cost of classifying x into $BND(F)$ when $\mu_F(x) > 0.5$ is reduced to 0.5. Also, as the elements in $POS(F)$, $NEG(F)$ and $BND(F)$ are assigned new membership grades of 1,0.5 and 0, respectively, the quantities $(1 - \mu_F(x))$,$(0.5 - \mu_F(x))$ and $(\mu_F(x) - 0)$ are the errors incurred when an element x is put in the positive, boundary and negative regions, respectively.

The decision to classify an element into $BND(F)$ is governed by the rule

iii) If $\lambda_{0.5\uparrow}(0.5 - \mu_F(x)) \leq \lambda_1(1 - \mu_F(x))$ and $\lambda_{0.5\uparrow}(0.5 - \mu_F(x)) \leq \lambda_0(\mu_F(x) - 0)$, then the membership grade of x, $T(x)$, in T equals 0.5. That is, $T(x) = 0.5$,

iv) If $\lambda_{0.5\downarrow}(\mu_F(x) - 0.5) \leq \lambda_1(1 - \mu_F(x))$ and $\lambda_{0.5\downarrow}(\mu_F(x) - 0.5) \leq \lambda_0(\mu_F(x) - 0)$, then the membership grade of x, $T(x)$, in T equals 0.5. That is, $T(x) = 0.5$, and comes with the cost $C(BND(F)) = \lambda_{0.5\downarrow}(\mu_F(x) - 0.5) + \lambda_{0.5\downarrow}(\mu_F(x) - 0.5)$

The decision to classify an element into $NEG(F)$ is governed by the rule

v) If $\lambda_0(\mu_F(x) - 0) \leq \lambda_{0.5\uparrow}(0.5 - \mu_F(x))$ and $\lambda_0(\mu_F(x)-0) \leq \lambda_1(1-\mu_F(x))$, then the membership grade of x, $T(x)$, in T equals 0. That is, $T(x)=0$.

vi) If $\lambda_0(\mu_F(x) - 0) \leq \lambda_{0.5\downarrow}(\mu_F(x) - 0.5)$ and $\lambda_0(\mu_F(x) - 0) \leq \lambda_1(1 - \mu_F(x))$ $\lambda_0(\mu_F(x) - 0) \leq \lambda 1(1-\mu_F(x))$, then the membership grade of x, $T(x)$, in T equals 0. That is, $T(x) = 0$.

The cost of taking this decision is given as $C(NEG(F)) = \lambda_0(\mu_F(x) - 0)$.

The total cost associated with the tri-partition T is the sum of the cost arising from each region:

$$C = C(NEG(F)) + C(BND(F)) + C(POS(F)) \tag{3.1}$$

The goal of decision-theoretic three-way approximation is to determine the optimum partition thresholds α and β which minimize C in Equation (3.1).

3.3 Theoretical Foundations of Shadowed Sets

In this section, induction of shadowed sets from different optimization-based principles is discussed. Two notions occupy a central position when constructing shadowed sets: *uncertainty balance* and *minimization of approximation error.*

a. **Uncertainty balance**
   When constructing shadowed sets, two types of actions are taken, viz., elevation and reduction actions. The elevation actions involve increase membership grades $\mu_F(x) \leq 0.5$ to 0.5 and also $0.5 < \mu_F(x \leq 1$ to 1. The elements with these membership grades are placed in *Shd*(*F*) and *Cor*(*F*), respectively. The reduction action decreases membership grades $\mu_F(x)$ within the range $0 \leq \mu_F(x) < 0.5$ to 0 and also those within the range $0.5 \leq \mu_F(x) < 1$ to 0.5. The elements with these membership grades are put in *Red*(*F*) and *Shd*(*F*), respectively. Due to the membership grades assigned to elements in *Red*(*F*) and *Cor*(*F*), these regions have no fuzziness. However, the fuzziness associated with *F* is relocated into *Shd*(*F*). In practical situations, it is difficult to account for the total amount of fuzziness of *F* in *Shd*(*F*). Therefore, an equivalent amount of the fuzziness in *F* is expected to be retained in *Shd*(*F*).

Remark 3. 1
A principle of uncertainty balance requires that when transforming a fuzzy set *F* into a shadowed set *S* induced by the pair of thresholds $(\alpha,1-\alpha)$, the total amount of fuzziness associated with *F* should be reasonably retained in *S*. This requires the determination of optimal pair of thresholds that guarantee an equivalent amount of fuzziness in *F* in the resultant shadowed set *S*.

b. **Error minimization:** When a shadowed set *S* is induced from a fuzzy set *F*, the original membership grades of elements, generically expressed as $x \in F$, are changed from $\mu_F(x)$ to a new membership grade $t \in \{0, 0.5, 1\}$. A principle of minimum error requires that *t* should be the closest number to $\mu_F(x)$. As any $x \in F$ can only be put into one of the three regions *Red*(*F*), *Shd*(*F*) or *Cor*(*F*) induced by $(\alpha,1-\alpha)$, the principle of minimum error searches for a pair of thresholds which minimize the total error incurred in the three regions of *S*.

3.3.1 Uncertainty Balance Models

Uncertainty balance plays a role of information preservation in the field of data mining and granular computing. Its usefulness in guaranteeing justifiable information granules has drawn the interest of researchers in three-way decision to apply this notion as a guiding principle for the induction of shadowed sets. The existing uncertainty balance models that captured our attentions include *Pedrycz's model, Tahayori-Sadeghian-Pedrycz model* and *Ibrahim-William-West-Kana-Singh (IWKS) model.*

3.3.1.1 Pedrycz's (Pd) Model

The origin of shadowed set could be attributed to the work in [19]that provides a practical approach for determining an optimal pair of partition thresholds (α,1−α) for making three-way decision with fuzzy sets. Given a shadowed set *S*, the elements in *Cor*(*F*) are *accepted* as objects which describe the concept of concern. The elements in *Red*(*F*) are *rejected* as describing the concept of concern. Subsequently, the elements in *Shd*(*F*) are regarded as doubtful. Hence, a non-commitment decision is taken on them. The central objective of shadowed set is to determine a suitable α which address the problem of uncertainty balance stated in the following way:

Uncertainty eliminated from *Red*(*F*) and *Cor*(*F*) should be accounted for in *Shd*(*F*). For discrete fuzzy sets, Pedrycz [19] captured this idea with the following equation:

$$V_1(\alpha)=\left|\sum_{\mu_F(x_i)<\alpha}\mu_F(x_i)+\sum_{\mu_F(x_i)>1-\alpha}(1-\mu_F(x_i))-Card\{x_i:\alpha\leq\mu_F(x_i)\leq 1-\alpha\}\right|\ldots \tag{3.2}$$

The value V_1 in Equation (3.2) is Pedrycz's mathematical model of uncertainty balance. The threshold $\alpha \in (0, 0.5]$ which minimizes V_1 is taken as the optimum α.

3.3.1.2 Tahayori-Sadeghian-Pedrycz (TSP) Model

The quantity of fuzziness associated with each $x \in F$ and its equivalent relocation in *Shd*(*S*), a fuzziness measure $\varphi(F)=\sum_{x\in X}[1-|2\mu_F(x)-1|]$ [10] on *F* and its fuzziness set $\varphi_F=\{(x,\mu_{\varphi_F}(x)):x\in X\}$ are exploited to determine the optimal partition thresholds α and β [20]. It was observed that the problem of uncertainty balance reduces to finding a threshold $\beta = 2\alpha$ in φ_F such that for al $x \in X$ l we have

$$\sum_{\mu_{\varphi F}(x)<\beta} \mu_{\varphi F}(x) = \sum_{\mu_{\varphi F}(x)\geq\beta} (1-\mu_{\varphi F}(x)),$$

which is equivalent to

$$V_2(\alpha) = \left|\sum_{\mu_{\varphi F}(x)<\beta} \mu_{\varphi F}(x) - \sum_{\mu_{\varphi F}(x)\geq\beta} (1-\mu_{\varphi F}(x))\right| \tag{3.3}$$

Equation (3.3) requires that elements with membership grades $\mu_{\varphi F}(x) < \beta$ should compensate for the elements with membership grades $\mu_{\varphi F}(x) \geq \beta$ to have full support in φ_F. In this way, the overall uncertainty in F will be adequately accounted for. Moreover, the amount of fuzziness in *Shd*(*S*) which is not accounted for by Equation (3.2), is duly considered by Equation (3.3). The optimum thresholds should minimize Equation (3.3).

Note also that other measures of fuzziness [3, 9, 11, etc.] could be used for the aforesaid calculation. The fuzziness measures satisfy the following properties.

A measure of fuzziness of F characterized by

$$\varphi(F) = h(\Sigma_{x \in X}\, g(\mu_F(x)\,))$$

is such that $h: \mathbb{R}^+ \rightarrow \mathbb{R}^+$ is a monotonically increasing function, and g is monotonically increasing and decreasing on [0, 0.5] and [0.5, 1], respectively. Moreover, g has a unique maximum at $g(0.5)$, and $g(0) - g(1)$.

3.3.1.3 *Ibrahim-William-West-Kana-Singh (IWKS) Model*

A shadowed set S induced from a fuzzy set F is the union of crisp sets and a set consisting of elements with the maximum degree of fuzziness. That is, $S = Cor(F) \cup Red(F) \cup Shd(F)$. Thus, the uncertainty reduced (UR) from F can be calculated as

$$UR = \Sigma_{x \in Red(F)}\, \varphi(x) + \Sigma_{x \in Cor(F)}\, \varphi(x) \tag{3.4}$$

To relocate the fuzziness of F into S, additional uncertainty induced by the elevation actions of S may be introduced. We refer to it as uncertainty introduced (UI); which is computed as

$$UI = Card(Shd(F)) - \sum_{x \in Shd(F)} \varphi(x) \tag{3.5}$$

When the uncertainty reduced from F is equivalent to the uncertainty introduced in it, we expect to have uncertainty balance. That is, $UR = UI$.

Along this line, IWKS model of shadowed set searches for an optimum threshold α which minimizes the following equation:

$$\sum_{x \in Red(F)} \varphi(x) + \sum_{x \in Cor(F)} \varphi(x) + \sum_{x \in Shd(F)} \varphi(x) = Card(Shd(F))$$

Or equivalently,

$$V_3(\alpha) = \left| \sum_{x \in Red(F)} \varphi(x) + \sum_{x \in Cor(F)} \varphi(x) + \sum_{x \in Shd(F)} \varphi(x) - Card(Shd(F)) \right| \tag{3.6}$$

The left-hand side of Equation (3.6) captures the overall fuzziness in F, whereas its right-hand side represents the total fuzziness in S. As noted in [8], for $\mu_F(x) = 0.5$, $\varphi(x) = 1$. Since the fuzziness of S is in $Shd(F)$ and any $x \in Shd(F)$ is assigned a membership grade 0.5. Therefore, $Card(Shd(F)) = \varphi(S)$.

3.3.2 Minimum Error or Deng-Yao (DY) Model

To provide error-based interpretation on induction of a shadowed set from a fuzzy set, Deng and Yao [4] observed that the elevation and reduction of membership grades from $\mu_F(x)$ to $t \in \{0, 0.5, 1\}$ induces an error of $|\mu_F(x) - t|$. Therefore, the optimum threshold α is computed by minimizing the overall error $E(\alpha)$ incurred by the shadowed set transformation actions under the threshold α. Here the error is determined by

$$E(\alpha) = E(Cor(S)) + E(Red(S)) + E(Shd(S)) \tag{3.7}$$

where the errors in the core, reduced and shadow regions are respectively calculated as

$$E(Cor(S)) = \Sigma_{x \in Cor(S)} (1 - \mu_F(x)),$$
$$E(Red(S)) = \Sigma_{x \in Red(S)} \mu_F(x),$$
$$E(Shd(S)) = \Sigma_{x \in Shd(S)} |0.5 - \mu_F(x)|.$$

3.3.3 Average Uncertainty or Ibrahim-West (IW) Model

Uncertainty is inherent in any fact or data. A larger number of uncertainties in data degrades the number of classified data. On the other hand, a smaller number of uncertainties in data upgrade the number of classified data. Consequently, by minimizing (respectively maximizing) the original uncertainty of a fuzzy set F, its classification ability is upgraded (respectively degraded). However, minimization of the uncertainty in F leads to information loss. At the same time, maximization of the uncertainty in F may not guarantee information gain. To strike a compromise between retention of the original uncertainty and its reduction, the average uncertainty associated with F is pursued when constructing a shadowed set S from F [7]. By modeling uncertainty balance with $J(\alpha)$ in the following Equation (3.8), a threshold which produces the average value of $J(\alpha)$ is sort as the optimal α -cut for constructing S. Here

$$J(\alpha) = \left| \Sigma_{x \in Red(F)} \mu_F(x) + \Sigma_{x \in Cor(F)} (1 - \mu_F(x)) \Sigma_{x \in Shd(F)} \mu_F(x) \right| \quad (3.8)$$

That is, for feasible thresholds $\alpha_i \in (0, 0.5]$, $1 \leq i \leq k \leq Card\ (F)$, we search for $\alpha \in \{\alpha_1, \alpha_2, \ldots, \alpha_k\}$ such that $J(\alpha)$ is the closest value to

$$A = \frac{J(\alpha_1) + J(\alpha_2) + \cdots + J(\alpha_k)}{k}$$

This is expressed as a minimization problem in the following way:
The optimal threshold α should minimize

$$V_4(\alpha) = |J(\alpha) - A| \quad (3.9)$$

We note that the third term in the absolute value on the right-hand side of Equation (3.8) is the scalar cardinality of the shadow region. Ibrahim and William-West [7] have used this quantity to modify Equation (3.2).

3.3.4 Nearest Quota of Uncertainty (WIK) Model

When modeling uncertainty balance in practical situations, where it is difficult to obtain a shadowed set with the same amount of fuzziness associated with *F*, decision makers may be satisfied with a shadowed set with the nearest quota of fuzziness as encountered in the original fuzzy set *F*. Further, decision makers may be much interested in the outcome of decision actions than the algorithmic procedure of elevation and reduction of membership grade. Thus, for feasible thresholds $\alpha_1, \alpha_2, \ldots, \alpha_k$ the overall amount of fuzziness in *F*, $\varphi(F)$, is compared against the total amount of fuzziness of each candidate shadowed set S_{aj}, $1 \leq j \leq k$. This direct approach eliminates the computational effort associated with verifying the most suitable approximation regions to assign each element *x*, and leads to the following closed-form formula for calculating the optimum partition threshold α [21]:

$$V_5(\alpha) = |\varphi(F) - \varphi(S_{\alpha j})| \tag{3.10}$$

The threshold which minimizes Equation (3.10) is embraced as optimum. Here, for $x \in F$, the elements with membership grades fulfilling $\alpha_j \leq \mu_F(x_r) \leq 1 - \alpha_j$, $1 \leq r \leq Card(S_{\alpha j})$ are assigned a new membership grade $\frac{1}{Card(S_{\alpha_j})}\sum_{r=1}^{Card(S_{\alpha_j})} \mu_F(x_r)$ and the fuzziness of each $S_{\alpha j}$ is computed and compared with $\varphi(F)$. Equation (3.10) asserts that the absolute difference between $\varphi(F)$ and $\varphi(S_{aj})$ be minimized.

We remark that besides the shadowed set approximation methods considered in this chapter, other methods dealing with constrained shadowed set method, game-theoretic shadowed sets, generalized Pedrycz's method and generalized entropy-based method can be found in [27, 30, 31].

3.3.5 Algorithm for Constructing Shadowed Sets

A shadowed set *S* approximation of a fuzzy set *F* is obtained by applying the following steps:

Input: Fuzzy set $F = \{(x, \mu_F(x_i)) : x \in F\}$.
Output: Shadowed set $S = Cor(S) \cup Shd(S)$ and optimal partition threshold
Step 1: Select $\mu_F(x_j) \leq 0.5$ as feasible partition threshold α_j.
Step 2: For each feasible threshold α, compute the quantity defined by Equations (3.2), (3.3), (3.6), (3.7), (3.9) and (3.10).

Step 3: Select the shadowed set producing the minimum values obtained in Step 2 as the optimal partition of each method.
Step 4: Print the threshold producing the shadowed set obtained in Step 3 as the optimum α-cut.

In the algorithm, Step 2 is aimed at calculating the candidate shadowed sets with their associated amount of fuzziness (or approximation error in the case of Equation (3.7)) and their absolute difference from $\varphi(F)$. Step 3 obtains the optimum shadowed set, whereas in Step 4 the optimal threshold is retrieved.

Throughout this chapter, we adopt the measure of fuzziness proposed in [10] to compute the underlying fuzziness of a fuzzy set. This measure is computed by

$$\varphi(F) = \Sigma_{x \in X}[1 - |2\mu_F(x) - 1|] \tag{3.11}$$

We note that no matter the choice of measure of fuzziness, the results obtained in this chapter holds.

3.3.6 Examples on Shadowed Set Approximation

In order to demonstrate how the shadowed set approximation methods considered so far are applied, let us consider a synthetic fuzzy set

$$F = \{(a, 0.1), (b, 0.25), (c, 0.3), (d, 0.55), (e, 0.6), (f, 0.75), (g, 0.8), (h, 1)\}.$$

The feasible thresholds for constructing a shadowed set lies in the interval (0, 0.5]. That is, we consider the values in {0.1, 0.25 0.3}.

Based on Equations (3.2), (3.3), (3.6), (3.7), (3.9) and (3.10), we have the computations to demonstrate each method:

Pedrycz's model:
The formula for uncertainty balance is used for the calculation:

$$V_1(\alpha) = \left| \sum_{\mu_F(x_i) < \alpha} \mu_F(x_i) + \sum_{\mu_F(x_i) > 1-\alpha} (1 - \mu_F(x_i)) - Card\{x_i : \alpha \le \mu_F(x_i) \le 1 - \alpha\} \right|$$

Iteration 1: When $\alpha = 0.1$

$$V_1(0.1) = |0 + 0 - 7| = 7,$$

Iteration 2: When $\alpha = 0.25$

$$V_1(0.25) = |0.1 + 0.2 - 5| = 4.7,$$

Iteration 3: When $\alpha = 0.3$

$$V_1(0.3) = |0.35 + 0.45 - 3| = 2.2.$$

Therefore, $\min_\alpha V_1(\alpha) = 2.2$, and it is obtained at $\alpha = 0.3$. Hence, the optimum threshold is $\alpha = 0.3$.

TSP model:

The fuzziness measure [6] on F is used to obtain the membership grades of elements in the fuzziness set $\varphi_F = \{(x, \mu_{\varphi F}(x)) : x \in X\}$ in order to evaluate $V_2(\alpha)$; $V_2(\alpha) = \left|\sum_{\mu_{\varphi F}(x)<\beta} \mu_{\varphi F}(x) - \sum_{\mu_{\varphi F}(x)\geq\beta} (1 - \mu_{\varphi_F}(x))\right|$.

Therefore, we have $\varphi(x) = 1 - |2\mu_F(x) - 1| = \mu_{\varphi_F}(x)$ as shown in the following Table 3.1:

Abstracting from Table 3.1, the fuzziness set of F is obtained as:

$\varphi_F = \{(a, 0.2), (b, 0.5), (c, 0.6), (d, 0.9), (e, 0.8), (f, 0.5), (g, 0.4), (h, 0.0)\}$, whereas the value β in Equation (3.3) is applied on $\mu_{\varphi F}(x)$.

Iteration 1: When $\alpha = 0.1, \beta = 2\alpha = 0.2$

$$V_2(0.1) = |0 - 3.1| = 3.1,$$

Iteration 2: When $\alpha = 0.25, \beta = 2\alpha = 0.5$

$$V_2(0.25) = |0.6 - 1.7| = 1.1,$$

Table 3.1 Fuzziness of elements in F.

x	a	b	c	d	e	f	g	h
$\varphi(x)$	0.2	0.5	0.6	0.9	0.8	0.5	0.4	0.0

Iteration 3: When $\alpha = 0.3, \beta = 2\alpha = 0.6$

$$V_2(0.3) = |1.6 - 0.7| = 0.9.$$

Therefore, $\min_\alpha V_2(\alpha) = 0.9$, and it is obtained at $\alpha = 0.3$. Hence, the optimum threshold is $\alpha = 0.3$.

IWKS model:

Iteration 1: When $\alpha = 0.1$

$$V_3(0.1) = |0 + 0 + 3.9 - 7| = 3.1,$$

Iteration 2: When $\alpha = 0.25$

$$V_3(0.25) = |0.2 + 0.4 + 3.3 - 5| = 1.1,$$

Iteration 3: When $\alpha = 0.3$

$$V_3(0.3) = |0.7 + 0.9 + 2.3 - 3| = 0.9.$$

Therefore, $\min_\alpha V_3(\alpha) = 0.9$, and it is obtained at $\alpha = 0.3$. Hence, the optimum threshold is $\alpha = 0.3$.

DY model:

Iteration 1: When $\alpha = 0.1$

$$E(0.1) = 0.0 + 0.0 + 1.55 = 1.15,$$

Iteration 2: When $\alpha = 0.25$

$$E(0.25) = 0.2 + 0.1 + 0.85 = 1.15,$$

Iteration 3: When $\alpha = 0.3$

$$E(0.3) = 0.45 + 0.35 + 0.35 = 1.15.$$

Therefore, $\min_\alpha E(\alpha) = 1.15$ and it is obtained at $\alpha = 0.25$ and $\alpha = 0.3$. Hence, indicating that the optimum threshold is $\alpha = 0.25$ and $\alpha = 0.3$. Usually, in situation where the optimum partition thresholds are not unique, a tie-breaking rule is deployed. Also, a decision maker may choose his preferred optimum threshold according to the size of its associate approximation regions.

IW model:
Before calculating the parameters involved, we give a quick description of the procedure involved.

First of all, the relationship between reduced and increased fuzziness in *Red(F)* ∪ *Cor(F)* and *Shd(F)*, respectively are computed for each feasible threshold $\alpha \in \{0.1, 0.25, 0.3\}$ by using

$$J(\alpha)=\left|\sum_{x\in Red(F)}\mu_F(x)+\sum_{x\in Cor(F)}(1-\mu_F(x))-\sum_{x\in Shd(F)}\mu_F(x)\right|:$$

$$J(0.1)=|0+0-3.35|=3.35,$$

$$J(0.25)=|0.1+0.2-2.45|=2.15,$$

$$J(0.3)=|0.35+0.45-1.45|=0.65.$$

Next, the average of the above values is determined with the aid of

$$A=\frac{J(\alpha_1)+J(\alpha_2)+\cdots+J(\alpha_k)}{k}=\frac{J(0.1)+J(0.25)+J(0.3)}{3}=2.05.$$

Finally, we determine the minimum absolute difference between $J(\alpha)$ and A. That is,

$$\min_\alpha |J(\alpha)-A|$$

$$V_4\,(0.1) = |J(0.1) - A| = |3.35 - 2.05| = 1.3,$$

$$V_4\,(0.25) = |J(0.25) - A| = |2.15 - 2.05| = 0.1,$$

$$V_4\,(0.3) = |J(0.3) - A| = |0.65 - 2.05| = 1.4,$$

$$\min_\alpha |J(\alpha)-A| = \min_\alpha\{1.3,0.1,1.4\} = 0.1.$$

The threshold value producing $J(\alpha)$ which keeps the minimum distance from A will be selected as optimum. That is, the entire procedure is summarized as follows.

Iteration 1: When $\alpha = 0.1$

$$J(\alpha) = |0 + 0 - 3.35| = 3.35,$$

Iteration 2: When $\alpha = 0.25$

$$J(\alpha) = |0.1 + 0.2 - 2.45| = 2.15,$$

Iteration 3: When $\alpha = 0.3$

$$J(\alpha) = |0.35 + 0.45 - 1.45| = 0.65.$$

The average $J(\alpha)$ is calculated as $A = 2.05$.

Therefore, $\min_\alpha V_4(\alpha) = \min_\alpha \{V_4(0.1), V_4(0.25), V_4(0.3)\} = V_4(0.25) = 0.1$, and it is obtained at $\alpha = 0.25$. Hence, the optimum threshold is $\alpha = 0.25$.

WIK model:

The formula for nearest quota of fuzziness is used for the calculation:

$$V_5(\alpha) = |\varphi(F) - \varphi(S_\alpha)|,\ \alpha \in \{0.1, 0.25, 0.3\}.$$

The fuzziness of F is computed by

$\varphi(F) = \Sigma_{x \in X}[1 - |2\mu_F(x) - 1| = 0.2 + 0.5 + 0.6 + 0.9 + 0.8 + 0.5 + 0.4 + 0 = 3.9.$

Iteration 1: When $\alpha = 0.1$, elements with membership grade $0.1 \le \mu_F(x) \le 0.9$ are assigned new membership grade of 0.48 in candidate shadowed set $S_{\alpha = 0.1}$. This new membership grade is computed by

$$\frac{1}{Card(Shd(S_{0.1}))} \sum_{r=1}^{Card(Shd(S_{0.1}))} \mu_F(x_r) = \frac{0.1 + 0.25 + 0.3 + 0.55 + 0.6 + 0.75 + 0.8}{7} = 0.48$$

The fuzziness of

$$\varphi(S_{0.1}) = \sum_{x \in X} \left[1 - |2\mu_{S_{0.1}}(x) - 1|\right] = \varphi(Shd(S_{0.1}))$$

$$\varphi(Shd(S_{0.1})) = \Sigma_{x\in X}[1-|2\times 0.48-1|]$$
$$= 0.96+0.96+0.96+0.96+0.96+0.96+0.96 = 6.72.$$

The proximity of the fuzziness in $S_{0.1}$ to the fuzziness in F is determined by

$$V5\ (0.1) = |\varphi(F) - \varphi(S_{0.1})|:$$

$$V5\ (0.1) = |3.9 - 6.72| = 2.82,$$

Iteration 2: When $\alpha = 0.25$, elements with membership grade $0.25 \le \mu_F(x) \le 0.75$ are assigned new membership grade of 0.49 in candidate shadowed set $S_{\alpha=0.25}$. This new membership grade is computed by

$$\frac{1}{Card(Shd(S_{0.25}))}\sum_{r=1}^{Card(Shd(S_{0.25}))}\mu_F(x_r) = \frac{0.25+0.3+0.55+0.6+0.75}{5} = 0.49.$$

The fuzziness of

$$\varphi(S_{0.25}) = \sum_{x\in X}\left[1-\left|2\mu_{S_{0.25}}(x)-1\right|\right] = \varphi(Shd(S_{0.25}))$$

$$\varphi(Shd(S_{0.25})) = \Sigma_{x\in X}[1-|2\times 0.49-1|] = 0.98+0.98+0.98+0.98+0.98 = 4.9.$$

The proximity of the fuzziness in $S_{0.25}$ to the fuzziness in F is determined by

$$V_5(0.25) = |\varphi(F) - \varphi(S_{0.25})|:$$

$$V_5(0.25) = |3.9 - 4.9| = 1,$$

Iteration 3: When $\alpha = 0.3$, elements with membership grade $0.3 \le \mu_F(x) \le 0.7$ are assigned new membership grade of 0.48 in candidate shadowed set $S_{\alpha=0.3}$. This new membership grade is computed by

$$\frac{1}{Card(Shd(S_{0.3}))}\sum_{r=1}^{Card(Shd(S_{0.3}))}\mu_F(x_r) = \frac{0.3+0.55+0.6}{3} = 0.48.$$

The fuzziness of

$$\varphi(S_{0.3}) = \sum_{x \in X} \left[1 - \left| 2\mu_{S_{0.3}}(x) - 1 \right| \right] = \varphi(Shd(S_{0.3}))$$

$$\varphi(Shd(S_{0.3})) = \Sigma_{x \in X}[1 - |2 \times 0.48 - 1|] = 0.96 + 0.96 + 0.96 = 2.88.$$

The proximity of the fuzziness in S0.3 to the fuzziness in F is determined by

$$V_5\,(0.3) = |\varphi(F) - \varphi(S_{0.3})|:$$

$$V_5\,(0.3) = |3.9 - 2.88| = 1.02.$$

Therefore, $\min_\alpha$, $V_5(\alpha) = \min_\alpha \{V_5(0.1), V_5(0.25), V_5(0.3)\} =$ and $\min_\alpha$ $\{2.82, 1, 1.02\} = 1$ and it is obtained at $\alpha = 0.25$. Hence, the optimum threshold is $\alpha = 0.25$.

From the threshold values obtained in the example, two types of shadowed sets are constructed as follows.

a. **Shadowed set with optimum cuts $\alpha = 0.3$ and $1 - \alpha = 0.7$:**

$$S_{\alpha = 0.3} = \{(c, 0.5), (d, 0.5), (e, 0.5), (f, 1), (g, 1), (h, 1)\}.$$

Noticeable, the elements in $S_{\alpha = 0.3}$ now assume new membership grades of 0.5 and 1 as a result of elevation and reduction actions on *F*.

We note that the elements with membership grades satisfying $\mu_F(x) < 0.3$ are excluded from $S_{\alpha = 0.3}$, and have been put into the reduced region. That is,

$$Red(S) = \{(a, 0), (b, 0)\}.$$

b. **Shadowed set with optimum cuts $\alpha = 0.25$ and $1 - \alpha = 0.75$:**

$$S_{\alpha = 0.25} = \{(b, 0.5), (c, 0.5), (d, 0.5), (e, 0.5), (f, 0.5), (g, 1), (h, 1)\}.$$

As previously done, observe that the elements in $S_{\alpha = 0.25}$ now assume new membership grades of 0.5 and 1 as a result of elevation and reduction actions.

We note that the elements with membership grades satisfying $\mu_F(x) < 0.25$ are excluded from $S_{\alpha = 0.25}$, and have been put into the reduced region. That is,

$$Red(S) = \{(a, 0)\}.$$

The two types of shadowed sets $S_{\alpha = 0.3}$ and $S_{\alpha = 0.25}$ are obtained from the method in [8, 19, 20] and [7, 21], respectively. The method in [4] produces any of these shadowed sets depending on the decision maker's preference or the applied tie-breaking rule. It is important to underline here that although the same threshold value emerge as optimum for two or three methods, in general this situation may be rare to come by.

3.4 Principles for Constructing Decision-Theoretic Approximation

Decision-theoretic three-way approximation, T, of a fuzzy set (DT3WA) assigns membership grades to $x \in T$ as shadowed set approximation does. The difference between the two frameworks is that while the shadowed set rely on a search algorithm for the optimum partition thresholds on T, the DT3WA on the other hand, depends on the user-defined decision cost parameters to obtain optimum but non-symmetrical thresholds, α and β on T. These costs satisfy the following conditions as provided by [4]:

i) non-negativity: $\lambda_0 > 0$, $\lambda_1 > 0$ and $\lambda_{0.5\downarrow} \geq 0$, $\lambda_{0.5\uparrow} \geq 0$,
ii) partial ordering: $\lambda_{0.5\uparrow} \leq \lambda_0$ and $\lambda_{0.5\downarrow} \leq \lambda_1$.

A general function $T{:}F \rightarrow \{n, m, p\}$ for membership assignment is given as

$$T(\mu F(x)) = \begin{cases} p, & x \in POS(T) \\ m, & x \in BND(T) \\ n, & x \in NEG(T) \end{cases} \tag{3.12}$$

The variables n, m and p are used to denote the membership grades of elements in $NEG(T)$, $BND(T)$ and $POS(T)$, respectively.

3.4.1 Deng and Yao Special Decision-Theoretic (DYSD) Model

Deng and Yao [4] considered special equidistant values of n, m and p to propose {0, 0.5, 1} -system by setting $n = 0$, $m = 0.5$ and $p = 1$ as the membership grades of elements in $NEG(T)$, $BND(T)$ and $POS(T)$, respectively. Two types of membership grades in F(*Membership grades satisfying* $\mu_F(x) \leq 0.5$ and *Membership grades satisfying* $\mu_F(x) > 0.5$) together with the decision rules discussed in Section 4 are considered to obtain a formula for α and β.

i. **Membership grades satisfying $\mu_F(x) \leq 0.5$:**
If $\mu_F(x) \leq 0.5$, then there are only two possible actions, viz.

a) reduce $\mu_F(x)$ to 0. That is, $\lambda_0 (\mu_F(x) - 0) \leq \lambda_{0.5\uparrow} (0.5 - \mu_F(x))$ and $\lambda_0 (\mu_F(x) - 0) \leq \lambda_1 (1 - \mu_F(x))$ are true. This action results in the assignment; $T(x) = 0$.

By simplifying the inequality $\lambda_0 (\mu_F(x) - 0) \leq \lambda_{0.5\uparrow} (0.5 - \mu_F(x))$ and $\lambda_0 (\mu_F(x) - 0) \leq \lambda_1 (1 - \mu_F(x))$, we have

$$\lambda_0 \mu_F(x) \leq \frac{1}{2}\lambda_{0.5\uparrow} - \lambda_{0.5\uparrow}\mu_F(x) \text{ and } \lambda_0\mu_F(x) \leq \lambda_1 - \lambda_1\mu_F(x)$$

$$(\lambda_0 + \lambda_{0.5\uparrow})\mu_F(x) \leq \frac{1}{2}\lambda_{0.5\uparrow} \text{ and } (\lambda_0 + \lambda_1)\, \mu_F(x) \leq \lambda_1$$

$$\mu_F(x) \leq \frac{\lambda_{0.5\uparrow}}{2(\lambda_0 + \lambda_{0.5\uparrow})} \text{ and } \mu_F(x) \leq \frac{\lambda_1}{\lambda_0 + \lambda_1}.$$

Since $\lambda_{0.5\uparrow} \leq \lambda_0$, the quantity $\frac{\lambda_{0.5\uparrow}}{2(\lambda_0 + \lambda_{0.5\uparrow})}$ is less than (or equal to) $\frac{\lambda_1}{\lambda_0 + \lambda_1}$
Hence, it is enough to bound the membership grades less than 0.5 by $\frac{\lambda_{0.5\uparrow}}{2(\lambda_0 + \lambda_{0.5\uparrow})}$. That is,

$$\mu_F(x) \leq \frac{\lambda_{0.5\uparrow}}{2(\lambda_0 + \lambda_{0.5\uparrow})} \tag{3.13}$$

b) elevate $\mu_F(x)$ to 0.5. That is, $\lambda_{0.5\uparrow}$ $(0.5 - \mu_F(x)) \le \lambda_1$ $(1 - \mu_F(x))$ and $\lambda_{0.5\uparrow}$ $(0.5 - \mu_F(x)) \le \lambda_0$ $(\mu_F(x) - 0)$ must be true. This action results in the assignment; $T(x) = 0.5$.

By simplifying the inequalities $\lambda_{0.5\uparrow}$ $(0.5 - \mu_F(x)) \le \lambda_1$ $(1 - \mu_F(x))$ and $\lambda_{0.5\uparrow}$ $(0.5 - \mu_F(x)) \le \lambda_0$ $(\mu_F(x) - 0)$, we have

$$\frac{1}{2}\lambda_{0.5\uparrow} - \lambda_{0.5\uparrow}\mu_F(x) \le \lambda_1 - \lambda_1\mu_F(x) \text{ and } \frac{1}{2}\lambda_{0.5\uparrow} - \lambda_{0.5\uparrow}\mu_F(x) \le \lambda_0\mu_F(x),$$

$$(\lambda_1 - \lambda_{0.5\uparrow})\mu_F(x) \le \frac{2\lambda_1 - \lambda_{0.5\uparrow}}{2} \text{ and } \frac{1}{2}\lambda_{0.5\uparrow} \le (\lambda_0 + \lambda_{0.5\uparrow})\mu_F(x)$$

$$\mu_F(x) \le \frac{2\lambda_1 - \lambda_{0.5\uparrow}}{2(\lambda_1 - \lambda_{0.5\uparrow})} \text{ and } \frac{\lambda_{0.5\uparrow}}{2(\lambda_0 + \lambda_{0.5\uparrow})} \le \mu_F(x)$$

By the initial assumption that $\mu_F(x) \le 0.5$, we can dismiss the last inequalities above. Hence, only Equation (3.13) is obtainable.

ii. **Membership grades satisfying $\mu_F(x) > 0.5$:**
If $\mu_F(x) > 0.5$, then there are only two possible actions viz.

1) reduce $\mu_F(x)$ to 0.5. That is, $\lambda_{0.5\uparrow}$ $(\mu_F(x) - 0.5) \le \lambda_1$ $(1 - \mu_F(x))$ and $\lambda_{0.5\downarrow}$ $(\mu_F(x) - 0.5) \le \lambda_0$ $(\mu_F(x) - 0)$ must hold. This action results in the assignment; $T(x) = 0.5$.

By simplifying $\lambda_{0.5\downarrow}$ $(\mu_F(x) - 0.5) \le \lambda_1$ $(1 - \mu_F(x))$ and $\lambda_{0.5\downarrow}$ $(\mu_F(x) - 0.5) \le \lambda_0$ $(\mu_F(x) - 0)$, we have

$$\lambda_{0.5\downarrow}\mu_F(x) - \frac{1}{2}\lambda_{0.5\downarrow} \le \lambda_1 - \lambda_1\mu_F(x) \text{ and } \lambda_{0.5\downarrow}\mu_F(x) - \frac{1}{2}\lambda_{0.5\downarrow} \le \lambda_0\mu_F(x),$$

$$(\lambda_{0.5\downarrow} + \lambda_1)\mu_F(x) \le \frac{1}{2}\lambda_{0.5\downarrow} \text{ and } (\lambda_{0.5\downarrow} - \lambda_0)\mu_F(x) \le \frac{1}{2}\lambda_{0.5\downarrow},$$

$$\mu_F(x) \le \frac{\lambda_{0.5\downarrow}}{2(\lambda_{0.5\downarrow} + \lambda_1)} \text{ and } \mu_F(x) \le \frac{\lambda_{0.5\downarrow}}{2(\lambda_{0.5\downarrow} - \lambda_0)}. \quad (3.14)$$

Again, since $\mu_F(x) \leq 0.5$ and Equation (3.14) may not always fulfill this assumption, we conclude that Equation is not the best choice.

2) elevate $\mu_F(x)$ to 1. That is, $\lambda_1 (1 - \mu_F(x)) \leq \lambda_{0.5\downarrow} (\mu_F(x) - 0.5)$ and $\lambda_1 (1 - \mu_F(x)) \leq \lambda_0 (\mu_F(x) - 0)$ hold. This action results in the assignment; $T(x) = 1$.

By simplifying $\lambda_1 (1 - \mu_F(x)) \leq \lambda_{0.5\downarrow} (\mu_F(x) - 0.5)$ and $\lambda_1 (1 - \mu_F(x)) \leq \lambda_0 (\mu_F(x) - 0)$, we have

$$\lambda_1 - \lambda_1\mu_F(x) \leq \lambda_{0.5\downarrow}\mu_F(x) - \frac{1}{2}\lambda_{0.5\downarrow} \text{ and } \lambda_1 - \lambda_1\mu_F(x) \leq \lambda_0\mu_F(x),$$

$$\frac{2\lambda_1 + \lambda_{0.5\downarrow}}{2} \leq (\lambda_{0.5\downarrow} + \lambda_1)\mu_F(x) \text{ and } \lambda_1 \leq (\lambda_0 + \lambda_1)\mu_F(x),$$

$$\frac{2\lambda_1 + \lambda_{0.5\downarrow}}{2(\lambda_{0.5\downarrow} + \lambda_1)} \leq \mu_F(x) \text{ and } \frac{\lambda_1}{(\lambda_0 + \lambda_1)} \leq \mu_F(x).$$

The quantity $\frac{2\lambda_1 + \lambda_{0.5\downarrow}}{2(\lambda_{0.5\downarrow} + \lambda_1)}$ is bigger than $\frac{\lambda_1}{(\lambda_0 + \lambda_1)}$. Therefore, $\frac{2\lambda_1 + \lambda_{0.5\downarrow}}{2(\lambda_{0.5\downarrow} + \lambda_1)} \leq \mu_F(x)$ is in line with the assumption that $\mu_F(x) > 0.5$. Moreover, $\frac{\lambda_1}{(\lambda_0 + \lambda_1)} \leq \mu_F(x)$ is fulfilled whenever $\frac{2\lambda_1 + \lambda_{0.5\downarrow}}{2(\lambda_{0.5\downarrow} + \lambda_1)} \leq \mu_F(x)$ holds. Hence, we choose the following inequality:

$$\frac{2\lambda_1 + \lambda_{0.5\downarrow}}{2(\lambda_{0.5\downarrow} + \lambda_1)} \leq \mu_F(x) \tag{3.15}$$

From the above computations, the quantities $\frac{\lambda_{0.5\uparrow}}{2(\lambda_0 + \lambda_{0.5\uparrow})}$ and $\frac{2\lambda_1 + \lambda_{0.5\downarrow}}{2(\lambda_{0.5\downarrow} + \lambda_1)}$ are obtained as the best cuts for partitioning F [4], with

$$\alpha = \frac{\lambda_{0.5\uparrow}}{2(\lambda_0 + \lambda_{0.5\uparrow})} \text{ and } \beta = \frac{2\lambda_1 + \lambda_{0.5\downarrow}}{2(\lambda_{0.5\downarrow} + \lambda_1)} \tag{3.16}$$

When $\lambda_0 = \lambda_{0.5\uparrow} = \lambda_{0.5\downarrow} = \lambda_1 = 1$, minimum distance-based three-way approximation of a fuzzy set is obtained from Equation (3.16) with optimum threshold values:

$$\alpha = \frac{1}{4} \text{ and } \beta = \frac{3}{4} \tag{3.17}$$

Remark 4.1
In decision-theoretic three-way approximation the optimum thresholds α and β with symmetric connotation are obtained under normal condition of $\lambda_0 = \lambda_{0.5\uparrow} = \lambda_{0.5\downarrow} = \lambda_1 = 1$.

3.4.2 Zhang, Xia, Liu and Wang (ZXLW) Generalized Decision-Theoretic Model

The thresholds determined by Equations (3.15) and (3.16) obtain minimum decision cost; however, in some situations, it may not produce the least overall cost. Such instances are pointed out in [28]. The case when the (non-zero) unit costs $\lambda_{0.5\downarrow}$ and $\lambda_{0.5\uparrow}$ are equal and the costs λ_1 and λ_0 are very large is an example. To address this issue, the intermediate membership value m is taken as a variable value in (0, 1), not necessarily equal to 0.5. A quick approach for determining $m \in (0, 1)$ has been proposed in [28] by imposing some restriction on the unit costs:

i) If $\lambda_1 \to \infty$, $\lambda_0 \to \infty$ and $\lambda_{m\downarrow} = \lambda_{m\uparrow}$, then m is the mean value of the membership grades in $F = \{\mu_F(x_1), \mu_F(x_2), \ldots, \mu_F(x_n)\}$. That is, $m = \frac{1}{n}\sum_{i=1}^{n} \mu_F(x_i)$.
ii) If $\lambda_{m\downarrow} \to 0$, then $m = \max\{\mu_F(x_1), \mu_F(x_2), \ldots, \mu_F(x_n)\}$.
iii) If $\lambda_{m\downarrow} \to 0$, then $m = \min\{\mu_F(x_1), \mu_F(x_2), \ldots, \mu_F(x_n)\}$.
iv) If m is such that

$$m = \frac{\sum_{\alpha \le \mu_F(x) < m} \lambda_{m\uparrow}\mu_F(x) + \sum_{m \le \mu_F(x) < \beta} \lambda_{m\downarrow}\mu_F(x)}{\sum_{\alpha \le \mu_F(x) < m} \lambda_{m\uparrow} + \sum_{m \le \mu_F(x) < \beta} \lambda_{m\downarrow}}$$
$$= \frac{\lambda_{m\downarrow}(\lambda_0 + \lambda_{m\uparrow})}{\lambda_{m\downarrow}(\lambda_0 + \lambda_{m\uparrow}) + \lambda_{m\uparrow}(\lambda_1 + \lambda_{m\downarrow})},$$

then the thresholds α and β are symmetric. That is, $\alpha + \beta = 1$.

By applying the rules simplification procedure deployed in Section 4.1, Zhang *et al.* [28] obtained the following closed-form formula for the optimum thresholds where m is not necessarily 0.5:

$$\alpha = \frac{m\lambda_{m\uparrow}}{\lambda_0 + \lambda_{m\uparrow}} \text{ and } \beta = \frac{\lambda_1 + m\lambda_{m\downarrow}}{\lambda_{m\downarrow} + \lambda_1} \tag{3.18}$$

When $\lambda 0 = \lambda_{m\uparrow} = \lambda_{m\downarrow} = \lambda_1 = 1$, minimum distance-based three-way approximation of a fuzzy set is obtained from Equation (3.17) with optimum threshold values:

$$\alpha = \frac{m}{2} \text{ and } \beta = \frac{1+m}{2} \tag{3.19}$$

3.4.3 A General Perspective to Decision-Theoretic Three-Way Approximation

A general assignment function on membership grades of elements in the three regions of decision-theoretic three-way approximation of fuzzy sets has been proposed as in Equation (3.12) in [15, 25]. A motivation of Equation (3.12) is to allow switching of a region's membership grade from 0, 0.5 or 1; where elements membership grades are normally distributed in F, to $n \neq 0$, $m \neq 0.5$ and $p \neq 1$; when elements' membership grades are in form of subnormal distribution.

Besides, the approximation obtained by Equation (3.12) is capable of minimizing approximation error, and may also facilitate uncertainty balance. Also, n, m and p, can be used to represent status of our knowledge about the elements classified into $NEG(T)$, $BND(T)$ and $POS(T)$. When $n \neq 0$ and $p \neq 1$, the three-way approximation emphasizes that $NEG(T)$ and $POS(T)$ do not reflect null and full support of membership to T. Subsequently, when $m \neq 0.5$, $BND(T)$ does not reflect the maximum degree of uncertainty.

In view of the aforesaid, determination of appropriate values of n, m and p is significant for interpreting the type of support associated with the approximation regions. That is, null support when $n = 0$, non-null support otherwise, maximum uncertainty when $m = 0.5$, non-maximum otherwise, full support when $p = 1$, non-full support otherwise. Existing studies always set $n = 0$ and $p = 1$, thereby assuming that objects in $NEG(T)$ and

$POS(T)$ have null and full support in T. These assumptions may lead to increase in approximation error and uncertainty imbalance. To address this issue, in this chapter an approach for determining appropriate values of n, m and p is proposed. These values are calculated from a given data that are to be partitioned with minimum approximation error and uncertainty imbalance.

3.4.3.1 Determination of n, m and p for Decision-Theoretic Three-Way Approximation

Yao *et al.* [25] suggested a possible way to follow in order to compute the values of n, m and p. Anchoring on two guiding principles: truth ordering $\prec_t$ and information ordering $\prec_i$ (see [25], for details) for ordering n, m and p, two standardizations were proposed:

a) $d(0, n) = d(1, p) = d(c, m)$,
b) $d(0, p) = d(1, n) > max\{d(0, m), d(1, m)\}$,

where d, is any distance measure and $c \in (0, 1)$ is an intermediate membership grade.

The above standardizations ensure that the distance between membership grades in [0, 1] are preserved (or measure uniformly) in $\{n, m, p\}$ with respect to $\prec_t$ and $\prec_i$. Simply put, $n \prec_t m \prec_t p$ holds if and only if $n < m < p$ and it is expected that as $n < m < p$, we should have $m \prec_i$ n and m $\prec_i p$ in $\{n, m, p\}$ system.

The standardization in (a) above requires that the distance between 0 and n should be equal to the distance between 1 and p, and that the distance should be maintained between 0.5 and m. We find this requirement to be too restrictive and impractical under subnormal fuzzy environment. In that case $d(1, p) \neq d(0, 1)$.

Hence, the distance imposed by standardization (a) may increase approximation error. Moreover, the two standardizations (a) and (b) contain four unknowns and can be reduced to a single equation. Hence, posing a challenge in determining n, m and p.

Taking insight from studies in [25], we propose a more flexible standardization which is membership-sensitive and require that:

i) $d(\mu_{F_{min}}, n) = d(\mu_{F_{max}}, p) = d(\overline{\mu_F(x)}, m)$,
ii) $(\mu_{F_{min}}, p) = d(\mu_{F_{max}}, n) > \max\{(\mu_{F_{min}}, m), d(\mu_{F_{max}}, m)\}$.

Here $\mu_{F_{min}}, \mu_{F_{max}}$ and $\overline{\mu_F(x)}$ are the minimum, maximum and mean value of $\{\mu_F(x_1), \mu_F(x_2), \ldots, \mu_F(x_n)\}$.

Standardization (i) produces the following equations:

$$p+n=\mu_{F_{max}}+\mu_{F_{min}},$$

$$p+m=\mu_{F_{max}}+\overline{\mu_F(x)},$$

$$m+n=\overline{\mu_F(x)}+\mu_{F_{min}},$$

Standardization (ii) leads to the following equation

$$p-\mu_{F_{min}}>m-\mu_{F_{min}},$$

$$p-\mu_{F_{min}}>\mu_{F_{max}}-m,$$

$$\mu_{F_{min}}-n>m-\mu_{F_{min}},$$

$$\mu_{F_{max}}-n>\mu_{F_{max}}-m.$$

Consequently, we have

$$p-m>0,$$

$$p+m>\mu_{F_{max}}+\mu_{F_{min}},$$

$$\mu_{F_{max}}+\mu_{F_{min}}>m+n,$$

$$m-n>0.$$

Given the above equations, we could perform direct calculations to obtain the values of n, m and p as shown in Equation (3.20), or ensure that the distance between elements in $\{0, 0.5, 1\}$ measures uniformly in between elements in $\{n, m, p\}$. That is, $m - n$ should be equal to $p - m$. Hence, $p + n = 2m$, and subsequently,

$$n=\mu_{F_{min}},m=\overline{\mu_F(x)},p=\mu_{F_{max}} \tag{3.20}$$

or

$$n = \frac{\mu_{F_{max}} + 3\mu_{F_{min}} - 2\overline{\mu_F(x)}}{2}, m = \frac{\mu_{F_{max}} + \mu_{F_{min}}}{2},$$

$$p = \frac{2\overline{\mu_F(x)} + \mu_{F_{max}} + \mu_{F_{min}}}{2} \tag{3.21}$$

3.4.3.2 *A General Decision-Theoretic Three-Way Approximation Partition Thresholds*

With a formula for computing appropriate values of n, m and p in place, general partition thresholds of decision-theoretic three-way approximation can be determined. The decision rules for obtaining well-guided three-way approximation are defined in the following way:

Membership grades satisfying $\mu_F(x) > 0.5$.
Rule for elevating $\mu_F(x)$ to p:

P1) If $\lambda_p(p - \mu_F(x)) \leq \lambda_{m\downarrow}(\mu_F(x) - m)$, then $T(x) = p$,

Rule for reducing $\mu_F(x)$ to m:

M1) If $\lambda_{m\downarrow}(\mu_F(x) - m) \leq \lambda_p(p - \mu_F(x))$, then $T(x) = m$.

Membership grades satisfying $\mu_F(x) \leq 0.5$.
Rule for reducing $\mu_F(x)$ to n:

N1) If $\lambda_n(\mu_F(x) - n) \leq \lambda_{m\uparrow}(m - \mu_F(x))$, then $T(x) = n$.

Rule for elevating $\mu_F(x)$ to m:

M2) If $\lambda_{m\uparrow}(m - \mu_F(x)) \leq \lambda_n(\mu_F(x) - n)$, then $T(x) = m$

We apply the same procedure employed in Section 4.1. Accordingly, we have

P1) $p\lambda_p + m\lambda_{m\downarrow} \leq (\lambda_p + \lambda_{m\downarrow})\, \mu_F(x)$

$$\frac{p\lambda_p + m\lambda_{m\downarrow}}{\lambda_p + \lambda_{m\downarrow}} \leq \mu_F(x) \tag{3.22}$$

M1) $(\lambda_{m\downarrow} + \lambda_p)\, \mu_F(x) \leq p\lambda_p + m\lambda_{m\downarrow}$

$$\mu_F(x) \leq \frac{p\lambda_p + m\lambda_{m\downarrow}}{\lambda_p + \lambda_{m\downarrow}} \tag{3.23}$$

N1) $(\lambda_n + \lambda_{m\uparrow})\,\mu_F(x) \le n\lambda_n + m\lambda_{m\uparrow}$

$$\mu_F(x) \le \frac{n\lambda_n + m\lambda_{m\uparrow}}{\lambda_n + \lambda_{m\uparrow}} \tag{3.24}$$

M2) $n\lambda_n + m\lambda_{m\uparrow} \le (\lambda_n + \lambda_{m\uparrow})\,\mu_F(x)$

$$\frac{n\lambda_n + m\lambda_{m\uparrow}}{\lambda_n + \lambda_{m\uparrow}} \le \mu_F(x) \tag{3.25}$$

When $\mu_F(x) > 0.5$:
Since $m < p \le 1$ and the conditions on the costs assume that $\lambda_{m\downarrow} \le \lambda_p$. It follows that $\frac{p\lambda_p + m\lambda_{m\downarrow}}{\lambda_p + \lambda_{m\downarrow}} < 1$ and $0.5 < \frac{p\lambda_p + m\lambda_{m\downarrow}}{\lambda_p + \lambda_{m\downarrow}}$. So, Equation (3.22) is not admissible. Therefore, we are left with Equation (3.23). Rules (P1) and (M1) can only be applied when $\beta = \frac{p\lambda_p + m\lambda_{m\downarrow}}{\lambda_p + \lambda_{m\downarrow}}$.

When $\mu_F(x) \le 0.5$:
Since $n < m < 1$ and the conditions on the costs assume that $\lambda_{m\uparrow} \le \lambda_n$. It follows that $\frac{n\lambda_n + m\lambda_{m\uparrow}}{\lambda_n + \lambda_{m\uparrow}} \le 0.5$ and $0 < \frac{n\lambda_n + m\lambda_{m\uparrow}}{\lambda_n + \lambda_{m\uparrow}}$. So, Equation (3.25) is not admissible. Therefore, we are left with Equation (3.24). Rules (N1) and (M2) can only be applied when $\alpha = \frac{n\lambda_n + m\lambda_{m\uparrow}}{\lambda_n + \lambda_{m\uparrow}}$.

From the above computations, we have the following three simplified rules:

P) If $\mu_F(x) > \frac{p\lambda_p + m\lambda_{m\downarrow}}{\lambda_p + \lambda_{m\downarrow}}$, then $T(x) = p$.

M) If $\frac{n\lambda_n + m\lambda_{m\uparrow}}{\lambda_n + \lambda_{m\uparrow}} \le \mu_F(x) \le \frac{p\lambda_p + m\lambda_{m\downarrow}}{\lambda_p + \lambda_{m\downarrow}}$, then $T(x) = m$.

N) If $\mu_F(x) < \frac{n\lambda_n + m\lambda_{m\uparrow}}{\lambda_n + \lambda_{m\uparrow}}$, then $T(x) = n$.

Here the optimum partition thresholds are

$$\alpha = \frac{n\lambda_n + m\lambda_{m\uparrow}}{\lambda_n + \lambda_{m\uparrow}},\ \beta = \frac{p\lambda_p + m\lambda_{m\downarrow}}{\lambda_p + \lambda_{m\downarrow}}. \tag{3.26}$$

When $\lambda_n = \lambda_{m\uparrow} = \lambda_{m\downarrow} = \lambda_p = 1$, minimum distance-based three-way approximation of a fuzzy set is obtained from Equation (3.26) with optimum threshold values:

$$\alpha = \frac{n+m}{2}, \beta = \frac{p+m}{2}. \tag{3.27}$$

3.4.4 Example on Decision-Theoretic Three-Way Approximation

Let us consider the fuzzy set given in Section 3.6:

$$F = \{(a, 0.1), (b, 0.25), (c, 0.3), (d, 0.55), (e, 0.6), (f, 0.75), (g, 0.8), (h, 1)\}.$$

To induce a T from F based on the models Equations (3.16), (3.17) and (3.18), (3.19), and also the one proposed in this chapter (i.e., Equations (3.26) and (3.27)), there are three main steps before calculating the optimum thresholds of each method:

Step 1: Compute the value of m for [28] model according to conditions (i)-(iv) of Section 4.2.

Step 2: Assign unit costs values to λ_n, $\lambda_{m\uparrow}$, $\lambda_{m\downarrow}$ and λ_p according to conditions (i) and (ii) of Section 4.

Here we assign the values $\lambda_n = 300$, $\lambda_{m\uparrow} = 8$, $\lambda_{m\downarrow} = 8$ and $\lambda_p = 250$ in line with [28].

Step 3: Compute the values of n, m and p for the proposed model according to Equations (3.20) and (3.21).

We note that all the decision-theoretic models require Step 2, only the model in [28] requires Step 1. Also, Step 3 is only required by the proposed model.

DY model:
Substituting $\lambda_n = 300$, $\lambda_{m\uparrow} = 8$, $\lambda_{m\downarrow} = 8$ and $\lambda_p = 250$ in Equations (3.16) and (3.17) as appropriate, we have:

Cost-sensitive model:

$$\alpha = \frac{\lambda_{0.5\uparrow}}{2(\lambda_0 + \lambda_{0.5\uparrow})} = \frac{8}{2(300+8)} = 0.013,$$

$$\beta = \frac{2\lambda_1 + \lambda_{0.5\downarrow}}{2(\lambda_{0.5\downarrow} + \lambda_1)} = \frac{2 \times 250 + 8}{2(250 + 8)} = 0.984.$$

$$T = POS(T) \cup BND(T),$$

where $POS(T) = \{(h, 1)\}$ and $BND(T) = \{(b, 0.5), (c, 0.5), (d, 0.5), (e, 0.5), (f, 0.5), (g, 0.5)\}$.

Based on the obtained thresholds, $NEG(T) = \varnothing$.

Minimum distance model:

$$\alpha = 0.25,$$

$$\beta = 0.75.$$

$$T = POS(T) \cup BND(T),$$

where $POS(T) = \{(h, 1), (g, 1)\}$ and $BND(T) = \{(b, 0.5), (c, 0.5), (d, 0.5), (e, 0.5), (f, 0.5)\}$.

Based on the obtained thresholds, $NEG(T) = \{(a, 0)\}$.

ZXLW model:

From the choice of unit cost, $m = mean(\{0.1, 0.25, 0.3, 0.55, 0.6, 0.75, 0.8, 1\}) = 0.54$.

Substituting $\lambda_n = 300, \lambda_{m\uparrow} = 8, \lambda_{m\downarrow} = 8, \lambda_p = 250$ and $m = 0.54$ in Equations (3.18) and (3.19) as appropriate, we have:

Cost-sensitive model:

$$\alpha = \frac{m\lambda_{m\uparrow}}{\lambda_0 + \lambda_{m\uparrow}} = \frac{0.54 \times 8}{300 + 8} = 0.014,$$

$$\beta = \frac{\lambda_1 + m\lambda_{m\downarrow}}{\lambda_{m\downarrow} + \lambda_1} = \frac{250 + 0.54 \times 8}{250 + 8} = 0.986.$$

$$T = POS(T) \cup BND(T),$$

where $POS(T) = \{(h, 1)\}$ and $BND(T) = \{(b, 0.54), (c, 0.54), (d, 0.54), (e, 0.54), (f, 0.54), (g, 0.54)\}$.

Based on the obtained thresholds, $NEG(T) = \emptyset$.

Minimum distance model:

$$\alpha = \frac{m}{2} = \frac{0.54}{2} = 0.27.$$

$$\beta = \frac{1+m}{2} = \frac{1+0.54}{2} = 0.77.$$

$$T = POS(T) \cup BND(T),$$

where $POS(T) = \{(h, 1), (g, 1)\}$ and $BND(T) = \{(c, 0.54), (d, 0.54), (e, 0.54), (f, 0.54)\}$.

Based on the obtained thresholds, $NEG(T) = \{(a, 0), (b, 0)\}$.

Proposed model:
From Equations (3.20) and (3.21), we have two types of region' membership grades:

1) $$n_1 = \frac{\mu_{F_{max}} + 3\mu_{F_{min}} - 2\overline{\mu_F(x)}}{2} = \frac{1+3\times 0.1 - 2\times 0.54}{2} = 0.11,$$

$$m_1 = \frac{\mu_{F_{max}} + \mu_{F_{min}}}{2} = \frac{1+0.1}{2} = 0.55 \text{ and}$$

$$p_1 = \frac{2\overline{\mu_F(x)} + \mu_{F_{max}} + \mu_{F_{min}}}{2} = \frac{2\times 0.54 + 1 - 0.1}{2} = 0.99$$

Substituting the values of $n_1, m_1, p_1, \lambda_n = 300, \lambda_{m\uparrow} = 8, \lambda_{m\downarrow} = 8, \lambda_p = 250$ in Equations (3.26) and (3.27) as appropriate, we have

Cost-sensitive model:

$$\alpha = \frac{n\lambda_n + m\lambda_{m\uparrow}}{\lambda_n + \lambda_{m\uparrow}} = \frac{0.11\times 300 + 0.55\times 8}{300+8} = 0.121,$$

$$\beta=\frac{p\lambda_p+m\lambda_{m\downarrow}}{\lambda_p+\lambda_{m\downarrow}}=\frac{0.99\times 250+0.55\times 8}{250+8}=0.976.$$

$$T = POS(T) \cup BND(T)$$

where $POS(T) = \{(h, 0.99)\}$ and $BND(T) = \{(b, 0.55), (c, 0.55), (d, 0.55), (e, 0.55), (f, 0.55), (g, 0.55)\}$.

Based on the obtained thresholds, $NEG(T) = \{(a, 0.11)\}$.

Minimum distance model:

$$\alpha=\frac{n+m}{2}=\frac{0.11+0.55}{2}=0.33.$$

$$\beta=\frac{p+m}{2}=\frac{0.99+0.55}{2}=0.77.$$

$$T = POS(T) \cup BND(T),$$

where $POS(T) = \{(h, 0.99),(g, 0.99)\}$ and $BND(T) = \{(d, 0.55), (e, 0.55), (f, 0.55)\}$.

Based on the obtained thresholds, $NEG(T) = \{(a, 0.11), (b, 0.11), (c, 0.11)\}$.

2) $n_2=\mu_{F_{min}}=0.1, m_2=\overline{\mu_F(x)}=0.54$ and $p_2=\mu_{F_{max}}=1$

Substituting the values of n_2, m_2, p_2, $\lambda_n = 300$, $\lambda_{m\uparrow} =, 8\lambda_{m\downarrow} = 8$, $\lambda_p = 250$ in Equations (3.26) and (3.27) as appropriate, we have

Cost-sensitive model:

$$\alpha=\frac{n\lambda_n+m\lambda_{m\uparrow}}{\lambda_n+\lambda_{m\uparrow}}=\frac{0.1\times 300+0.54\times 8}{300+8}=0.111,$$

$$\beta=\frac{p\lambda_p+m\lambda_{m\downarrow}}{\lambda_p+\lambda_{m\downarrow}}=\frac{1\times 250+0.54\times 8}{250+8}=0.986.$$

$$T = POS(T) \cup BND(T),$$

where $POS(T)$={(h,1)} and $BND(T)$={(b,0.54),(c,0.54),(d,0.54),(e,0.54),(f,0.54),(g,0.5)4}.

Based on the obtained thresholds, $NEG(T)$ = {(a,0.1)}.

Minimum distance model:

$$\alpha = \frac{n+m}{2} = \frac{0.1+0.54}{2} = 0.32.$$

$$\beta = \frac{p+m}{2} = \frac{1+0.54}{2} = 0.77.$$

$$T = POS(T) \cup BND(T),$$

where $POS(T)$ = {(h,1),(g,1)} and $BND(T)$ = {(d,0.54),(e,0.54),(f,0.54)}.

Based on the obtained thresholds, $NEG(T)$ = {(a,0.1),(b,0.1),(c,0.1)}.

3.5 Concluding Remarks and Future Directions

This chapter has considered several three-way approximation methods in simplied form. The methods are grouped into two categories, *viz.*, shadowed set approximation and decision-theoretic three-way approximation. The shadowed set approximation methods rely on a search algorithm to determine optimum partition thresholds. Hence, some additional computational time is required as compared to decision-theoretic approximation methods, which require no search algorithm. Unlike shadowed set approximation methods, decision-theoretic methods heavily depend on the unit decision costs to produce optimum thresholds.

The influence of variable values of n, m and p on decision-theoretic three-way approximations in real applications and a comparative study on the performance of these models on synthetic and some benchmark datasets from UCI Machine Learning repository are opened for further considerations.

By subjecting the SSA and DTA models to relevant case studies and experimental studies from benchmark datasets in UCI Machine Learning repository, appreciable plus and promising advantages of the models can easily be characterized. Finally, due to lack of detection of optimum threshold from nature of data, decision-theoretic approximation techniques may not exhibit better performance compared to shadowed set methods when applied to unsupervised learning. As a remedy, when the values of n, m and p are determined before the learning process, the performance of the model may be enhanced.

References

1. Barman B., Mitra S., and Pedrycz W., "Shadowed Clustering for Speech Data and Medical Image Segmentation", RSCTC '08: *Proceedings of the 6th International Conference on Rough Sets and Current Trends in Computing,* 475–484, 2008.
2. Cattaneo G., and Ciucci D., "Shadowed Sets and Related Algebraic Structures", *Fundamenta Informaticae* 55: 255–284, 2003.
3. De Luca A. and Termini S., "A Definition of a Nonprobabilistic Entropy in the Setting of Fuzzy Sets Theory", *Inform. and Control* (1972), 301-312.
4. Deng X. F. and Yao Y. Y., "Decision-Theoretic Three-Way Approximations of Fuzzy Sets", *InformationSciences*, 79: 702–715, 2014.
5. Gao M., Zhang Q., Zhao F., and Wang G., "Mean-Entropy-Based Shadowed Sets: A Novel Three-Way Approximation of Fuzzy Sets". *Approximate Reasoning*, 120: 102 –124, 2020.
6. Higashi M., and Klir G. J., "On Measures of Fuzziness and Fuzzy Complements", *International Journal ofGeneral Systems* 8(3):169–180, 2008.
7. Ibrahim M. A. and William-West T.O., "Induction of Shadowed Sets from Fuzzy Sets", *GranularComputing*, 4:27–38, 2019.
8. Ibrahim M. A., William-West T. O., Kana A. F. D., and Singh D., "Shadowed Sets with Higher Approximation Regions", *SoftComputing*, Springer, Heidelberg, doi.org/10.1007/s00500-020-04992-8, 2020.
9. Kaufmann A., *Introduction to the Theory of Fuzzy Subsets - Fundamental Theoretical Elements*, Vol. I, Academic Press, New York, 1975.
10. Klir G. J., "A Principle of Uncertainty and Information Invariance", *International Journal of GeneralSystems* 17(2): 249–275, 1990.
11. Knopfmacher. J., "On measures of fuzziness", *Journal of Mathematical Analysis and Applications* 49 (1975) 529–534.
12. Liu. D., and Liang D. C., "Three-Way Decisions in Ordered Decision System", *Knowledge BasedSystems*, 137:182–195, 2017, 2017.
13. Mitra S., Pedrycz W. and Barman B., "Shadowed C-Means: Interpreting Fuzzy and Rough Clustering", *Pattern Recognition*, 43: 1282–1291, 2010.
14. Mitra S. and Kundu P. P. "Satellite Image Segmentation with Shadowed C-Means", *Information Sciences*, 181: 3601–3613, 2011.

15. Nguyen H. T., Pedrycz W. and Kreinovich V., "On Approximation of Fuzzy Sets by Crisp Sets: From Continuous Control-Oriented Defuzzification to Discrete Decision Making", *Proceeding of InternationalConference on Intelligent Technologies, Bangkok, Thailand*, 254–260, 2000.
16. Pauker, S. G., Kassirer, J. P., "The threshold approach to clinical decision making". *New England Journal of Medicine* 302, 1109-1117, 1980.
17. Pedrycz W., and Vukovich G., "Granular computing with shadowed sets", *International Journal of Intelligent Systems*, (17) 173–197, 2002.
18. Pedrycz. W., "Interpretation of Clusters in the Framework of Shadowed sets", *Pattern Recognition Letters*, 26: 2439– 2449, 2005.
19. Pedrycz W., "Shadowed sets: Representing and processing fuzzy sets", *IEEE Transactions on System,* Man *and Cybernetics*, 28:103–109, 1998.
20. Tahayori H., Sadeghian A., and Pedrycz W., "Induction of shadowed sets based on the gradual grade of fuzziness", *IEEE Transactions on Fuzzy System* 21: 937–949, 2013.
21. William-West T. O., M. A. Ibrahim and Kana A. F. D., "Shadowed sets approximation of fuzzy sets based on nearest quota of fuzziness", *Annals of Fuzzy Mathematics and Informatics*, 17:133–145, 2019.
22. Yao. Y. Y., "An Outline of a Theory of Three-Way Decisions", In: Yao J., Yang Y., Slowinski R., Greco S., Li H., Mitra S., and Polkowski L., (eds.) *RSCTC 2012.Lecture Notes in Computer Sciences*, Springer, Heidelberg (LNAI), 7413: 1–17, 2012.
23. Yao. Y. Y., "Three-Way Decisions and Cognitive Computing", *Cognition Computing*, 8: 543–554, 2016.
24. Yao J., and Azam N., "Web-Based Medical Decision Support Systems for Three-Way Medical Decision Making with Game-Theoretic Rough Sets", *IEEE Transactions on Fuzzy Systems* 23(1):3-15, 2015.
25. Yao, Y. Y., Wang S., and Deng X. F., "Constructing Shadowed Sets and Three-Way Approximations of fuzzy sets", *Information Sciences*, 413: 132–153, 2017.
26. Zadeh. A., "Fuzzy Sets", *Information and Control* 8: 338–353, 1965.
27. Zhang. Y., and Yao J. T., "Game Theoretic Approach to Shadowed Sets: A Three-Way Tradeoff Perspective", *Information Sciences*, doi:10.1016/j.ins.2018.07.058, 2020
28. Zhang. Q., Xia D., Liu K.., and Wang G., "A General Model of Decision-Theoretic Three-Way Approximations of Fuzzy Sets Based on a Heuristic Algorithm", *InformationSciences* 507: 522–539, 2020.
29. Zhou. j., Pedrycz W., and Miao D., "Shadowed Sets in the Characterization of Rough-Fuzzy Clustering", *Pattern Recognition*, 44: 1738–1749, 2011.
30. Zhou. J., C. Gao., Pedrycz W., and Lai Z., "Constrained Shadowed Sets and Fast Optimization Algorithm". *International Journal of Intelligent Systems* 34:2655–2675. https://doi.org/10.1002/int.22170, 2019.
31. Zhou. J., W. Pedrycz., C. Gao., Z. Lai., X. Yue., "Principles for Constructing Three-Way Approximation of Fuzzy Sets: A Comparative Evaluation Based

on Unsupervised Learning". *Fuzzy Sets and Systems,* doi.org/10.1016/j.fss.2020.0619, 2020.

32. Zhou, B., Yao, Y., & Luo, J. (2010). *A Three-Way Decision Approach to Email Spam Filtering. Lecture Notes in Computer Science, 28–39.* doi:10.1007/978-3-642-13059-5_
33. LIU, D., LI, T., & LIANG, D. (2012). *three-way government decision analysis with decision-theoretic rough sets. International Journal of Uncertainty, Fuzziness and Knowledge-Based Systems, 20(supp01), 119–132.* doi:10.1142/s0218488512400090

4

Intuitionistic Fuzzy Rough Sets: Theory to Practice

Shivani Singh[1] and Tanmoy Som[2*]

[1]DST-Centre for Interdisciplinary Mathematical Sciences, Institute of Science, Banaras Hindu University, Varanasi, India
[2]Department of Mathematical Sciences, Indian Institute of Technology (Banaras Hindu University), Varanasi, India

Abstract

The Rough set theory is a very successful tool to deal with vague, inconsistent, imprecise and uncertain knowledge. In recent years, rough set theory and its applications have drawn many researchers' interest progressively in one of its hot issues, viz. the field of artificial intelligence. An intuitionistic fuzzy (IF) set, which is a generalization of fuzzy set, has more practical and flexible real-world proficiency to characterize a complex information and provide a better glimpse to confront uncertainty and ambiguity when compared with those of the fuzzy set. Moreover, rough sets and intuitionistic fuzzy sets deal with the specific aspects of the same problem – imprecision, and their combination IF rough set has been studied by many researchers in the past few years. The present chapter deals with a review of IF rough set theory, their basic concepts, properties, topological structures, logic operators, approximation operators and similarity relations on the basis of axiomatic and constructive approaches. The characterization of IF rough sets based on various operators, similarity relations, distances, IF cut sets, IF coverings and inclusion degrees is also discussed. Moreover, several extensions of IF rough sets and their hybridization with other extended rough set theories are thoroughly surveyed. The applications of IF rough sets in different real-world problems are also discussed in detail.

Keywords: Rough set, intuitionistic fuzzy set, intuitionistic fuzzy rough set, feature selection, decision making

**Corresponding author*: tsom.apm@iitbhu.ac.in

Ramakant Bhardwaj, Jyoti Mishra, Satyendra Narayan and Gopalakrishnan Suseendran (eds.) Mathematics in Computational Science and Engineering, (91–134) © 2022 Scrivener Publishing LLC

4.1 Introduction

Fuzzy sets, introduced by Zadeh [1], provide an effective method of representing uncertainty and vagueness in order to describe the behavior of systems that are too inaccurate or too complex to process certain mathematical analysis with classical tools and methods. From that time onward, fuzzy set theory has become a powerful research field in several disciplines, such as computer networks, artificial intelligence, decision making, signal processing, medical and life sciences, social sciences, pattern recognition, graph theory, robotics, management sciences, expert systems, engineering, automata theory and multi-agent systems.

The notion of rough set theory (RST) was first proposed by Pawlak [2], as a useful mathematical tool to deal with intelligent systems described by incomplete, noisy or inexact information. The fundamental structure of rough set theory is an approximation space comprising a universe of discourse with an imposed binary relation. In an information system, hidden knowledge can be expressed and unravelled as decision rules by incorporating the concept of lower approximation and upper approximation of all decision classes regarding the approximation space induced from the set of conditional attributes. RST does not need any additional or prior information for most of the knowledge discovery process in a database and deals with vagueness after probability theory, fuzzy set theory (FST), and evidence theory. Nevertheless, the major drawback of this theory is its discretization process before pre-processing of data as it can only be implemented to discrete information systems, which leads to some information loss in the system.

To handle the above drawback, rough set and fuzzy set are integrated to build a fuzzy rough set (FRS) based technique introduced by Dubois and Prade [3, 4]. Fuzzy rough set idea has been realized to outperform the shortcomings of the traditional rough set technique from different perspectives. FRS has been used to preserve a strategic distance from the loss of data due to discretization.

Though FST has a great ability to handle imprecision, it does have some limitations. FST has an inability of handling several decision-making issues; for instance, in a group leader selection problem of 25 persons, assume that 12 persons vote in favour of a particular person with a conclusion "agree", eight of them "oppose" and the remaining persons "abstain" from any decision. Such a situation can be efficiently processed by providing a non-membership degree for "oppose" and hesitancy degree for "abstain". Furthermore, a data structure in which data values are vaguely specified is prevalent these days in various real-life applicable problems, like sensor information, medical diagnosis, fault prediction, etc. FST is applied to cope with such vagueness by extending the idea of membership of any object in a set. In this theory, any element belongs to a universe with a single value between 0 and 1 as the membership degree, but that single value may not encompass all the information of the system since uncertainty lies in both judgment as well as identification. Hence, some advancements of FST are needed to manage such a situation.

Considering the above, IF set (proposed by Atanassov) [5] as an extension of fuzzy set is another significant mathematical framework to deal with imprecise and/or incomplete information. It takes into account positive, negative as well as hesitancy degree of an object to associate with a set. Consequently, it has a powerful ability to better illustrate the vagueness in the case where the introduction of non-membership degree is rather easier than that of membership degree. Hence, it is foreseen that IF set concept can be used to simulate the inevitably imprecise or not completely reliable activities that require human knowledge, expertise and decision-making process. In a larger sense intuitionistic FST-based methods have been effectively applied in decision-making techniques and pattern identification.

The hybridization of the theory of IF set and rough set leads to a new mathematical structure for the prerequisite of knowledge handling techniques. A number of authors have researched on this subject. In this paper, we survey the IF rough set theory from theory to practice. The structure of the rest of the chapter is organized as follows. Some preliminaries on Rough set, fuzzy set, IF sets and IF rough set are discussed in Section 4.2. In Section 4.3, a detailed survey on IF rough sets, their construction and properties is presented. The study of hybridization and extension theory of IF rough sets is presented in Section 4.4. In Section 4.5, a systematic survey on applications of IF rough sets is added. We present the work distribution of IF rough set country-wise and year-wise with the limitation of IF rough set theory in Section 4.6. In Section 4.7, we conclude our work.

4.2 Preliminaries

In this section, we discuss the concepts and basic definitions related to rough set theory, IF set theory and IF rough set theory.

4.2.1 Rough Set Theory

Rough set theory (RST) provides great convenience as a mathematical tool when used to tackle with uncertainty. It can be used to extract the knowledge from any domain in a concise manner as it reduces the size of the system by maintaining the original information content.

Definition 2.1 (*Information System*) [6]. An information system is a quadruple expressed as (U, P, V, f), where U (non-empty) is a set of finite objects, P (non-empty) is finite set of attributes, $V = \bigcup_{b\hat{\imath}P} V_b$ is the set of values V_b of attribute b, and $f{:}U \rightarrow V$ is an information or description function. If C and D are set of conditional attributes and decision attributes, respectively and $P = C \cup D$ with condition $C \cap D = \varnothing$, then (U, P, V, f) is referred to as a decision system.

Definition 2.2 (*Indiscernible Relation*) [2]. For any subset of attributes $B \subseteq P$, an associated equivalence relation R_B is defined as

$$R_B = \{(x, y) \in U \times U \,|\forall\, b \in B, b(x) = b(y)\}$$

x and y are said to be indiscernible by attributes from B if $(x, y) \in R_B$. Here, the equivalence relation is a B-indiscernible relation with equivalence classes as $[x]_B$. The pair $(U, [x]_B)$ is said to be an approximation space.

Definition 2.3 (*Upper and lower approximation*) [2]. Let (U, P, V, f) is an information system. For $X \subseteq U$; X can be approximated by B-lower approximation $R_B \downarrow X$ and B-upper approximation $R_B \uparrow X$ which are defined as below:

$$R_B \downarrow X = \{x \in U \mid [x]_B \subseteq X\}$$

$$R_B \uparrow X = \{x \in U \mid [x]_B \cap X \neq \varnothing\}$$

The pair $(R_B \downarrow X, R_B \uparrow X)$ is called a rough set. Figure 4.1 demonstrates the lower and upper approximation of the set X.

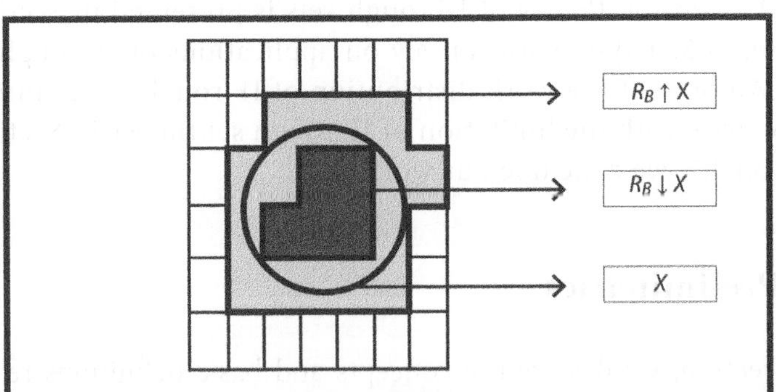

Figure 4.1 Lower and upper approximation of set X.

4.2.2 Intuitionistic Fuzzy Set Theory

Intuitionistic fuzzy sets, as a further development of fuzzy sets, can well simulate and describe the fragile natural vagueness of the real world. It consists of non-membership degree along with membership degree, which gives extra parameter to better describe the characteristics of things and handle the uncertain information in a better way.

Definition 2.4 (*Intuitionistic Fuzzy Set*) [5]. Given a non-empty finite universe of discourse U. A set A on U having the form $A = \{\langle x, \mu_A(x), \vartheta_A(x)\rangle | x \in U\}$ is said to be an intuitionistic fuzzy (IF) set, where $\mu_A : U \to [0,1]$ and $\vartheta_A : U \to [0,1]$ with $0 \leq \mu_A(x) + \vartheta_A(x) \leq 1$ for all x in U. $\mu_A(x)$, $\vartheta_A(x)$ and $\pi_A(x) = 1 - \mu_A(x) - \vartheta_A(x)$ are the membership, non-membership and hesitancy degree of the element x in A, respectively.

Any fuzzy set $A = \{\langle x, \mu_A(x)\rangle | x \in U\}$ is an IF set as it can be characterized by an IF set of the form $\{\langle x, \mu_A(x), 1 - \mu_A(x)\rangle | x \in U\}$. The relation between crisp set, fuzzy set and IF set can be seen in Figure 4.2.

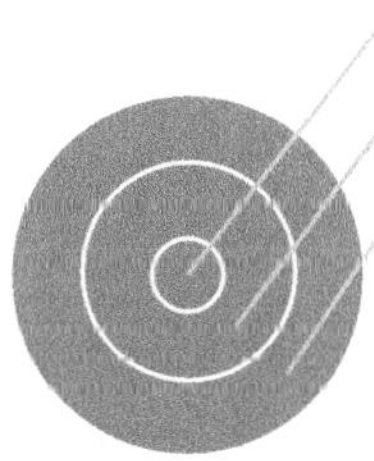

Figure 4.2 Intuitionistic fuzzy set as a generalization of fuzzy set.

Properties. Let A_1 and A_2 are two IF sets in X,

1. $A_1 \subseteq A_2$ iff $(\forall x \in X)(\mu_{A_1}(x)) \leq \mu_{A2}(x), \vartheta_{A_1}(x) \geq \vartheta_{A_2}(x))$
2. $A_1 \supseteq A_2$ iff $A_2 \subseteq A_1$
3. $A_1^c = \{x, \vartheta_{A_1}(x), \mu_{A_1}(x), x \in X\}$
4. $A_1 \cup A_2 = \{(x, \max(\mu_{A_1}(x), \mu_{A2}(x)), \min(\vartheta_{A_1}(x), \vartheta_{A2}(x))\}$
5. $A_1 \cup A_2 = \{(x, \min(\mu_{A_1}(x), \mu_{A2}(x)), \max(\vartheta_{A_1}(x), \vartheta_{A2}(x))\}$
6. If X is finite, the cardinality of the IF set A_1, is calculated as:

$$|A_1| = \sum_{x \in A_1} \frac{1 + \mu_{A_1}(x) - \vartheta_{A_1}(x)}{2}.$$

4.2.3 Intuitionistic Fuzzy-Rough Set Theory

The theory of Rough set has been widely used by researchers for many problems such as decision making, feature selection and rule acquisition, though, the traditional rough set is not successful for handling the attributes with continuous or real values, which occur more frequently than the attributes having nominal values in the physical world. Thus, IF rough set, shown in Figure 4.3, has gained popularity in recent years to solve the issues created by rough sets and fuzzy sets as discretization of nominal values and better deals with information loss in several real-world problem with the presence of extra parameter, i.e., non-membership degree. IF rough sets use IF similarity or IF tolerance relations for IF data values of objects in place of equivalence relations in use by the RST. IF lower approximation and upper approximation are defined with the usage of similarity relation, IF t-norm and IF implicator.

Definition 2.5 (*IF tolerance relation*) [7]. An IF binary relation $R(x_p, x_q) = < \mu_A (x_p, x_q), \vartheta_A (x_p, x_q) >$ between objects $x_p, x_q \in U$ is said to be an IF tolerance relation if it is reflexive (i.e. $\mu_A(x_p, x_p) = 1$ and $\vartheta_A(x_p, x_p) = 0, \forall x_p \in X$) and symmetric (i.e., $\mu_A(x_p, x_q) = \mu_A(x_q, x_p)$ and $\vartheta_A(x_p, x_q) = \vartheta_A(x_q, x_p), \forall x_p, x_q \in X$).

Definition 2.6 (*IF triangular-norm*) [8]. An IF triangular norm or IF *t*-norm T is a mapping from $[0,1] \times [0,1] \to [0,1]$ which is increasing, associative and commutative and satisfies $T(1, x) = x, \forall x \in [0,1]$.

Example 2.1. For IF sets $x = (x_p, x_q), y = (y_p, y_q)$ in [0,1], few widely used IF t-norms are given as:

$$T_M(x, y) = \langle \min(x_p, y_p), \max(x_q, y_q) \rangle$$

$$T_W(x, y) = \Box\max(0, x_p + yp - 1), \min(1, x_q + y_q) \rangle$$

Figure 4.3 Intuitionistic fuzzy rough set.

Definition 2.7 (*IF Implicator*) [8]. An IF implicator is a mapping I: $[0,1] \times [0,1] \rightarrow [0,1]$, which is decreasing in its first component, and increasing in second with condition $I(0,0) = 1$ and $I(1,x) = x, \forall x \in [0,1]$.

Example 2.2. For IF sets $x = (x_p, x_q)$, $y = (y_p, y_q)$ in $[0,1]$, few widely used IF implicators corresponding to t-norms given in Example 2.2 are as follows:

$$I_{TM}(x,y) = \left\{ \begin{array}{ll} 1 & \text{if } x_p \leq y_p \text{ and } x_q \geq y_q \\ (1-y_q, y_q) & \text{if } x_p \leq y_p \text{ and } x_q < y_q \\ (y_p, 0) & \text{if } x_p > y_p \text{ and } x_q \geq y_q \\ (y_p, y_q) & \text{if } x_p > y_p \text{ and } x_q < y_q \end{array} \right\}$$

$$I_{TW}(x, y) = \langle \min(1, 1 + \mu_2 - \mu_1, 1 + \vartheta_1 - \vartheta_2), \max(0, \vartheta_2 - \vartheta_1) \rangle$$

Definition 2.8 (*IF Upper and Lower Approximation*) [8]. Given an IF set $X \subseteq U$ and $R(x_p, x_q)$ is an IF similarity relation from $U \times U \rightarrow [0,1]$ which assigns degree of similarity to each distinct pair of objects. The lower and upper approximation of X by R can be computed in many ways. A general definition is given as:

$$\underline{R_I}X(x_p) = \inf_{xq \in U} I(R(x_p, x_q), X(x_q), \forall x_p, x_q \in U$$

$$\overline{R_T}X(x_i) = \sup_{x_j \in U} T(R(x_p, x_q), X(x_q), \forall x_p, x_q \in U$$

Here, I is an IF implicator and T is an IF t-norm and $X(x_q) = 1$, for $x_q \in X$, otherwise $X(x_q) = 0$. The pair $(\underline{R_I}X(x_p), \overline{R_T}X(x_q))$ is called as IF rough set.

4.3 Intuitionistic Fuzzy Rough Sets

As rough sets and IF sets both encapsulate specific facets of the same notion – imprecision, the work intended for the integration of IF set theory and RST has become a topic of interest for many researchers in order to deal with imprecision in a better way. This section, discusses a detailed survey on IF rough sets and rough IF sets introduced by many researchers.

In 1998, Chakrabarty *et al.* introduced a framework to construct an IFRS (U, V) of a rough set (A, B), where U and V are both IF sets in X such that

$U \subseteq V$ [9]. The lower approximation U and upper approximation V introduced here are both IF sets. Samanta and Mondal also proposed such notion and termed it as rough IF rough in 2001 [10]. They generalized the concept proposed by Nanda and Majumdar [11] and defined a couple (U, V) IF rough set such that membership function U and non-membership function V are both fuzzy rough sets are no longer fuzzy sets and $U \subseteq$ Complement (V). In 2002, Jena *et al.* independently proposed the notion of IF rough sets, where the lower and upper approximations are both IF sets [12].

However, the above introduced IF rough sets and rough IF sets are not characterized by means of an approximation space. In contrast, the above methods as well as the lines of Pawlak's rough sets, Rizvi *et al.* [13] advanced this idea to rough IF sets by using a Pawlak approximation space, but in that case, the lower and upper approximations are IF sets in the family of equivalence classes generated by the equivalence relation R instead of IF sets in the universe of discourse. To conquer this drawback, in 2003, an approximation space determined by an IF triangular norm T, an IF implicator I, and an IF T-equivalence relation on the universe of discourse is used by Cornelis *et al.* to propose the notion of IF rough sets with the lower and upper approximations as IF sets again in the universe [8].

In 2006, Radzikowska presented rough approximation operations based on IF sets and discussed some properties of IF rough approximation operations [14]. It is shown in the paper that their methodology allows them to demonstrate weak/strong certainties and possibilities of membership as well as non-membership of elements to a set. In 2007, Zhou and Wu discussed rough approximations in IF sets and introduced the notion of intuitionistic rough fuzzy sets and IF rough sets also examined their basic operations and properties [15].

In 2008, Zhou and Wu proposed a structure for relation-based IF rough approximation operators by using both constructive and axiomatic approaches and by introducing cut sets of IF sets, they presented classical representations of IF rough approximation operators. An operator-oriented characterization of IF rough sets is also proposed by them [16]. Wu and Zhou investigated the topological structures of IF rough sets and showed that an IF topological space could be induced by IF rough approximation space if and only if the IF relation in the approximation space is reflexive and transitive [17]. Lu *et al.* proposed IF set, IF inclusion degree and lower and upper approximation based on a special Lattice and constructed the model of IF rough sets based on inclusion degree [18]. A model of IF rough set by introducing t-norm, t-conorm, S-implicator and R-implicator based on IF similarity relation is presented by Lu *et al.*, and their properties is examined by them [19]. Xu *et al.* introduced IF rough sets determined by IF triangular norm [20].

In 2009, Lin and Wang investigated the relationship between IF rough sets and IF topologies on finite universes and proposed IFT condition for IF topology and proved that IF topology induced by IF approximation space satisfies IFT condition [21]. Liu and Liu studied IF rough sets by means of axiomatic characterization and characterized IF approximations with two simple axioms. They proved that the collection of all definable IF sets is a completely distributive lattice [22]. An integration between the theories of IF descriptive logics and rough descriptive logics by providing IF rough descriptive logics based on IF rough set theory is presented by Jiang *et al.* [23]. Zhou *et al.* presented a general scenario for the study of relation-based (I, T)-IF rough sets with both constructive and axiomatic approaches and established the connection between special types of IF relations and IF operators [24]. The notion of pseudo-closure operator is introduced by Zhou *et al.*, IF topological spaces and IF rough sets in an infinite universe of discourse are studied by them [25].

In 2010, Lin and Wang proposed a general scenario for the study of IF rough approximation operators determined by IF triangle norm, defined lower and upper approximations generated from IF relation, and examined the basic properties of IF rough operators [26]. Xu and Wu defined the concept and properties of IF rough *set al*gebras (IFRSA) by axiomatic approach and proved that the derived system of an IFRSA is IFRSA [27]. A general kind of relation-based $(I, \daleth)$-IF rough sets determined by two IF implicators I and $\daleth$ is proposed, and the properties of IF rough approximation operators are presented by Wu [28]. Zhang *et al.* introduced the ideas of an IF set, an IF covering and IF logical operators. This article structured two pairs of generalized lower and upper IF rough approximation operators with the help of an IF triangular norm and an IF implicator based on an IF covering of a universe of discourse. The primary properties of the generalized IF rough approximation operators and the nature of duality among the operators are viewed in this paper [29]. Two operations of IF sets and an axiomatic system of IF rough sets using the operations and the axiomatization mode are proposed by Yang *et al.* [30]. Tang defined an IF rough set based on k-step fuzzy relations and studied the basic properties of the IF rough operators and discussed the relationship between IF relations and IFRSA [31].

In 2011, Thomas and Nair presented basic notions of rough IF set theory and IF rough sets theory and its relationship with lattices. They proved that an IF rough set in $Apr(X)$ is an IF rough lattice if its level rough set $\left(\underline{A}_{(\alpha,\beta)}, \overline{A}_{(\alpha,\beta)}\right)$ is a rough sub lattice of $Apr(X)$ [32]. Based on the Hamming distance, Fan *et al.* proposed an improved measure of similarity between IF rough sets and by adding the hesitancy degree and weight, solved the problem of inaccurate similarity measure [33]. Zhou and Wu discussed the rough approximations of IF sets in crisp and fuzzy approximation spaces by constructive as well as axiomatic approaches and defined IF rough approximation operators and rough IF approximation operators by axioms [34]. The basic theory of IF rough sets over two universes by constructive approach is proposed by Sun *et al.* They introduced the concept of level cut sets for IF rough sets and established the decomposition theorem for IF rough sets over two universes [35]. Feng *et al.* discussed about IF Rough Sets and IF Topology Spaces [36]. Further, the IF rough set models based on the dual triangular norms are investigated by Gong and Zhao [37].

In 2012, Zhang constructed two pairs of generalized lower and upper IF rough approximation operators determined by an IF triangular norm, an IF implicator and an IF covering, and discussed the duality of the operators. To remove the redundancy in an IF covering, two notions, namely "reduct" and "exclusion" are introduced in the paper [38]. Acharjya and Tripathy introduced IF rough set on two universal sets and studied its properties. They developed the proposed concepts for the application in knowledge extraction and in designing knowledge bases [39]. Moreover, a new kind of IF rough sets by analysing its basic properties based on the concept of two universes are proposed by Zhang *et al.*, and IF rough approximation operators by axioms are defined by them [40]. Xu *et al.* introduced a pair of dual IF operators (θ, ϕ), constructed a novel IF rough set model based on an IF equivalence relation, and gave some properties of IF approximation operators [41]. Yang *et al.*, 2012, proposed a different kind of definition of IF rough set and examined the similarity measure between IF rough sets. A strong similarity measure between IF rough sets along with "upper rough degree" and "lower rough degree" for the characterization of the components in every IF rough value is proposed by them. They also developed a measure function to overcome the problem of market prediction [42]. A new believe factor related to an IF value in IF rough environment, a new kind of uncertain rules, i.e., believable rules, for improved performance in the field of decision-making are introduced by Chai *et al.* They implemented the proposed technique in multicriteria sorting [43]. The concept of incomplete multigranulation on rough IF sets (MGRIFS) proposed by Tripathy *et al.*, and topological properties of MGRIFS were studied in the paper [44]. Xu and Luo presented IF rough set model over two different universes and studied properties of lower/upper approximation operators [45].

In 2013, Weihua and Shuqun proposed IF rough set model over two different universes and studied some properties of proposed approximation operators [46]. Mukherjee and Das introduced IF rough relation on a set and demonstrated that the collection of such relations is closed under several binary compositions like algebraic sum, algebraic product, etc. They gave the definitions of various IF similarity relations and investigated their properties [47]. Yang and Yang improved IF RST by evaluating the independence of axioms in theory and presented the independent axiom set characterizing IF rough approximation operators [48]. Dual IF rough approximation operators characterised by an IF implication operator I in infinite universes of discourse are presented, and properties of such operators are investigated by Wu *et al.* [49]. Zhang and He presented a generalized IF rough sets by a constructive approach. IF approximation operators are defined by employing an IF residual implicator and its dual operator and some properties are examined by them. The connection between special types of IF relations and proposed IF rough approximation operators are established in this paper [50]. Huang *et al.* introduced a ranking technique for constructing the neighborhood of every object, obtained by means of IF values of condition attribute and proposed a dominance IF decision tables (DIFDT). Lower/upper approximation set of an object and crisp decision classes is discovered by the comparison between their relation. To obtain knowledge from proposed dominance IF decision tables, a lower and upper approximation reduction and rule extraction algorithm is developed with the help of discernibility matrix and discernibility function [51].

In 2014, Wu *et al.* investigated relation-based IF rough approximation operators determined by an IF triangular norm T and its dual IF triangular co-norm. Lower and upper approximation of IF sets corresponding to IF approximation space, relationships between special types of IF relations and related properties of the T- IF operators are discussed by them [52]. The topological structures of rough IF sets were investigated by Xu *et al.* They demonstrated that an IF Alexandrov space could be generated by a reflexive crisp relation and an IF clopen topological space could be produced by a similarity crisp approximation space [53].

In 2015, Liu and Lin constructed a unique IF rough set model by defined the conflict distance in IF environment. They developed a method to solve the problem of decision-making in IF information system by employing the notion of measuring IF similarity [54]. The basic properties of IF rough approximation operators and IF relations by means of topology are investigated by Yun and Lee. They proved that if the IF relation is reflexive, the upper approximation of any set is the set itself if the set is a lower set

[55]. Wu *et al.* proposed an axiomatic characterization of relation-based IF rough approximation operators determined by IF-implicator operator *I*. They proved that IF rough approximation operators produced by reflexive and *T*- transitive IF relations could be characterized by single axiom [56].

In 2016, Das developed a different approach of partially included IF rough (PIFR) relation on a set and determined sufficient conditions under which an IF rough relation becomes a PIFR-relation. They investigated semi-properties of IF rough relations such as reflexivity, irreflexivity, asymmetry, transitivity, etc. [57]. Zhong and Yan introduced lower and upper approximations of Intuitionistic *L*-fuzzy with respect to intuitionistic *L*-fuzzy approximation space determined by intuitionistic *L*-fuzzy *t*-norm and implicators and investigated their properties. They presented a general framework for the analysis of IF *L*- fuzzy rough sets, intuitionistic *L*-fuzzy pre-orders and IF *L*-fuzzy topologies, and proved that there is one-to-one correspondence between the set of all intuitionistic *L*-fuzzy pre-orders and the set of all intuitionistic *L*- fuzzy Alexandrov topologies [58].

In 2017, Wu and Yuan proposed IF rough set based on triple valued fuzzy sets used as cut sets of IF sets, constructed four methods for upper and lower approximation of IF sets using the representation theorem [59]. Bandyopadhyay *et al.* discovered the applicability of IF rough sets in a problem related to granular computing with game theoretic IF rough sets (GTIFRS). The so-constructed GTIFRS is defined over compatibility-based IF relation and the obtained imprecise information is found to be composed of IF granules [60]. A new definition of IF rough set as an extension of fuzzy rough sets (introduced by A. Nakamura [61]) is proposed, and the properties of the IF rough lower and upper approximations are explored by Sheeja and Sunny. A decomposition theorem is also presented for these approximations [62]. Bai *et al.* presented a study on IF rough set with axiomatic approach, researched correlative conclusion as well as equivalence conditions. They introduced two operators for binary order number based on fuzzy lattice and IF set, and briefly discussed relevant properties [63].

In 2019, Wu *et al.* presented axiomatic and operation-oriented characterization of relation-based IF rough approximation operators determined by an IF t-norm *T* and its dual IF t-conorm *S*. They introduced the concept of *T*-inner and *S*- outer products of IF sets [64]. Recently, in 2020, Bashir *et al.* studied the topological properties of IF rough sets in detail in presence of many conditions like serial, strongly serial, left continuity, transitivity on IF relations, t-norms and implicators. They explored special IF relations identified as *T*-similarity class algebraically and for better applicability of IF rough sets, discovered lattices to model

Table 4.1 Intuitionistic fuzzy rough sets.

S. no.	Authors	Proposed work
1	Chakarbarty *et al.*, 1998	Proposed a framework to construct an IFRS (U, V) of a rough set (A, B), where U (lower approximation) and V (upper approximation) are both IF sets in X such that $U \subseteq V$.
2	Samanta and Mondal, 2001	Introduced rough IF rough and defined the pair (U, V) as IF rough set such that U and V are both fuzzy rough sets and $U \subseteq$ Complement (V).
3	Jena *et al.*, 2002	Independently proposed the idea of IF rough sets in which the lower and upper approximations are both IF sets.
4	Rizvi *et al.*, 2002	Defined the concept of rough IF sets and studied their properties.
5	Cornelis *et al.*, 2003	Proposed the notion of IF rough sets in which the lower and upper approximations are both IF sets in the universe.
6	Radzikowska, 2006	Presented rough approximation operations based on IF sets, discussed some properties of IF rough approximation operations and demonstrated weak/ strong and possibilities of membership as well as non-membership of elements to a set.
7	Zhou and Wu, 2007	Discussed rough approximations in IF sets and introduced the notion of intuitionistic rough fuzzy sets and IF rough sets.
8	Zhou and Wu, 2008	Proposed a structure for relation-based IF rough approximation operators by using both constructive and axiomatic approaches and by introducing cut sets of IF sets, they presented classical representations of IF rough approximation operators.

(Continued)

Table 4.1 Intuitionistic fuzzy rough sets. (*Continued*)

S. no.	Authors	Proposed work
9	Wu and Zhou, 2008	Investigated the topological structures of IF rough sets and showed that an IF topological space could be induced by IF rough approximation space if and only if the IF relation in the approximation space is reflexive and transitive.
10	Lu *et al.*, 2008	Proposed IF set, IF inclusion degree and lower and upper approximation based on a special Lattice and constructed the model of IF rough sets based on inclusion degree.
11	Lu *et al.*, 2008	Presented a model of IF rough set by introducing t-norm, t-conorm, S-implicator and R-implicator based on IF similarity relation.
12	Xu *et al.*, 2008	Introduced IF rough sets determined by IF triangular norm.
13	Lin and Wang, 2009	Investigated the relationship between IF rough sets and IF topologies on finite universes and proposed IFT condition for IF topology.
14	Liu and Liu, 2009	Studied IF rough sets by means of axiomatic characterization and characterized IF approximations with two simple axioms.
15	Jiang *et al.*, 2009	Presented an integration between the theories of IF descriptive logics and rough descriptive logics by providing IF rough descriptive logics based on IF rough set theory.

(*Continued*)

Table 4.1 Intuitionistic fuzzy rough sets. (*Continued*)

S. no.	Authors	Proposed work
16	Zhou *et al.*, 2009	Presented a general scenario for the study of relation-based (*I*, *T*)-IF rough sets with both constructive and axiomatic approaches and established the connection between special types of IF relations and IF operators.
17	Zhou *et al.*, 2009	Introduced the notion of pseudo-closure operator and studied IF topological spaces and IF rough sets in an infinite universe of discourse.
18	Lin and Wang, 2010	Proposed a general scenario for the study of IF rough approximation operators determined by IF triangle norm, defined lower and upper approximations generated from IF relation.
19	Xu and Wu, 2010	Defined the concept and properties of IF rough *set al*gebras (IFRSA) by axiomatic approach and proved that derived system of an IFRSA is IFRSA.
20	Wu, 2010	Proposed a general kind of relation-based (*I*, ℸ)-IF rough sets determined by two IF implicators *I* and ℸ and presented the properties of IF rough approximation operators.
21	Zhang *et al.*, 2010	Structured two pairs of generalized lower and upper IF rough approximation operators by means of an IF triangular norm and an IF implicator based on an IF covering of a universe of discourse.
22	Yang *et al.*, 2010	Proposed two operations of IF sets and an axiomatic system of IF rough sets using the operations and the axiomatization mode.

(*Continued*)

Table 4.1 Intuitionistic fuzzy rough sets. (*Continued*)

S. no.	Authors	Proposed work
23	Tang, 2010	Defined an IF rough set based on k-step fuzzy relations and studied the basic properties of the IF rough operators and discussed the relationship between IF relations and IFRSA.
24	Thomas and Nair, 2011	Presented basic notions of rough IF set theory and IF rough sets theory and its relationship with lattices.
25	Fan *et al.*, 2011	Proposed an improved measure of similarity between IF rough sets and by adding the hesitancy degree and weight, solved the problem of inaccurate similarity measure.
26	Zhou and Wu, 2011	Discussed the rough approximations of IF sets in crisp and fuzzy approximation spaces by constructive as well as axiomatic approaches and defined IF rough approximation operators and rough IF approximation operators by axioms.
27	Sun *et al.*, 2011	Proposed the basic theory of IF rough sets over two universes by constructive approach. Introduced the concept of the level cut sets for IF rough sets and established the decomposition theorem for IF rough sets over two universes.
28	Feng *et al.*, 2011	Discussed about IF Rough Sets and IF Topology Spaces.
29	Gong and Zhao, 2011	Investigated the IF rough sets models based on the dual triangular norms.
30	Zhang, 2012	Constructed two pairs of generalized lower and upper IF rough approximation operators determined by an IF triangular norm, an IF implicator and an IF covering, and discussed the duality of the operators.

(*Continued*)

Table 4.1 Intuitionistic fuzzy rough sets. (*Continued*)

S. no.	Authors	Proposed work
31	Acharjya and Tripathy, 2012	Introduced IF rough set on two universal sets and studied its properties and proposed concepts for the application in knowledge extraction and in designing knowledge bases.
32	Zhang *et al.*, 2012	Proposed a new kind of IF rough sets by analyzing its basic properties based on the concept of two universes and defined IF rough approximation operators by axioms.
33	Xu *et al.*, 2012	Introduced a pair of dual IF operators (θ, ϕ), constructed a novel IF rough set model based on an IF equivalence relation.
34	Li *et al.*, 2012	Proposed a different kind definition of IF rough set and examined the similarity measure between IF rough sets, and developed a measure function to overcome the problem of market prediction.
35	Chai *et al.*, 2012	Introduced a new believe factor related to an IF value in IF rough environment, a new kind of uncertain rule, i.e., believable rules, for improved performance in the field of decision-making.
36	Tripathy *et al.*, 2012	Proposed the concept of incomplete MGRIFS proposed and studied some of it's topological properties.
37	Yang *et al.*, 2012	Proposed a different kind definition of IF rough set and examined the similarity measure between IF rough sets. Developed a measure function to overcome the problem of market prediction.

(*Continued*)

Table 4.1 Intuitionistic fuzzy rough sets. (*Continued*)

S. no.	Authors	Proposed work
38	Xu and Luo, 2012	Presented IF rough set model over two different universes and studied properties of lower/upper approximation operators.
39	Mukherjee and Das, 2012	Introduced IF rough relation on a set and demonstrated that the collection of such relations is closed under several binary compositions like algebraic sum, algebraic product, etc.
40	Yang and Yang, 2012	Improved IF rough set theory by evaluating the independence of axioms in the theory and presented the independent axiom set characterizing IF rough approximation operators.
41	Weihua and Shuqun, 2013	Proposed IF rough set model over two different universes and studied some properties of proposed approximation operators.
42	Wu *et al.*, 2013	Presented dual IF rough approximation operators determined by an IF implication operator *I* in infinite universes of discourse.
43	Zhang and He, 2013	Presented a generalized IF rough sets by a constructive approach and defined IF approximation operators by employing an IF residual implicator and its dual operator.
44	Huang *et al.*, 2013	Introduced a ranking technique for constructing the neighborhood of every object in IF environment and proposed a DIFDT. Developed rule extraction algorithm with the help of discernibility matrix approach.

(*Continued*)

Table 4.1 Intuitionistic fuzzy rough sets. (*Continued*)

S. no.	Authors	Proposed work
45	Wu *et al.*, 2014	Investigated relation-based IF rough approximation operators determined by an IF triangular norm *T* and its dual IF triangular co-norm.
46	Xu *et al.*, 2014	Studied and investigated the topological structures of rough IF sets.
47	Liu and Lin, 2015	Constructed a unique IF rough set model by defined the conflict distance in IF environment, and developed method to solve the problem of decision-making in IF information system.
48	Yun and Lee, 2015	Introduced and investigated basic properties of IF rough approximation operators and IF relations by means of topology.
49	Wu *et al.*, 2015	Proposed an axiomatic characterization of relation-based IF rough approximation operators determined by IF-implicator operator *I*.
50	Das, 2016	Developed a different approach of PIFR relation on a set and determined sufficient conditions under which an IF rough relation becomes a PIFR-relation.
51	Zhong and Yan, 2016	Introduced lower and upper approximations of Intuitionistic *L*-fuzzy with respect to intuitionistic *L*-fuzzy approximation space determined by intuitionistic *L*-fuzzy *t*-norm and implicators.
52	Wu and Yuan, 2016	Proposed IF rough set based on triple valued fuzzy sets used as cut sets of IF sets, constructed four methods for upper and lower approximation of IF sets using the representation theorem.

(*Continued*)

Table 4.1 Intuitionistic fuzzy rough sets. (*Continued*)

S. no.	Authors	Proposed work
53	Bandyopadhyay *et al.*, 2017	Discovered the applicability of IF rough sets in a problem related to granular computing with GTIFRS.
54	Sheeja and Sunny, 2017	Proposed a new definition of IF rough set as an extension of fuzzy rough sets introduced by A. Nakamura and decomposition theorem for the operators.
55	Bai *et al.*, 2017	Presented a study on IF rough set with axiomatic approach, researched correlative conclusion as well as equivalence conditions.
56	Wu *et al.*, 2019	Presented axiomatic and operation-oriented characterization of relation-based IF rough approximation operators determined by an IF t-norm T and its dual IF t-conorm S.
57	Bashir *et al.*, 2020	Studied the topological properties of IF rough Sets in detail in presence of many conditions like serial, strongly serial, left continuity, transitivity on IF relations, t-norms and implicators.

the real-life problems [65]. The summary of work done by the researchers in the theoretical background of IF rough sets is presented in Table 4.1.

4.4 Extension and Hybridization of Intuitionistic Fuzzy Rough Sets

In this section, extension of IF rough set theory and hybridization of IF rough sets with other extended rough set theories have been discussed.

4.4.1 Extension

IF rough set theory is generalized based on a few concepts that are earlier proposed in fuzzy rough set theory in order to better tackle the problems concerned with uncertainty exist in real life.

4.4.1.1 Dominance-Based Intuitionistic Fuzzy Rough Sets

To solve the problem of discovering inconsistencies that come from consideration of criteria, i.e., attributes having preference-ordered domains like debt ratio, product quality and market share, Greco *et al.* proposed an extension of rough sets, named as dominance-based rough set approach (DRSA) [66]. Zhang and Yang generalized the dominance-based rough set model to IF environment and defined the IF dominance-based lower and upper approximations with constructive approach. They examined the basic properties of the IF dominance-based rough approximations, and attribute reduction is performed with the model [67]. A DIFDT is proposed by Huang *et al.* To obtain knowledge from proposed DIFDTs, a lower and upper approximation reduction along with rule extraction algorithm are developed using discernibility function [51].

4.4.1.2 Covering-Based Intuitionistic Fuzzy Rough Sets

The notion of IF covering is introduced by Zhang *et al.* The article structured two pairs of generalized lower and upper IF rough approximation operators through an IF triangular norm and an IF implicator based on an IF covering of a universe of discourse [29, 38]. Huang *et al.* developed a new generalized rough set model, a combination of the β-neighborhood fuzzy covering rough sets and IF rough sets. An approach of matrix computation for the upper and lower approximations of the proposed IF rough sets based on the IF graded neighborhood is presented by them. They also examined the IF graded approximation space, IF rough sets, uncertainty measures and computation methods for its reducts from the viewpoint of multi-granulation [68]. Further, Wang and Zhang investigated some properties of IF β-covering approximation spaces and the IF covering rough set model, and solved the issue of multiple criteria group decision-making using the proposed concept [69]. Zhang *et al.* proposed numerous types of covering-based general multigranulation IF rough set (CGMIFRS) models by using four types of IF-neighborhoods. A multi-attribute group decision-making (MAGDM) model with IF information built on CGMIFRS models is constructed by the alternative to the principle of the PROMETHEE II method [70]. Three classes of coverings based on IF rough set models through β-neighborhoods and IF complementary β-neighborhood (IFCβ-neighborhood) by combining the theories of covering based rough sets, IF sets and fuzzy rough sets are introduced by Zhan and Sun. IF TOPSIS technique for multi-attribute decision-making (MADM) problem with an assessment of IF information is discussed by them [71].

4.4.1.3 Kernel Intuitionistic Fuzzy Rough Sets

The kernel IF rough set (KIFRS) model encloses new kernel IF clustering (KIFCM) to improve the performance of rough set theory. Lin *et al.* developed a hybrid model as KIFRS to get the better effects of rule generation based on rough set theory. For the purpose, they first proposed a hybrid method that accepts KIFCM for clustering raw data into similarity groups and then, based on the results of KIFCM, the rough set theory can acquire superior performance in producing rules [72].

4.4.1.4 Tolerance-Based Intuitionistic Fuzzy Rough Sets

The tolerance degree is a similarity threshold that provides the required level of similarity for inclusion inside tolerance classes. Tiwari *et al.* proposed a new method for feature selection using tolerance-based IF RST. They proposed tolerance-based IF lower and upper approximations and devised a degree of dependency of decision attributes over the set of conditional attributes [73].

4.4.1.5 Interval-Valued Intuitionistic Fuzzy Rough Sets

In an interval-valued IF set, the membership and non-membership degrees take an interval number of the unit interval as their values. The concept of an interval-valued IF set, an extension of Zadeh's interval-valued fuzzy sets, was proposed by Atanassov and Gargov and Atanassov [74, 75]. Zhang proposed an interval-valued IF rough model and rough IF models, defined the lower and upper interval-valued IF rough approximation operators generated from an interval-valued IF relation, and investigated some concepts and properties of model and operators. Classical representations of interval-valued IF rough approximation operators are presented in this paper by introducing cut sets of interval-valued IF sets [76, 77]. Zhang and Tian proposed an interval-valued IF rough set model based on an interval-valued IF implicator. The upper approximations and lower approximations of interval-valued IF sets are defined by them with the help of interval-valued IF t-norm and implicator, respectively, and examined their basic properties [78]. Lin and Yang defined interval-valued IF by employing two interval-valued IF implicators I and J and investigated their properties [79]. Interval-valued IF rough set based on the interval-valued fuzzy compatible relation is presented, some of its properties are discussed, and its application are illustrated by Yang [80]. Xu *et al.* presented the study of (*I*, *T*)-interval-valued IF rough sets with axiomatic approach and proposed an operator-oriented characterization of interval-valued IF rough sets [81]. Wang and Wang defined a new interval-valued IF upper and lower approximate operators by generalizing the definition of fuzzy rough membership function and entropy of interval-valued IF rough sets. They solved the problem of uncertainty measurement on interval-valued IF rough sets [82]. Moreover, Xu *et al.* also defined interval-valued IF rough sets and developed relative properties. They discussed the knowledge reduction of interval-valued IF information system by using discernibility matrix approach [83].

4.4.2 Hybridization

Due to the uncertainty and complexity of the real world, as well as the inadequacy of the human ability to comprehend, it is very tough for a single method of uncertainty to efficiently tackle the physical world problems. So, it is instinctive to think about integrating the advantages of different theories of uncertainty for the development of a more powerful hybrid method.

4.4.2.1 Variable Precision Intuitionistic Fuzzy Rough Sets

By combining the theory of variable precision rough sets (VPRS) (proposed by Ziarko [84]) and the IF rough sets, Variable precision IF rough set model (VPIFRS) was proposed by Gong and Zhang [85], which is based on the IF inclusion sets and IF inclusion ratio. They employed the IF implicator and the IF t-norm for the construction of new model. The approximation quality and attribute reduction of the variable precision IF rough sets are also presented by them. Liu *et al.* also constructed a hybrid model VPIFRS on the idea of conflict distance. Their model retains the ability to tolerate faults up to some extent on adjustment of the confidence threshold parameter β. They implemented the proposed model on the problem of decision-making, having decision attribute values consist of IF numbers [86]. Gong and Zhang applied the model of VPIFRS to process the data in decision table and get the weight of all conditional attributes and the weighted T-equivalence IF relation. A new T-equivalence IF partition is discovered based on the weight set of conditional attributes that leads to the conclusion the obtained partition is more suitable and less sensitive to perturbation and misclassification [87]. Zhang *et al.* presented an efficient model based on the covering-based VPIFRS i.e., CVPIFRS. To effectively settle down the complex and changeable bone transplant selections (a typical MADM problem) they introduced a decision-making method by incorporating the CVPIFRS models along with the concept of TOPSIS [88].

4.4.2.2 Intuitionistic Fuzzy Neighbourhood Rough Sets

A ranking technique for constructing the neighborhood of every object, found by means of IF values of condition attributes, is introduced by Huang *et al.* A DIFDT is proposed in the article, and lower/upper approximation set of an object and crisp decision classes is discovered by the comparison between their relations [51]. Shreevastava *et al.* introduced an IF neighborhood rough set model by an amalgamation of the theories of IF sets and neighborhood rough set, and defined IF neighbourhood-based lower and upper approximations [89]. The articles [68, 70, 71] described in 4.1.2 had also worked on IF neighborhood rough sets.

4.4.2.3 Intuitionistic Fuzzy Multigranulation Rough Sets

A new multigranulation rough set model by the combination of multigranulation rough sets and IF rough sets, called an IF multigranulation rough set (IFMGRS) is developed by Huang *et al.* They proposed three kinds of IFMGRSs and concluded that they are extensions of three existing IF rough sets on the basis of their basic properties. They examined the reduction techniques of IFMGRS [90]. Bandyopadhyay *et al.* discovered the applicability of IF rough sets in a problem related to granular computing with GTIFRS [60]. Tan *et al.* introduced numerous types of fuzzy granules for addressing the lower and upper approximations of IF set with suitable employment of granular structures. Significance measures are developed by them for the evaluation of approximation quality and classification ability of IF relations. A reduction algorithm was also established by them to reduce redundant IF relations and addressed the knowledge reduction of IF decision systems from the perspective of retaining the IF approximations of decision classes [91]. The articles [68, 70] as described in section 4.1.2 are also discovered IF multigranulation rough set theory.

4.4.2.4 Intuitionistic Fuzzy Decision-Theoretic Rough Sets

A simple model of IF decision-theoretic rough sets (IFDTRSs) based upon DTRSs and IFNs is proposed by Liang and Liu. Then, from the single period to the multi-period decision-making situation, three-way decisions are derived, and their properties are studied by them. They designed three strategies with respect to single-period decision-making situation and obtained their area of applications and three aggregation operations of IFDTRSs, i.e., DIFWA, DIFPA and DIFOA with respect to multi-period decision-making scenario [92]. Liang *et al.* reconstructed the above basic model of IFDTRSs by analyzing the impact of IFPO on the loss functions and examined the new derivation of three-way decisions. They proved that the prerequisites amid loss functions hold in every stage under the variation setup of loss function, which leads to the implication that Bayesian decision theory could be directly utilized to deduce three-way decisions [93]. A novel similarity measure by the amalgamation of the information contained by hesitancy degree with the endpoint distance of membership as well as non-membership is presented by Liang *et al.* IF-DTRS model and multigranulation IF-DTRS (MG-IF-DTRS) model are constructed by using similarity measure, and a methodology for optimal granulation selection is devised by them [94].

4.4.2.5 *Intuitionistic Fuzzy Rough Sets and Soft Intuitionistic Fuzzy Rough Sets*

The notions of soft rough IF sets and IF soft rough sets is proposed, and a few properties of the proposed concept is investigated in detail by Zhang *et al.* A methodology is developed by them in the application of decision making based on IF soft rough sets [95]. Shao and Shao presented a generic approach towards the IF soft sets. They obtained a broad collection of soft IF rough sets, where ($\mathcal{E}$, $\mathcal{F}$)- soft IF rough set is structured by a pair of intuitionistic implicators ($\mathcal{E}$, $\mathcal{F}$). They also proved that proposed ($\mathcal{E}$, $\mathcal{F}$)-soft IF rough sets are equivalent to ($\mathcal{E}$, $\mathcal{F}$)-IF rough sets of Zhou *et al.* [24] by employing a K-equivalence IF relation [96]. Moreover, Zhang *et al.* proposed the notion of generalized IF soft rough set, investigated its properties and developed an approach for decision making with the proposed approach [97].

4.4.2.6 *Multi-Adjoint Intuitionistic Fuzzy Rough Sets*

A multi-adjoint IF rough set by employing adjoint triples under IF information system is constructed by Liang *et al.* IF indiscernibility relation and multi-adjoint approximation operators are proposed by them. They generalised the basic results from the model of multi-adjoint fuzzy rough set to multi-adjoint IF rough set. A new technique to feature selection is proposed. An approximate reduction to preserve the dependence of the positive region to a degree α is constructed, and on basis of that, a new approach for attribute reduction is presented by them [98].

4.4.2.7 Intuitionistic Fuzzy Quantified Rough Sets

The hybridization procedure by incorporating IF quantifiers into the concept of upper and lower approximations of IF rough sets is reconsidered by Singh *et al.* The supremacy of IF quantifier over VPRS and VQRS is shown. IF quantified rough set effectively dealt with the problem of radically changing the output of the approximations by ignoring or adding just one element. Feature selection with IF quantifier-based lower approximation is also performed [99].

4.4.2.8 Genetic Algorithm and IF Rough Sets

A classification rule base mining algorithm with the combination of the genetic algorithm and IF rough set for large-scale IF information system is proposed by Zhang. This article discovered a multi-objective optimization methodology to optimize the population size of the data sample, and to decreased the attribute collection of fuzzy information system using IF rough set. An optimal rule to acquire large-scale IF information system base is obtained in the article [100].

4.5 Applications of Intuitionistic Fuzzy Rough Sets

IF rough set theory is successfully implemented in many issues inbuilt with imprecision and vagueness of the real-world, which are surveyed in this section.

4.5.1 Attribute Reduction

Attribute reduction or feature selection is a pre-processing technique, devotes in the process of reducing the size of the high-dimensional input dataset in a will to extract the most significant attributes from the dataset for the analysis and further processing. Attribute reduction is one of the most useful applications of rough set theory. IF rough set-based feature selection is proposed by many researchers as follows.

In 2009, Lu *et al.* extended IF approximation space by proposing an improved definition of lower approximation described with fuzzy equivalent classes. They generalized the notion of positive region and relative reduct for IF information system and proposed a heuristic algorithm for attribute reduction based on IF rough set [101]. In 2011, Chen and Yang hybridized the theory of information entropy and IF rough set and gave a new kind of algorithm for attribute reduction using mutual information-based IF rough set [102].

In 2012, Zhang and Yang generalized the dominance-based rough set model to IF environment and attribute reduction is performed with the model [67]. Zhang and Tian systematically analyzed attribute reduction based on IF rough sets and introduced a few notions and theorems of attribute reduction based on IF rough sets. To find all attribute reduct using the discernibility matrix technique, an algorithm is presented by them [103]. Wang and Shu investigated a methodology of rough sets under IF similarity relations, and gave the definitions of positive field, dependence degree and non-dependence degree. Then, an algorithm for attribute reduction based on IF rough sets is analyzed [104].

In 2013, Huang *et al.* introduced a ranking technique for constructing the neighborhood of every object, and proposed a DIFDT. To extract knowledge from proposed dominant IF decision tables with the help of discernibility matrix and discernibility function, the lower and upper approximating reduction and rule extraction method are developed [51].

In 2014, Huang *et al.* developed the model IFMGRS. They examined the reduction techniques of IFMGRS to eliminate redundant IF granulations by defining the reducts of the three types of IFMGRSs [90]. In 2016, Zhang thoroughly examined the structure of attribute reduction and introduced a few ideas and results of feature selection using IF rough sets. An algorithm to find all the feature selected is also presented by using discernibility matrix technique [105].

In 2018, Tiwari *et al.* proposed the method for feature selection using tolerance-based IF rough set theory. They devised a degree of dependency method for feature selection [73]. Tiwari *et al.* also proposed the approach for feature selection based on IF rough set theory using membership grade, score function and cardinality of IF numbers [106]. Further, the attribute reduction technique using IF neighborhood rough set is introduced by Shreevastava *et al.* [89]. Tan *et al.* proposed numerous types of fuzzy granules to address the lower and upper approximations of IF set. A reduction algorithm for decision system was established by them for addressing the knowledge reduction and to lessen redundant IF relations [91].

In 2019, Shreevastava *et al.* proposed a unique attribute reduction technique for partially labelled data set using IF rough set theory [107]. Singh *et al.* performed feature selection with IF quantifier-based lower approximation using the degree of dependency-based methodology [99]. Tiwari *et al.* proposed attribute reduction technique assisted by IF rough set that can be implemented on high-dimensional data sets with the purpose of getting more reduced attribute data sets along with higher classification accuracy [108].

In 2020, Jain *et al.* established IF decision of an object relies on neighborhood concept, and introduced IF lower and upper approximations using IF decision accompanied by parameterized IF granule. A heuristic greedy forward algorithm for attribute reduction based on the proposed IF dependency function is also given by them [109]. By combining the genetic algorithm and IF rough set for large-scale IF information system, Zhang proposed an algorithm for classification rule base mining and to optimize the population size of the data sample, discovered a multi-objective optimization methodology using IF rough set. An optimal rule to acquire large-scale IF information system base with the minimal size, generation time, configuration and storage space is obtained in the article [100].

4.5.2 Decision Making

Decision making is the process of recognising and picking alternatives by the decision makers on the basis of preferences, values and beliefs. In past few years, the decision making on RST has been advanced by amalgamation of the theory of fuzzy sets, IF sets and soft sets. IF rough set-based decision making has been proposed by many researchers as follows.

In 2012, Chai *et al.* introduced the believe factor and believable rules related to an IF value for improved performance in the field of decision making and implemented the proposed technique in multicriteria sorting [43]. Acharjya and Tripathy developed the concepts of IF rough set on two universal sets for the application in knowledge extraction and in designing knowledge bases [39].

In 2013, Sun *et al.* presented constructive approach for the IF rough set model over two universes and discussed the elementary properties of the model. A new technique of decision making in uncertainty environment is given by them with the proposed model [110]. In 2014, a method for decision making based on IF soft rough sets is developed by Zhang *et al.* [95].

In 2015, Das *et al.* further extended the concepts of IF rough sets on two universal sets by developing approximation of classifications. The approximation accuracy and quality of classifications on two universal sets is defined by them using IF relation. Multi-criterion decision making based on IF rough set on two universal sets is accentuated by them [111]. Liu and Lin developed a method to solve the problem of decision making in IF information system by measuring IF similarity relation and conflict distances in IF environment [54].

In 2017, Das defined the notions of fuzzy rough set, IF rough sets and IF rough set on two universal sets, and studied their properties. Thereafter, a method based on IF rough set on two universal sets for decision-making problems is established by them [112]. In 2018, Wang and Zhang presented a novel technique to solve the issue of multiple criteria group decision making using IF covering rough set model [69].

In 2019, Zhang *et al.* proposed CGMIFRS models using IF-neighborhoods. A MAGDM model with IF information built on CGMIFRS models is constructed by the alternative to the principle of the PROMETHEE II method [70]. In 2020, Zhan and Sun discussed MADM problem with the evaluation of IF information using IF TOPSIS technique and introduced three classes of coverings based on IF rough set models through β-neighborhoods and IFCβ-neighborhood [71].

4.5.3 Other Applications

The major domain of applications of IF rough set theory is in the field of attribute reduction and decision making. There are few more areas of applications in which IF rough sets are implemented.

In 2012, Yang *et al.* developed a measure function to overcome the problem of market prediction by defining IF rough sets, IF similarity measure along with upper and lower rough degree [42]. In 2014, a method for image segmentation with the use of multiscale IF roughness measure is proposed by Nehare *et al.* They measured the IF roughness under multiple scales by incorporating the concepts of scale-space and employing IF representation for images [113].

In 2015, Liu applied IF rough sets to assess for grid disasters by exemplifying the grid disaster's assessment of snow disaster in Liaoning province, and accomplished the visualization system of disasters evaluation using software design. Attribute reduction is employed to reduce redundancy of the disaster's attributes, and value reduction is employed to extract the evaluation rules in the process [114].

Table 4.2 Applications of Intuitionistic fuzzy rough sets.

S. no.	Authors	Application field	Description/study contribution
1	Lu *et al.*, 2009	Attribute reduction	Degree of dependency
2	Chen and Yang, 2011	Attribute reduction	Mutual information
3	Zhang and Yang, 2012	Attribute reduction	Discernibility matrix
4	Zhang and Tian, 2012	Attribute reduction	Discernibility matrix
5	Wang and Shu, 2012	Attribute reduction	Degree of dependency
6	Chai *et al.*, 2012	Decision making	Believe factor and believable rules
7	Acharjya and Tripathy, 2012	Knowledge representation/ decision making	IFRS on two universal sets
8	Yang *et al.*, 2012	Market prediction problem	Similarity measure function
9	Sun *et al.*, 2013	Decision making	IFRS on two universal sets
10	Huang *et al.*, 2013	Rule extraction, attribute reduction	Discernibility matrix
11	Zhang *et al.*, 2014	Decision making	IF soft rough sets
12	Huang *et al.*, 2014	Attribute reduction	Granular computing
13	Nehare *et al.*, 2014	Image segmentation	Theory of scale-space
14	Das *et al.*, 2015	Decision making	IFRS on two universal sets

(Continued)

Table 4.2 Applications of Intuitionistic fuzzy rough sets. (*Continued*)

S. no.	Authors	Application field	Description/study contribution
15	Liu and Lin, 2015	Decision making	Conflict analysis model
16	Liu, 2015	Power grid disaster assessment	Evaluation index system
17	Zhang, 2016	Attribute reduction	Discernibility matrix
18	Chowdhary and Acharjya, 2016	Breast cancer detection, image segmentation, feature extraction	Gray-level co-occurrence matrix
19	Das, 2017	Decision making	IF rough set on two universes
20	Bandyopadhyay *et al.*, 2017	Granular computing	Game theoretic IF rough set
21	Tiwari *et al.*, 2018	Attribute reduction	Degree of dependency
22	Shreevastava *et al.*, 2018	Attribute reduction	Degree of dependency
23	Wang and Zhang, 2018	Decision making	IF β-covering
24	Tan *et al.*, 2018	Attribute reduction	IF rough set-based granular structures
25	Shreevastava *et al.*, 2019	Attribute reduction	Degree of dependency
26	Singh et. al., 2019	Attribute reduction	Degree of dependency
27	Tiwari *et al.*, 2019	Attribute reduction	Degree of dependency
28	Chuanchao, 2019	Dynamic fault feature extraction	Discernibility matrix

(*Continued*)

Table 4.2 Applications of Intuitionistic fuzzy rough sets. (*Continued*)

S. no.	Authors	Application field	Description/study contribution
29	Liang *et al.*, 2019	Granular selection	Multigranulation IF decision-theoretic rough sets
30	Zhang *et al.*, 2019	Multi-attribute group decision making	CGMIFRS
31	Jain *et al.*, 2020	Attribute reduction	Degree of dependency
32	Zhang, 2020	Rule mining, attribute reduction	Genetic algorithm
33	Zhan and Sun, 2020	Multi-attribute group decision making	Covering based IFRS and IF-TOPSIS

In 2016, Chowdhary and Acharjya proposed statistical feature extraction methodology in IF rough set environment. Their hybrid scheme begins with image segmentation employing IF set for extracting the zone of interest, which leads to enhance the edges at its surrounding. Feature extraction with gray-level co-occurrence matrix is presented by them and rough set is incorporated to engender all minimal reducts and rules. To categorize different zones of interest and to make sure these points contain decision class value as either cancer or not, they entered the above rules into a classifier [115]. In 2017, Bandyopadhyay *et al.* discovered the applicability of IF rough sets in a problem related to granular computing with GTIFRS [60].

In 2019, Liang *et al.* presented a methodology for optimal granulation selection by constricting models IF-DTRS and MG-IF-DTRS with the help of proposed similarity measure [94]. Chuanchao proposed a generalized algorithm for dynamic feature extraction based on the discernibility matrix approach of IF rough set theory and dynamic reduction [116]. The summary of work done by the researchers in the area of applications of IF rough set theory is presented in Table 4.2.

The Figure 4.4 depicts the various application domain of IF rough sets and their frequencies.

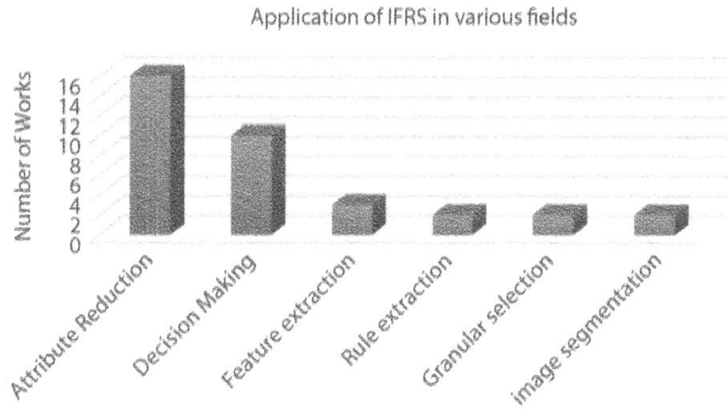

Figure 4.4 Application of IF rough sets in various fields.

4.6 Work Distribution of IFRS Country-Wise and Year-Wise

In this section, we present the work distribution of IF rough set theory country-wise as well as year-wise. Limitations of the concerned theory are also discussed here.

4.6.1 Country-Wise Work Distribution

The model of IF rough sets was first introduced in India, and many researchers in India are effectively working in the area of IF rough set theory. Figure 4.5 shows the country-wise distribution of work in the field of IF rough sets. It shows that most of the published research work in our area of interest for this survey is from China. India is the second most country for the usage of this theory. Belgium and Pakistan have some research articles in this domain, which shows in the other section in the figure.

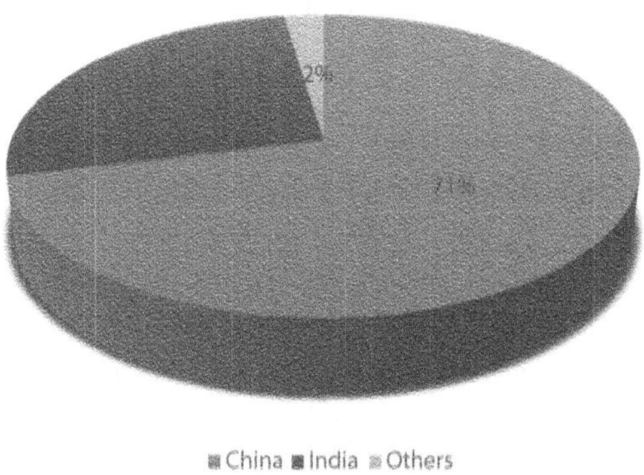

Figure 4.5 The country-wise distribution for the number of works in the field of IFRSs.

4.6.2 Year-Wise Work Distribution

Figure 4.6 shows the year-wise distribution of work in the field of IF rough sets. IF rough set theory was proposed in 1998. From Figure 4.6, It is clear that most research was done in this theory in 2012, then in 2019, and in 2020 work is still going, which shows that this topic is hot and trending in terms of theory and application in various application fields where uncertainty is concerned.

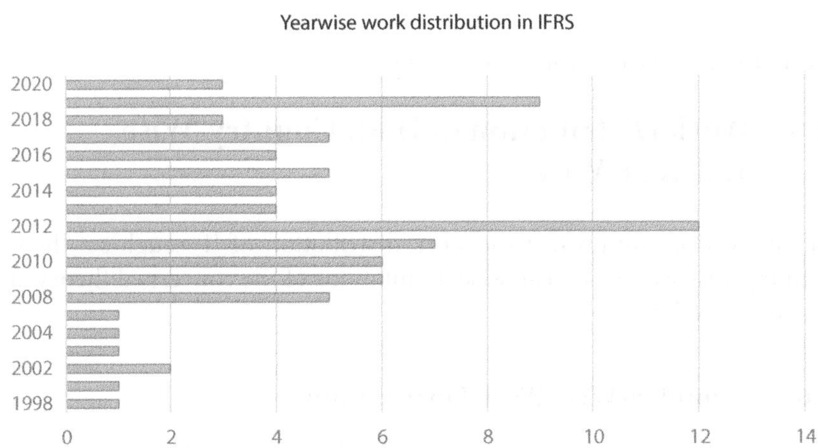

Figure 4.6 The year-wise distribution for the number of works in the field of IFRS.

4.6.3 Limitations of Intuitionistic Fuzzy Rough Set Theory

Hybridization of rough set with IF set successfully dealt with the drawback of rough set in which discretization process is needed before application in problem of attribute reduction, rule generation, decision making, etc. But there are some limitations of IF rough sets as follows.

- In this theory there may be a substantial variation in the result of the approximations due to ignoring or adding just one element. To cope with this situation, the hybridization process is reconsidered by incorporating IF quantifiers into the concept of upper and lower approximations.
- To discover inconsistencies coming from consideration of criteria, which is a drawback of rough sets, IF rough sets also face the same, that's why dominance-based IF rough set theory is developed.

4.7 Conclusion

In this chapter, we have presented a broad survey on techniques of IF rough set theory with discussion of their basic properties, characterization, topological structures, connection with lattices, various similarity relations, logic operators like t-norms, co-norms, implicators, construction of their approximation operators-lower and upper approximations based on axiomatic and constructive approaches. Rough IF set theory, IF rough sets on two universes, and Interval-valued IF rough set theory are also surveyed. There are some limitations of IF rough set theories; therefore, several extensions of this theory have taken place based on dominance, covering, tolerance, kernel etc., which have been widely discussed. Hybridization of IF rough set theory with other theories dealing with uncertainties like VPRS, multi-adjoint theory, soft sets have been studied in detail. Last, we have presented a detailed discussion of IF rough set theory in various application fields, namely, attribute reduction, topological structures, rule-generation and image processing. The analysis results show that IF rough set theory provides a significant contribution in the real-world problems dealing with vagueness and imprecision, and this theory can be applied to many other application fields in the future.

Acknowledgement

First author would like to thank the Research Foundation-CSIR for funding her research.

References

1. Zadeh, L. A. (1965). Fuzzy sets. *Information and control, 8*(3), 338-353.
2. Pawlak, Z. (2012). *Rough sets: Theoretical aspects of reasoning about data* (Vol. 9). Springer Science & Business Media.
3. Dubois, D., & Prade, H. (1990). Rough fuzzy sets and fuzzy rough sets. *International Journal of General System, 17*(2-3), 191-209.
4. Dubois, D., & Prade, H. (1992). Putting rough sets and fuzzy sets together. In *Intelligent Decision Support* (pp. 203-232). Springer, Dordrecht.
5. Atanassov, K. T. (1999). Intuitionistic fuzzy sets. In *Intuitionistic fuzzy sets* (pp. 1-137). Physica, Heidelberg.
6. Huang, S. Y. (Ed.). (1992). *Intelligent decision support: handbook of applications and advances of the rough sets theory* (Vol. 11). Springer Science & Business Media.

7. Bustince, H., & Burillo, P. (1996). Structures on intuitionistic fuzzy relations. *Fuzzy sets and systems, 78*(3), 293-303.
8. Cornelis, C., De Cock, M., & Kerre, E. E. (2003). Intuitionistic fuzzy rough sets: at the crossroads of imperfect knowledge. *Expert systems, 20*(5), 260-270.
9. Chakrabarty, K., Gedeon, T., & Koczy, L. (1998). Intuitionistic fuzzy rough set. In *Proceedings of 4th joint conference on information sciences (JCIS), Durham, NC* (pp. 211-214).
10. Samanta, S. K., & Mondal, T. K. (2001). Intuitionistic fuzzy rough sets and rough intuitionistic fuzzy sets. *Journal of Fuzzy Mathematics, 9*(3), 561-582.
11. Nanda, S., & Majumdar, S. (1992). Fuzzy rough sets. *Fuzzy sets and systems, 45*(2), 157-160.
12. Jena, S. P., Ghosh, S. K., & Tripathy, B. K. (2002). Intuitionistic fuzzy rough sets. *Notes on Intuitionistic Fuzzy Sets*, 8(1), 1-18.
13. Rizvi, S., Naqvi, H. J., & Nadeem, D. (2002, March). Rough Intuitionistic Fuzzy Sets. In *JCIS* (pp. 101-104).
14. Radzikowska, A. M. (2006, June). Rough approximation operations based on IF sets. In *International Conference on Artificial Intelligence and Soft Computing* (pp. 528-537). Springer, Berlin, Heidelberg.
15. Zhou, L., & Wu, W. Z. (2007, August). Rough Approximations of Intuitionistic Fuzzy Sets. In *2007 International Conference on Machine Learning and Cybernetics* (Vol. 7, pp. 3713-3718). IEEE.
16. Zhou, L., & Wu, W. Z. (2008). On generalized intuitionistic fuzzy rough approximation operators. *Information Sciences, 178*(11), 2448-2465.
17. Wu, W. Z., & Zhou, L. (2008, July). Topological structures of intuitionistic fuzzy rough sets. In *2008 International Conference on Machine Learning and Cybernetics* (Vol. 1, pp. 618-623). IEEE.
18. LU, Y. L., LEI, Y. J., & ZHOU, W. (2008). Intuitionistic fuzzy rough set based on inclusion degree [J]. *Journal of Computer Applications, 8.*
19. Lu, Y., Lei, Y., & Lei, Y. (2008, July). Intuitionistic fuzzy rough set based on intuitionistic similarity relation. In *2008 Chinese Control and Decision Conference* (pp. 794-799). IEEE.
20. XU, X. L., LEI, Y. J., & TAN, Q. Y. (2008). Intuitionistic fuzzy rough sets based on intuitionistic fuzzy triangle norm [J]. *Control and Decision, 8.*
21. Lin, R., & Wang, J. (2009, August). On the topological properties of intuitionistic fuzzy rough sets. In *2009 IEEE International Conference on Granular Computing* (pp. 404-408). IEEE.
22. Liu, G., & Liu, J. (2009, August). On structure of generalized intuitionistic fuzzy rough sets. In *2009 IEEE International Conference on Granular Computing* (pp. 419-423). IEEE.
23. Jiang, Y., Tang, Y., Wang, J., & Tang, S. (2009). Reasoning within intuitionistic fuzzy rough description logics. *Information Sciences, 179*(14), 2362-2378.

24. Zhou, L., Wu, W. Z., & Zhang, W. X. (2009). On characterization of intuitionistic fuzzy rough sets based on intuitionistic fuzzy implicators. *Information Sciences, 179*(7), 883-898.
25. Zhou, L., Wu, W. Z., & Zhang, W. X. (2009). On intuitionistic fuzzy rough sets and their topological structures. *International Journal of General Systems, 38*(6), 589-616.
26. Lin, R., & Wang, J. (2010, August). Intuitionistic fuzzy rough approximation operators based on intuitionistic fuzzy triangle norm. In *2010 IEEE International Conference on Granular Computing* (pp. 308-313). IEEE.
27. Xu, Y. H., & Wu, W. Z. (2010, July). On intuitionistic fuzzy rough *set al*gebras. In *2010 International Conference on Machine Learning and Cybernetics* (Vol. 2, pp. 581-586). IEEE.
28. Wu, W. Z. (2010, August). Intuitionistic fuzzy rough sets determined by intuitionistic fuzzy implicators. In *2010 IEEE International Conference on Granular Computing* (pp. 536-540). IEEE.
29. Zhi-mingla, Z. H. A. N. G., Yun-chaolb, B. A. I., & Jing-Feng, T. (2010). Intuitionistic fuzzy rough sets based on intuitionistic fuzzy coverings. *Control and Decision*, (9), 19.
30. Yang, Y., Zhu, X. Z., & Li, L. (2010). Axiomatization of intuitionistic fuzzy rough sets. *Journal of Hefei University of Technology, 33*(4), 590-592.
31. Xiao-jiang, T. A. N. G. (2010). Intuitionistic Fuzzy Rough Sets Based on the k-step Fuzzy Relations. *Journal of Zhangzhou Normal University (Natural Science)*, (4), 5.
32. Thomas, K. V., & Nair, L. S. (2011). Rough intuitionistic fuzzy sets in a lattice. In *International Mathematics Forum* (Vol. 6, No. 27, pp. 1327-1335).
33. Fan, C. L., Lei, Y. J., & Zhang, C. (2011). Improved measure of similarity between intuitionistic fuzzy rough sets. *Jisuanji Yingyong/ Journal of Computer Applications, 31*(5), 1344-1347.
34. Zhou, L., & Wu, W. Z. (2011). Characterization of rough set approximations in Atanassov intuitionistic fuzzy set theory. *Computers & Mathematics with Applications, 62*(1), 282-296.
35. Sun, B. Z., Ma, W. M., & Liu, Q. (2011, July). Theory for intuitionistic fuzzy rough sets of two universes. In *2011 International Conference on Machine Learning and Cybernetics* (Vol. 1, pp. 307-312). IEEE.
36. Feng, T., Zhang, S. P., & Mi, J. S. (2011). Intuitionistic fuzzy rough sets and intuitionistic fuzzy topology spaces. *Information-an international interdisciplinary journal, 14*(8), 2553-2562.
37. Z.T. Gong, W. Zhao, (2011) Intuitionistic fuzzy rough sets models based dual triangular norms, Journal of Lanzhou University (Natural) 6 (47) 1–9.
38. Zhang, Z. (2012). Generalized intuitionistic fuzzy rough sets based on intuitionistic fuzzy coverings. *Information Sciences, 198*, 186-206.
39. Acharjya, D. P., & Tripathy, B. K. (2012). Intuitionistic fuzzy rough set on two universal sets and knowledge representation. *Mathematical Sciences International Research Journal, 1*(2), 584-598.

40. Zhang, X., Zhou, B., & Li, P. (2012). A general frame for intuitionistic fuzzy rough sets. *Information Sciences, 216*, 34-49.
41. Xu, W., Liu, Y., & Sun, W. (2012, May). Intuitionistic fuzzy rough sets model based on (θ, φ)-operators. In *2012 9th International Conference on Fuzzy Systems and Knowledge Discovery* (pp. 234-238). IEEE.
42. Yang, H., Li, W., Zhang, C., & Li, J. (2012, August). A new definition of intuitionistic fuzzy rough set and its similarity measure. In *2012 International Conference on Computer Science and Service System* (pp. 1852-1855). IEEE.
43. Chai, J., Liu, J. N., & Li, A. (2012, August). A new intuitionistic fuzzy rough set approach for decision support. In *International Conference on Rough Sets and Knowledge Technology* (pp. 71-80). Springer, Berlin, Heidelberg.
44. Tripathy, B. K., Panda, G. K., & Mitra, A. (2012). Some Concepts of Incomplete Multigranulation based on Rough Intuitionistic Fuzzy Sets. In *Advances in Computer Science, Engineering & Applications* (pp. 683-693). Springer, Berlin, Heidelberg.
45. Xu, W., & Luo, S. (2012, December). The intuitionistic fuzzy rough sets model over two different universes. In *2012 Fourth International Symposium on Information Science and Engineering* (pp. 295-298). IEEE.
46. Weihua, X., & Shuqun, L. (2013). The intuitionistic fuzzy rough sets model over two different universes. *Global Journal on Technology, 3*.
47. Mukherjee, A., & Das, A. K. (2013). Intuitionistic fuzzy rough relations. *Ann. Fuzzy Math. Inform, 6*(1), 115-126.
48. Yang, X., & Yang, Y. (2013). Independence of axiom sets on intuitionistic fuzzy rough approximation operators. *International Journal of Machine Learning and Cybernetics, 4*(5), 505-513.
49. Wu, W. Z., Gao, C. J., Li, T. J., & Xu, Y. H. (2013, October). On dual intuitionistic fuzzy rough approximation operators determined by an intuitionistic fuzzy implicator. In *International Workshop on Rough Sets, Fuzzy Sets, Data Mining, and Granular-Soft Computing* (pp. 138-146). Springer, Berlin, Heidelberg.
50. Zhang, H., & He, Y. (2013, July). Generalized intuitionistic fuzzy rough sets based on an intuitionistic fuzzy residual implicator. In *2013 10th International Conference on Fuzzy Systems and Knowledge Discovery (FSKD)* (pp. 140-145). IEEE.
51. Huang, Bing, Yu-Liang Zhuang, Hua-Xiong Li, and Da-Kuan Wei. (2013) A dominance intuitionistic fuzzy-rough set approach and its applications. *Applied Mathematical Modelling* 37, no. 12-13 (2013): 7128-7141.
52. Wu, W. Z., Gu, S. M., Li, T. J., & Xu, Y. H. (2014, October). Intuitionistic fuzzy rough approximation operators determined by intuitionistic fuzzy triangular norms. In *International Conference on Rough Sets and Knowledge Technology* (pp. 653-662). Springer, Cham.
53. Xu, Y. H., Wu, W. Z., & Wang, G. (2014). On the Intuitionistic fuzzy topological structures of Rough Intuitionistic fuzzy sets. In *Transactions on Rough Sets XVIII* (pp. 1-22). Springer, Berlin, Heidelberg.

54. Liu, Y., & Lin, Y. (2015). Intuitionistic fuzzy rough set model based on conflict distance and applications. *Applied Soft Computing, 31*, 266-273.
55. Yun, S. M., & Lee, S. J. (2015). Intuitionistic fuzzy rough approximation operators. *International Journal of Fuzzy Logic and Intelligent Systems, 15*(3), 208-215.
56. Wu, W. Z., Xu, Y. H., Li, T. J., & Wang, X. (2015, November). Axiomatic Characterizations of Reflexive and T-Transitive I-Intuitionistic Fuzzy Rough Approximation Operators. In *Rough Sets, Fuzzy Sets, Data Mining, and Granular Computing: 15th International Conference, RSFDGrC 2015, Tianjin, China, November 20-23, 2015, Proceedings* (Vol. 9437, p. 218). Springer.
57. Das, A. K. (2016). On partially included intuitionistic fuzzy rough relations. *Afrika Matematika, 27*(5-6), 993-1001.
58. Zhong, Y., & Yan, C. H. (2016). Intuitionistic L-fuzzy Rough Sets, Intuitionistic L-fuzzy Preorders and Intuitionistic L-fuzzy Topologies. *Fuzzy Information and Engineering, 8*(3), 255-279.
59. Wu, L. T., & Yuan, X. H. (2016, September). Intuitionistic Fuzzy Rough Set Based on the Cut Sets of Intuitionistic Fuzzy Set. In *International workshop on Mathematics and Decision Science* (pp. 37-45). Springer, Cham.
60. Bandyopadhyay, S., Yao, J., & Zhang, Y. (2017, December). Granular Computing with Compatibility Based Intuitionistic Fuzzy Rough Sets. In *2017 16th IEEE International Conference on Machine Learning and Applications (ICMLA)* (pp. 378-383). IEEE.
61. Nakamura, A. (1988). Fuzzy rough sets. *Note on Multiple-Valued Logic in Japan, 9*(8), 1-8.
62. Sheeja, T. K., & Sunny, K. A. (2017). On Intuitionistic Fuzzy Rough Sets. *International Journal of Advanced Research in Computer Science, 8*(5).
63. Bai, Y., Shu, L., & Yao, B. S. (2017, July). A Study of Intuitionistic Fuzzy Rough Sets. In *International Conference on Fuzzy Information & Engineering* (pp. 245-253). Springer, Cham.
64. Wu, W. Z., Shao, M. W., & Wang, X. (2019). Using single axioms to characterize (S, T)-intuitionistic fuzzy rough approximation operators. *International Journal of Machine Learning and Cybernetics, 10*(1), 27-42.
65. Bashir, Z., Abbas Malik, M. G., Asif, S., & Rashid, T. (2020). The topological properties of intuitionistic fuzzy rough sets. *Journal of Intelligent & Fuzzy Systems, 38*(1), 795-807.
66. Greco, S., Matarazzo, B., & Słowiński, R. (2008). Fuzzy set extensions of the dominance-based rough set approach. In *Fuzzy sets and their extensions: representation, aggregation and models* (pp. 239-261). Springer, Berlin, Heidelberg.
67. Zhang, Y., & Yang, X. (2012). Intuitionistic Fuzzy Dominance-based Rough Set Approach: Model and Attribute Reductions. *JSW, 7*(3), 551-563.
68. Huang, B., Guo, C. X., Li, H. X., Feng, G. F., & Zhou, X. Z. (2016). An intuitionistic fuzzy graded covering rough set. *Knowledge-Based Systems, 107*, 155-178.

69. Wang, J., & Zhang, X. (2018). Two types of intuitionistic fuzzy covering rough sets and an application to multiple criteria group decision making. *Symmetry, 10*(10), 462.
70. Zhang, L., Zhan, J., Xu, Z., & Alcantud, J. C. R. (2019). Covering-based general multigranulation intuitionistic fuzzy rough sets and corresponding applications to multi-attribute group decision-making. *Information Sciences, 494*, 114-140.
71. Zhan, J., & Sun, B. (2020). Covering-based intuitionistic fuzzy rough sets and applications in multi-attribute decision-making. *Artificial Intelligence Review, 53*(1), 671-701.
72. Lin, K. P., Hung, K. C., & Lin, C. L. (2018). Rule generation based on novel kernel intuitionistic fuzzy rough set model. *IEEE Access, 6*, 11953-11958.
73. Tiwari, A. K., Shreevastava, S., Som, T., & Shukla, K. K. (2018). Tolerance-based intuitionistic fuzzy-rough set approach for attribute reduction. *Expert Systems with Applications, 101*, 205-212.
74. Atanassov, K. T. (1999). Interval valued intuitionistic fuzzy sets. In *Intuitionistic Fuzzy Sets* (pp. 139-177). Physica, Heidelberg.
75. Atanassov, K., Gargov, G. (1989) Interval valued intuitionistic fuzzy sets, *Fuzzy Sets and Systems*, 31(3), 343–349.
76. Zhang, Z. (2009). An interval-valued intuitionistic fuzzy rough set model. *Fundamenta Informaticae, 97*(4), 471-498.
77. Zhang, Z. (2010). An interval-valued rough intuitionistic fuzzy set model. *International Journal of General Systems, 39*(2), 135-164.
78. Zhang, Z. M., & Tian, J. F. (2010). Interval-valued intuitionistic fuzzy rough sets based on implicators. *Control and Decision, 25*(4), 614-618.
79. Lin, M. L., & Yang, W. P. (2011). Properties of interval-valued intuitionistic fuzzy rough sets with implicators. *Journal of Shandong University (Natural Science), 46*, 104-109.
80. Yang, H. L. (2012, August). Interval valued fuzzy rough set model on two different universes and its application. In *International Conference on Rough Sets and Current Trends in Computing* (pp. 66-72). Springer, Berlin, Heidelberg.
81. Xu, F., Yin, H., & Wu, Q. (2012, April). An axiomatic approch of interval-valued intuitionistic fuzzy rough sets based on interval-valued intuitionistic fuzzy approximation operators. In *2012 2nd International Conference on Consumer Electronics, Communications and Networks (CECNet)* (pp. 3098-3102). IEEE.
82. WANG, Y. P., & WANG, J. Y. (2014). Uncertainty Measurements of Inteval-Valued Intuitionistic Fuzzy Rough Sets. *Journal of North University of China (Natural Science Edition)*, (6), 5.
83. Xu, F., Xing, Z. Y., & Yin, H. D. (2016). Attribute reductions and concept lattices in interval-valued intuitionistic fuzzy rough set theory: Construction and properties. *Journal of Intelligent & Fuzzy Systems, 30*(2), 1231-1242.
84. Ziarko, W. (1993). Variable precision rough set model. *Journal of Computer and System Sciences, 46*(1), 39-59.

85. Gong, Z., & Zhang, X. (2014). Variable precision intuitionistic fuzzy rough sets model and its application. *International Journal of Machine Learning and Cybernetics, 5*(2), 263-280.
86. Liu, Y., Lin, Y., & Zhao, H. H. (2015). Variable precision intuitionistic fuzzy rough set model and applications based on conflict distance. *Expert systems, 32*(2), 220-227.
87. Gong, Z., & Zhang, X. (2017). The further investigation of variable precision intuitionistic fuzzy rough set model. *International Journal of Machine Learning and Cybernetics, 8*(5), 1565-1584.
88. Zhang, L., Zhan, J., & Yao, Y. (2020). Intuitionistic fuzzy TOPSIS method based on CVPIFRS models: an application to biomedical problems. *Information Sciences, 517*, 315-339.
89. Shreevastava, S., Tiwari, A. K., & Som, T. (2018). Intuitionistic fuzzy neighborhood rough set model for feature selection. *International Journal of Fuzzy System Applications (IJFSA), 7*(2), 75-84.
90. Huang, B., Guo, C. X., Zhuang, Y. L., Li, H. X., & Zhou, X. Z. (2014). Intuitionistic fuzzy multigranulation rough sets. *Information Sciences, 277*, 299-320.
91. Tan, A., Wu, W. Z., Qian, Y., Liang, J., Chen, J., & Li, J. (2018). Intuitionistic fuzzy rough set-based granular structures and attribute subset selection. *IEEE Transactions on Fuzzy Systems, 27*(3), 527-539.
92. Liang, D., & Liu, D. (2015). Deriving three-way decisions from intuitionistic fuzzy decision-theoretic rough sets. *Information Sciences, 300*, 28-48.
93. Liang, D., Xu, Z., & Liu, D. (2017). Three-way decisions with intuitionistic fuzzy decision-theoretic rough sets based on point operators. *Information Sciences, 375*, 183-201.
94. Liang, M., Mi, J., & Feng, T. (2019). Optimal granulation selection for similarity measure-based multigranulation intuitionistic fuzzy decision-theoretic rough sets. *Journal of Intelligent & Fuzzy Systems, 36*(3), 2495-2509.
95. Zhang, H., Shu, L., & Liao, S. (2014, January). Intuitionistic fuzzy soft rough set and its application in decision making. In *Abstract and Applied Analysis* (Vol. 2014). Hindawi.
96. Shao, W., & Shao, Y. (2015, August). Generalized soft intuitionistic fuzzy rough sets determined by a pair of intuitionistic fuzzy implicators. In *2015 12th International Conference on Fuzzy Systems and Knowledge Discovery (FSKD)* (pp. 226-230). IEEE.
97. Zhang, H., Xiong, L., & Ma, W. (2016). Generalized intuitionistic fuzzy soft rough set and its application in decision making. *J. Comput. Anal. Appl, 20*(4), 750-766.
98. Liang, M., Mi, J., Feng, T., & Zhao, T. (2018). Multi-adjoint intuitionistic fuzzy rough sets. *The Journal of Engineering, 2018*(16), 1637-1644.
99. Singh, S., Shreevastava, S., Som, T., & Jain, P. (2019). Intuitionistic fuzzy quantifier and its application in feature selection. *International Journal of Fuzzy Systems, 21*(2), 441-453.

100. Zhang, C. (2020). Classification rule mining algorithm combining intuitionistic fuzzy rough sets and genetic algorithm. *International Journal of Fuzzy Systems, 22,* 1694-1715.
101. Lu, Y. L., Lei, Y. J., & Hua, J. X. (2009). Attribute reduction based on intuitionistic fuzzy rough set [J]. *Control and Decision, 3,* 003.
102. Chen, H., & Yang, H. C. (2011). One new algorithm for intuitiontistic fuzzy-rough attribute reduction. *Journal of Chinese Computer Systems, 32*(3), 506-510.
103. Zhang, Z., & Tian, J. (2012). On attribute reduction with intuitionistic fuzzy rough sets. *International Journal of Uncertainty, Fuzziness and Knowledge-Based Systems, 20*(01), 59-76.
104. Wang, X. M., & Shu, L. (2012). An attribute reduction algorithm based on similarity measure of intuitionistic fuzzy rough sets. *Fuzzy Systems and Mathematics, 26,* 185-190.
105. Zhang, Z. (2016). Attributes reduction based on intuitionistic fuzzy rough sets. *Journal of Intelligent & Fuzzy Systems, 30*(2), 1127-1137.
106. Tiwari, A. K., Shreevastava, S., Shukla, K. K., & Subbiah, K. (2018). New approaches to intuitionistic fuzzy-rough attribute reduction. *Journal of Intelligent & Fuzzy Systems, 34*(5), 3385-3394.
107. Shreevastava, S., Tiwari, A., & Som, T. (2019). Feature Subset Selection of Semi-supervised Data: An Intuitionistic Fuzzy-Rough Set-Based Concept. In *Proceedings of International Ethical Hacking Conference 2018* (pp. 303-315). Springer, Singapore.
108. Tiwari, A. K., Shreevastava, S., Subbiah, K., & Som, T. (2019). An intuitionistic fuzzy-rough set model and its application to feature selection. *Journal of Intelligent & Fuzzy Systems, 36*(5), 4969-4979.
109. Jain, P., Tiwari, A. K., & Som, T. (2020). A fitting model based intuitionistic fuzzy rough feature selection. *Engineering Applications of Artificial Intelligence, 89,* 103421.
110. Sun, B., Ma, W., & Liu, Q. (2013). An approach to decision making based on intuitionistic fuzzy rough sets over two universes. *Journal of the Operational Research Society, 64*(7), 1079-1089.
111. Das, T. K., Acharjya, D. P., & Patra, M. R. (2015). Multi criterion decision making using intuitionistic fuzzy rough set on two universal sets. *International Journal of Intelligent Systems and Applications, 7*(4), 26-33.
112. Das, T. K. (2017). Decision Making by Using Intuitionistic Fuzzy Rough Set. In *Emerging Research on Applied Fuzzy Sets and Intuitionistic Fuzzy Matrices* (pp. 268-286). IGI Global.
113. Nehare, Prajakta R., Yogita K. Dubey, and Milind M. Mushrif. "Multiscale intuitionistic fuzzy roughness measure for image segmentation." *2014 International Conference on Communication and Signal Processing.* IEEE, 2014.

114. Liu X. (2015). The application of intuition fuzzy rough set theory in power grid disaster assessment. *Editorial Department, China Journal of Electrical Engineering.*
115. Chowdhary, C. L., & Acharjya, D. P. (2016). A hybrid scheme for breast cancer detection using intuitionistic fuzzy rough set technique. *International Journal of Healthcare Information Systems and Informatics (IJHISI), 11*(2), 38-61.
116. Chuanchao, Z. (2019). Large Data Generalized Dynamic Fault Feature Extraction Algorithm Based on Intuitionistic Fuzzy-Rough Set Discernibility Matrix. *JCP, 14*(1), 1-24.

5

Satellite-Based Estimation of Ambient Particulate Matters ($PM_{2.5}$) Over a Metropolitan City in Eastern India

Tamanna Nasrin[1], Sharadia Dey[2*] and Sabyasachi Mondal[1]

[1]*Department of Mathematics, Amity University Kolkata, Newtown, West Bengal, India*

[2]*Department of Environmental Studies, St. Xavier's College (Autonomous), Kolkata, West Bengal, India*

Abstract

Air quality of different metropolitan cities of India has worsened over the last decade. Kolkata is among the most polluted urban areas of the country. Particulate matter smaller than 2.5μm ($PM_{2.5}$) is considered as one of the significant parameters for indicating air quality. Ground-based monitoring stations for $PM_{2.5}$ are limited over Kolkata. So, Aerosol optical depth (AOD) obtained by Aqua satellites and Moderate Resolution Imaging Spectroradiometer (MODIS) onboard EOS Terra are used to evaluate the local $PM_{2.5}$ concentration over Kolkata. This work attempts to develop a statistical model to estimate $PM_{2.5}$ concentration using AOD_{MODIS} and meteorological parameters (Temperature, Relative Humidity, Planetary Boundary Layer Height, Total Cloud Cover, Wind Speed). The concentration of $PM_{2.5}$ is found to be influenced by various meteorological parameters. It is found that 52% of the variability of the dependent variable $PM_{2.5}$ is explained by the six explanatory variables (i.e., AOD_{MODIS}, temperature, relative humidity, average total cloud, planetary boundary layer height and wind speed) whereas only 3.9% of the variability of the dependent variable $PM_{2.5}$ is explained by AOD alone as explanatory variable.

***Keywords*:** $PM_{2.5}$, AOD_{MODIS}, statistical model, meteorological parameters

**Corresponding author*: sharadiadey1985@gmail.com

Ramakant Bhardwaj, Jyoti Mishra, Satyendra Narayan and Gopalakrishnan Suseendran (eds.) Mathematics in Computational Science and Engineering, (135–146) © 2022 Scrivener Publishing LLC

5.1 Introduction

Escalating particulate pollution and deteriorating air quality status is a matter of concern worldwide. Exposure to ambient $PM_{2.5}$ leads to lung cancer, cardiovascular disease, diabetes, pregnancy-related complications, chronic obstructive pulmonary disease (COPD), low birth weight and premature death across the globe [1–5]. The relationship between increased mortality or morbidity and ambient $PM_{2.5}$ concentrations has been confirmed by many epidemiological studies [6, 7]. Many investigations were performed for understanding the relationship between surface $PM_{2.5}$ measurements and total-column AOD. Most of these studies have developed simple empirical relationships between surface $PM_{2.5}$ measurements and total-column AOD [8, 9].

Many researchers have highlighted the use of regression model for estimating the concentration of ambient particulate matters by utilizing the satellite derived AOD [9, 10]. The predictive power of the regression models was improved by addition of different meteorological and other parameters [11–15] examined a methodology for generating the regional PM_{10} concentration maps for performing health impact studies. A satellite estimation of PM_{10} was performed over Agra by developing a statistical model [12]. In that study, meteorological parameters were found to influence the concentration of ambient $PM_{2.5}$. Wang and Christopher [9] carried out a study over Alabama, Jefferson County, and observed the correlation between the AOD obtained from seven locations within 100 km from both Aqua and Terra satellites and $PM_{2.5}$ (both hourly and 24-h averaged). This study concluded that the coefficient of correlation(R) for linearly related hourly $PM_{2.5}$ and AOD_{MODIS} for all the data of all the sites was 0.70 and R for 24-h average of $PM_{2.5}$ and AOD_{MODIS} for all the data of all the sites was 0.98.

AOD portrays total columnar aerosol optical properties. Different researchers have used AOD as an input parameter in statistical models for the estimation of the ambient PM concentration ($PM_{10}/PM_{2.5}$) at ground level [16–19]. Meteorological parameters and vertical distribution of aerosol influence the relation between AOD and ambient PM concentration. Therefore, estimation of PM concentrations simply from AOD as the only regressor would incorporate large uncertainties in the developed model. Atmospheric boundary layer height and different meteorological parameters should be introduced as various dependent parameters into the correlative models to reduce these uncertainties and increase the predictability.

India experiences some of the worst particulate air pollution in the world, with mean $PM_{2.5}$ concentrations consistently above World Health Organization guidelines [20, 21]. Ground-based monitoring of ambient $PM_{2.5}$ are limited. Satellite-based estimation of $PM_{2.5}$ is helpful for exploring the health impact of particulate matters to overcome the limitation of unavailability of ground-based $PM_{2.5}$ data.

The present study aims to perform satellite-based estimation of $PM_{2.5}$ over Kolkata by using statistical modelling. The in-situ $PM_{2.5}$ data collected from the WBPCB monitoring station at Victoria Memorial, the AOD data collected from MODIS and meteorological data are utilized to develop the statistical model for the estimation of $PM_{2.5}$. This model will help in estimating the concentration of ambient $PM_{2.5}$ over regions where there is no in-situ $PM_{2.5}$ monitoring station.

5.2 Methodology

Kolkata (22.5726° N, 88.3639° E) is chosen as the study area. It is a metropolitan city located in the eastern part of India; the estimated population of Kolkata is 5.4 million. It is located on the bank of Hooghly River and has an average elevation of 6 metres.

The monitoring station at Victoria Memorial established by West Bengal Pollution Control Board (WBPCB) is used for obtaining the $PM_{2.5}$ in-situ data for this study. The hourly concentration of $PM_{2.5}$ is collected for the duration of May 2018 to April 2019.

The Moderate Resolution Imaging Spectroradiometer (MODIS) onboard EOS Terra and Aqua satellites acquire aerosol optical depth (AOD) and other aerosol properties routinely over land and ocean [22]. AOD_{MODIS} values are reported at the wavelength of 0.55. In this study, the AOD data from the MODIS sensors onboard Terra and Aqua Satellites are retrieved from the Level 1 atmosphere archive and distribution system (LAADS) website of National Aeronautics and Space Administration (NASA). The MODIS onboard Terra and Aqua satellites cross the equator at about 10:30 a.m. and 1:30 p.m. (at local sun times) respectively, with a checking area of 2,330 km (cross track) by 10 km (along track at nadir) [22]. AOD_{MODIS} is reported at the wavelength of 0.55 μm at 10 km×10 km spatial resolution as level 2 information [23, 24]. Levy *et al.* [25] has already shown the worldwide approval of AOD_{MODIS} product.

The data of meteorological parameters (i.e., relative humidity, temperature, average total cloud, wind speed and planetary boundary layer height) coincident to AOD_{MODIS} retrieval are collected from NOAA's website (ready.noaa.gov/READY amet.php). In the proposed statistical model, AOD_{MODIS} and meteorological parameters are used as the regressors. The data of May 2018 to October 2018 were used for the development of regression model while the model testing and/or validation studies are performed with the data of November 2018 to April 2019.

5.3 Result and Discussions

The present study attempts to estimate $PM_{2.5}$ by utilizing satellite-derived AOD and different meteorological parameters like temperature (T), average total cloud (ATC), relative humidity (RH), wind speed (WS) and planetary boundary layer height (PBLH). Various regression models are used for estimating $PM_{2.5}$ where AOD_{MODIS} and meteorological parameters are used as independent regressors. Minimum, maximum, mean and standard deviation (i.e., statistical summary) of these dependent and independent parameters are shown in Table 5.1.

Correlation matrix of dependent ($PM_{2.5}$) and independent parameters (AOD_{MODIS} and meteorological parameters) in Table 5.2 shows that $PM_{2.5}$

Table 5.1 Statistical summary.

Variable	Minimum	Maximum	Mean	Std. deviation
PM2.5	20.820	288.828	106.227	54.666
AOD	0.238	0.990	0.709	0.198
TEMPERATURE	23.000	40.800	28.731	4.195
REL HUMID	22.400	55.500	37.112	9.057
AVG TTL CLD	0.000	98.600	16.584	25.039
CV PBL HEIGHT	873.500	1858.100	1230.217	223.607
WIND SPEED	0.200	12.600	4.781	2.239

Table 5.2 Correlation matrix of dependent ($PM_{2.5}$) and independent parameters.

	AOD	Temperature	REL humid	AVG TTL CLD	CV PBL height	Wind speed	PM2.5
AOD	1.000	-0.101	-0.284	0.164	0.023	0.086	0.199
TEMPERATURE	-0.101	1	0.521	0.397	0.478	0.115	-0.539
REL HUMID	-0.284	0.521	1	0.271	0.050	0.200	-0.593
AVG TTL CLD	0.164	0.397	0.271	1	0.202	0.177	-0.249
CV PBL HEIGHT	0.023	0.478	0.050	0.202	1	0.021	-0.276
WIND SPEED	0.086	0.115	0.200	0.177	0.021	1	-0.384
PM2.5	0.199	-0.539	-0.593	-0.249	-0.276	-0.384	1

holds inverse relationship with all the meteorological parameters. AOD shows very weak positive correlation with ambient $PM_{2.5}$. Temperature and relative humidity show significant inverse relationship with $PM_{2.5}$. It shows that with rise in temperature, the height of convective planetary boundary layer increases, thereby there is a decrease in the concentration of $PM_{2.5}$ near ground.

Table 5.3 shows the equations obtained from different regression models, root mean square error (RMSE), and coefficient of variation (R^2). This study shows that there is substantial in the R^2 value when meteorological parameters are gradually added along with AOD_{MODIS}. 4% of the variation of $PM_{2.5}$ is explained by the regression model when only AOD_{MODIS} is used as the regressor. In general, the R^2 value increases with gradual increase in the number of regressors and 52% of the variability of the dependent variable $PM_{2.5}$ is explained by the AOD_{MODIS} and six explanatory variables (i.e., temperature, relative humidity, average total cloud, wind speed and planetary boundary layer height). The improvement in regression model is also shown by gradual decrease in the value of RMSE with subsequent addition of meteorological parameters as independent regressors.

Ambient temperature is found to be an effective predictor for estimation of $PM_{2.5}$. The negative sign of its estimator shows that AOD_{MODIS} estimate lower $PM_{2.5}$ concentration at higher temperature. Similar results were shown for $PM_{2.5}$ and PM_{10} in other studies [15, 26]. Relative humidity changes the optical properties of aerosol thereby affecting the association

Table 5.3 Details of regression models.

Set no.	Equation of the model	R^2	RMSE
I	PM2.5 = 67.45+54.73*AOD	0.039	54.054
II	PM2.5 = 274.27+40.12*AOD-6.84*T	0.312	46.162
III	PM2.5 = 308.69+13.61*AOD-4.17*T-2.49*RH	0.427	42.524
IV	PM2.5 = 304.63+15.18*AOD-4.07*T-2.47*RH-5.08E-02*ATC	0.427	42.908
V	PM2.5 = 323.9+15.20*AOD-2.93*T-2.71*RH-3.96-02*ATC-3.52E-02*CV PBLH	0.442	42.753
VI	PM2.5 = 338.08+25.37*AOD-3.05*T-2.32*RH+2.80E-02*ATC-3.52E-02*CV PBLH-7.01*WS	0.519	40.095

of $PM_{2.5}$ - AOD_{MODIS}. Surface level wind speed is seen as a huge for estimation of $PM_{2.5}$ and negative sign of the estimator shows that AOD_{MODIS} estimates lower concentration of $PM_{2.5}$ ant higher wind speed. The impact of average total cloud and PBL height on estimation of $PM_{2.5}$ is found to be very low. Many studies have shown that meteorological parameters such as temperature, relative humidity, wind velocity, and atmospheric pressure can influence the AOD–PM association [11, 27, 28].

Model adequacy checks were applied for all the six linear regressor models. The residual error vs estimated $PM_{2.5}$ scatter plots of Set I to Set VI are shown in Figure 5.1 to Figure 5.6. All the six plots in this figure obviously demonstrated an expanding pattern of lingering error (positive/negative)

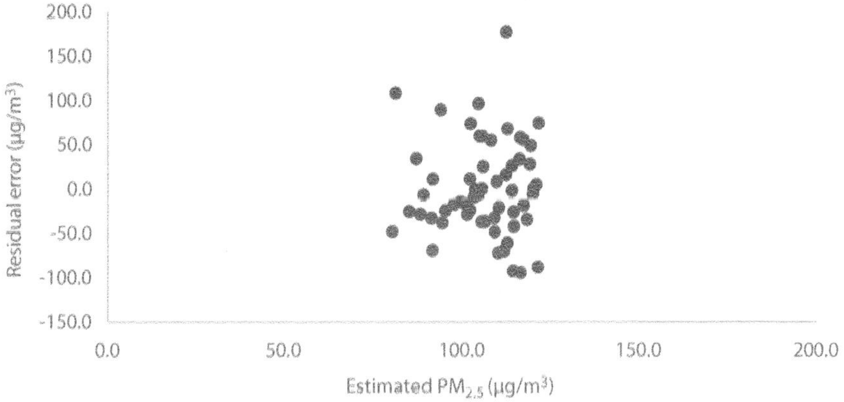

Figure 5.1 Residual plot for linear regression model of Set I.

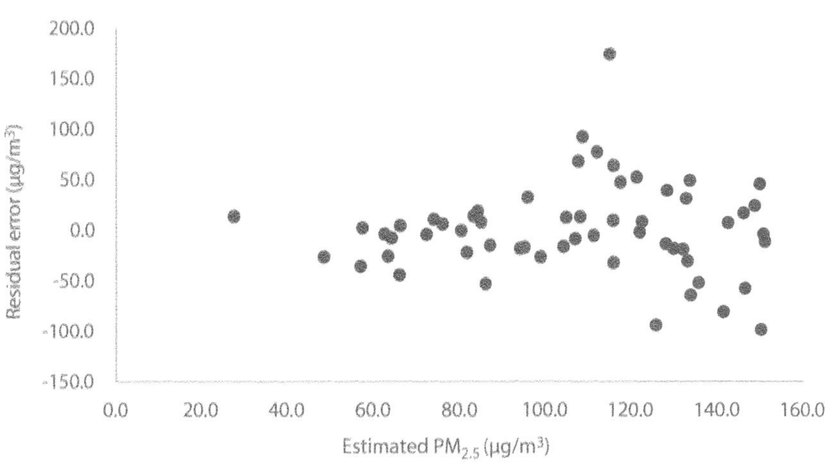

Figure 5.2 Residual plot for linear regression model of Set II.

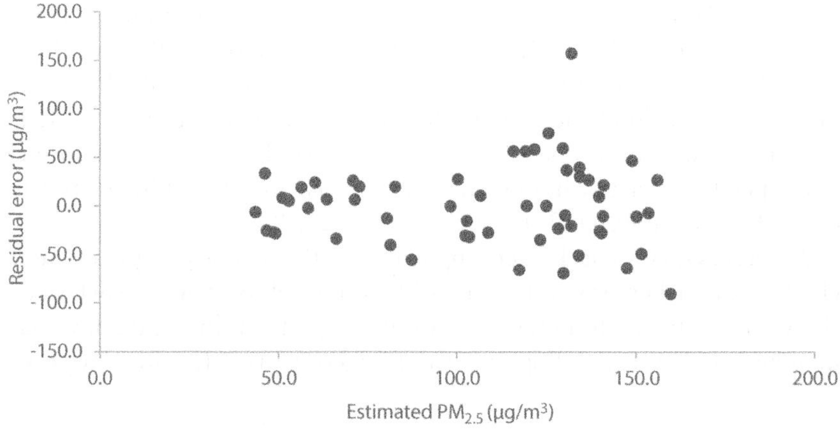

Figure 5.3 Residual plot for linear regression model of Set III.

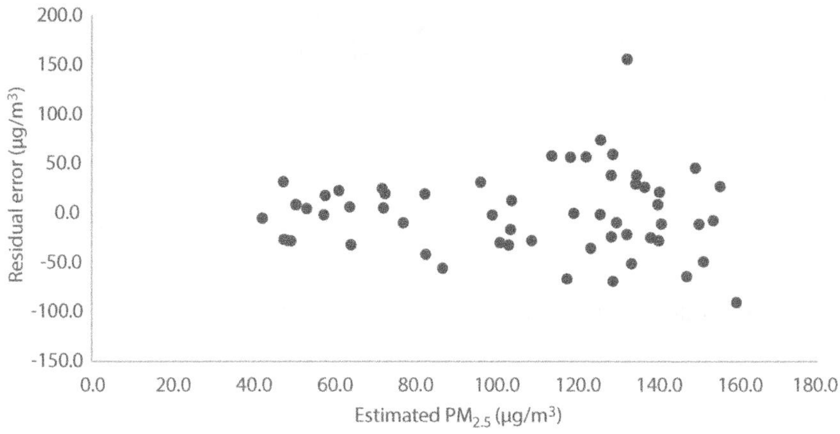

Figure 5.4 Residual plot for linear regression model of Set IV.

with expanding $PM_{2.5}$ values. These expanding patterns showed the need of model change for improved estimation of encompassing $PM_{2.5}$ fixation. Similar result was shown over Agra, India for $PM_{2.5}$ estimation [12]. However, with increase in the number of regressors the residual errors of estimation show a constant band over a considerable range of $PM_{2.5}$ and then become divergent in nature.

The observation of the results obtained in this study depicts that gradual addition of meteorological parameters along with AOD_{MODIS} as regressors in the linear regressor models improves estimation of ambient $PM_{2.5}$.

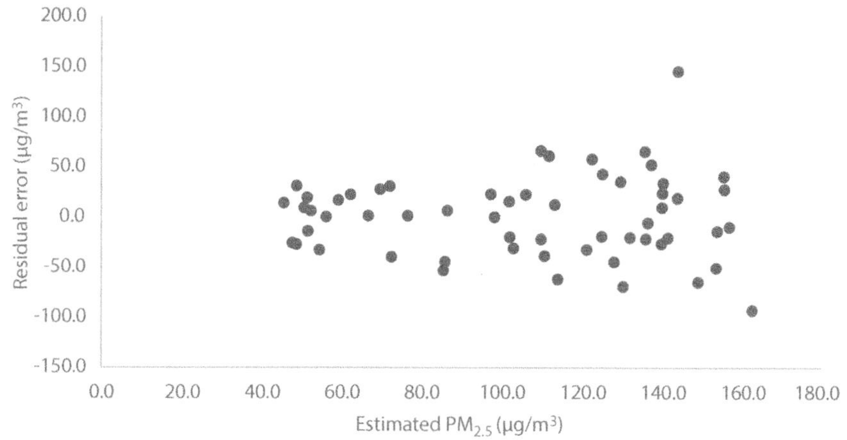

Figure 5.5 Residual plot for linear regression model of Set V.

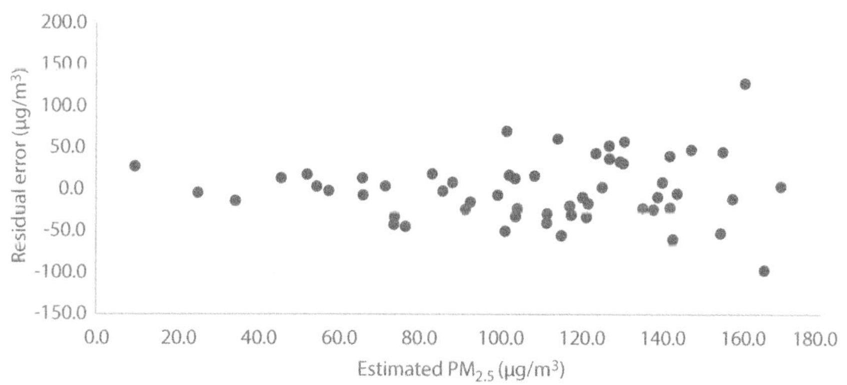

Figure 5.6 Residual plot for linear regression model of Set VI.

The coefficient of variation increases and RMSE decreases with subsequent addition of regressors. Surface based $PM_{2.5}$ is still limited to few selective sites over Kolkata. So, this method can be used for estimation of surface $PM_{2.5}$ by utilizing satellite derived AOD and available meteorological data thereby understanding the air quality of the study area.

5.4 Conclusion

The present study attempts to find a method for satellite-based estimation of ambient $PM_{2.5}$ over Kolkata. The proposed method can be used for exploring the $PM_{2.5}$ climatology and spatio-temporal variability of $PM_{2.5}$ over this region. This statistical model can also be used for air quality management and health management in different parts of the country. $PM_{2.5}$ and meteorological parameters are measured at a particular point in Kolkata, but AOD_{MODIS} was deduced as the average value of the pixel (10 × 10 km) covering the ground monitoring station of WBPCB. So very high performance of the model should not be expected. But this study shows that the ambient air quality is influenced by the local meteorology and gradual addition of meteorological parameters along with AOD_{MODIS} as regressors in the linear regressor models improves estimation of ambient $PM_{2.5}$.

References

1. Pope, C.A., Burnett, R.T., Thun, M.J., Calle, E.E., Krewski, D., Ito, K., Thurston, G.D., (2002). Lung cancer, cardiopulmonary mortality, and long-term exposure to fine particulate air pollution. *JAMA* 287, 1132–1141. https://doi.org/10.1016/j.scitotenv.2011.03.017.
2. Chen H, Goldberg MS, Villeneuve PJ (2008) A systematic review of the relation between long term exposure to ambient air pollution and chronic diseases. *Rev Environ Health* 23:243–297. doi:10.1515/reveh.2008.23.4.24
3. Krewski, D., Jerrett, M., Burnett, R.T., Ma, R., Hughes, E., Shi, Y., Turner, M.C., Pope III,C.A., Thurston, G., Calle, E.E., Thun, M.J. (2009). Extended follow-up and spatial analysis of the American cancer society study linking particulate air pollution and mortality. *Res.Rep. Health Eff. Inst.* 140, 5–114.
4. Anenberg, S.C., Horowitz, L.W., Tong, D.Q., West, J.J., (2010). An estimate of the global burden of anthropogenic ozone and fine particulate matter on premature human mortality using atmospheric modeling. *Environ. Health Perspect.* 118, 1189–1195.https://doi.org/10.1289/ehp.0901220
5. Pedersen, M., Giorgis-Allemand, L., Bernard, C., Aguilera, I., Andersen, A.-M.N., Ballester, F., Beelen, R.M.J., Chatzi, L., Cirach, M., Danileviciute, A., Dedele, A., Eijsden, M. van, Estarlich, M., Fernández-Somoano, A., Fernández, M.F., Forastiere, F., Gehring, U., Grazuleviciene, R., Gruzieva, O., Heude, B., Hoek, G., de Hoogh, K., van den Hooven, E.H., Håberg, S.E., Jaddoe, V.W.V., Klümper, C., Korek, M., Krämer, U., Lerchundi, A., Lepeule, J., Nafstad, P., Nystad, W., Patelarou, E., Porta, D., Postma, D., Raaschou-Nielsen, O., Rudnai, P., Sunyer, J., Stephanou, E., Sørensen, M., Thiering, E., Tuffnell, D., Varró, M.J., Vrijkotte, T.G.M., Wijga, A., Wilhelm, M., Wright, J., Nieuwenhuijsen, M.J., Pershagen, G., Brunekreef, B., Kogevinas, M., Slama, R. (2013). Ambient air pollution and low birthweight: a European cohort study (ESCAPE). *Lancet. Respir. Med.* 1, 695–704. https://doi.org/10.1016/S22132600(13)70192-9

6. Dominici, F.; Peng, R.D.; Bell, M.L.; Pham, L.; Dermott, A.M.; Zeger, S.L.; Samet, J.M. (2006). Fine particulate air pollution and hospital admission for cardiovascular and respiratory diseases. *J. Am.Med. Assoc.*, 295, 1127–1134.
7. Li, P.; Xin, J.; Wang, Y.; Wang, S.; Li, G.; Pan, X.; Liu, Z.; Wang, L. (2013). The acute effects of fine particles on respiratory mortality and morbidity in Beijing, 2004–2009. *Environ. Sci. Pollut. Res.Int.*, 20, 6433–6444.
8. Engel-Cox JA, Hoff RM, Haymet ADJ. (2004). Recommendations on the use of satellite remote-sensing data for urban air quality. *J Air Waste Manage Assoc* 54(11):1360–1371.
9. Wang J, Christopher SA (2003). Intercomparison between satellite derived aerosol optical thickness and $PM_{2.5}$ mass: implications for air quality studies. *Geophysical Research Letter* 30: 2095.
10. Chu DA, Kaufman YJ, Zibordi G, Chern JD, Mao JM, Li C, Holben HB (2003) Global monitoring of air pollution over land from EOS Terra MODIS. *J Geophys Res* 108:4661.
11. Kumar N, Chu A, Foster A (2007) An empirical relationship between $PM_{2.5}$ and aerosol optical depth in Delhi Metropolitan. *Atmos Environ* 41:4492–4503.
12. Chitranshi S., Sharma S. P., Dey S. (2015). Satellite-based estimates of outdoor particulate pollution (PM10) for Agra City in northern India. *Air Quality,Atmosphere and Health* 8:55–65. DOI 10.1007/s11869-014-0271-x
13. Kumar N, Chu A, Foster A (2008) Remote sensing of ambient particles in Delhi and its environs: estimation and validation. *Int J Remote Sens* 29:3383 3405.
14. Gupta P, Christopher SA, Wang J, Gehrig R, Lee Y, Kumar N (2006). Satellite remote sensing of particulate matter and air quality assessment over global cities. *Atmos Environ* 40:5880–5892.
15. Chitranshi S., Sharma S. P., Dey S. (2015).Spatio-temporal variations in the estimation of PM_{10} from MODIS-derived aerosol optical depth for the urban areasin the Central Indo-Gangetic Plain. *Meteorol Atmos Phys* 127:107–121DOI 10.1007/s00703-014-0347-z
16. Othman N, Jafri MZM, Lim HS, Abdullah K (2009) Retrieval of aerosol optical thickness (AOT) and its relationship to air pollution particulate matter (PM10). *Sixth Int Conf Comput Graph Imaging Vis.* doi:10.1109/CGIV.2009.22
17. Yap XQ, Hashim M (2013) A robust calibration approach for PM10 prediction from MODIS aerosol optical depth. *Atmos Chem Phys* 13:3517–3526.
18. Chu A, Szykman J, Kondragunta S (2006) Remote sensing of aerosol and chemical gases. Model simulation/assimilation and applications to air quality. *Proceedings of SPIE 6299.* San Diego, CA: SPIE Digital Library.
19. Lee HJ, Liu Y, Coull BA, Schwartz J, Koutrakis P (2011) A novel calibration approach of MODIS AOD data to predict PM2.5 concentration. *Atmos Chem Phys* 11:7991–8002. doi:10.5194/acp-11-7991

20. Apte JS, Marshall JD, Cohen AJ, Brauer M. (2015). Addressing global mortality from ambient $PM_{2.5}$. *Environ Sci Technol.*;49(13):8057–66.
21. Dey S, Girolamo LD, Donkelaar AV, Tripathi SN, Gupta T, Mohan M (2012) Decadal exposure to fine particulate matters ($PM_{2.5}$) in the Indian subcontinent using remote sensing data. *Remote SensEnviron* 127:153–161.
22. Remer L. A, Kaufman Y. J, Tanre D, Mattoo S, Chu D. A, Martins J. V, Li R. R, Ichoku C, Levy R. C, Kleidman R. G, Eck T. F, Vermote E, Holben B. N (2005). The MODIS aerosol algorithm, products, and validation. *Journal of Atmospheric Science* 62:947–973.
23. Levy RC, Remer L, Dubovik O (2007a) Global aerosol optical properties and application to moderate resolution imaging spectroradiometer aerosol retrieval over land. *Journal of Geophysical Research* Atmosphere 112(D13), D13210. doi:10.1029/2006JD0078151
24. Levy RC, Remer L, Mattoo S, Vermote E, Kaufman YJ (2007b) Second generation operational algorithm: retrieval of aerosol properties over land from inversion of moderate resolution imaging spectroradiometer spectral reflectance. *Journal of Geophysical Research-Atmosphere* 112(D13), D13211. doi:10.1029/2006JD007811
25. Levy RC, Remer LA, Kleidman RG, Mattoo S, Ichoku C, Kahn R, Eck TF (2010). Global evaluation of the collection 5 MODIS dark- target aerosol products over land. *Atmospheric Chemistry & Physics.* doi:10, 5194/acp-10-10399
26. Liu Y, Franklin M, Kahn R, Koutrakis P (2006). Using aerosol optical thickness to predict ground level PM2.5 concentrations in the St. Louis area: a comparison between MISR and MODIS. *RemoteSens Environ* 107:33–44.
27. Gao J, Zha Y (2010) Meteorological influence on predicting air pollution from MODIS-derived aerosol optical thickness: a case study in Nanjing, China. *Remote Sens* 2:2136–2147.
28. Grguric S, Krizan J, Gasparae G, Antonic O, Spiric Z, Mamouri ER, Christodoulou A, Nisantzi A, Agapiou A, Themistocleous K, Fedra K, Panaylotou C, Hadjilmitsis D (2013) Relationship between MODIS based aerosol optical depth and PM_{10} over Croatia. *Central European Journal of Geosciences.*doi: 10.2478/s13533012-0135-6

6

Computational Simulation Techniques in Inventory Management

Dr. Abhijit Pandit* and Dr. Pulak Konar

Department of Mathematics, Amity University, Kolkata, West Bengal, India

Abstract
In this paper we will see inventory systems and will learn to manage it with simulation technique. We will study simulation which is performed manually for better understanding of the topic. Then we will discuss some merits and demerits of it and study Monte Carlo simulation technique and how this method is used to solve a real life problem. We will use Excel software to generate random numbers and we will perform simulation in it. We will find the inventory sales for a confectionary bakery shop and will simulate and predict the sales for next few days. At the end we will compare and plot a graph so as to see and understand simulation better.

***Keywords*:** Simulation technique, Monte Carlo simulation technique, Excel software, visual inventory sales, simulation

6.1 Introduction

6.1.1 Inventory Management

Inventory management refers to the process of ordering, storing and using a company's inventory. This includes the processing of raw materials, components and finished products, warehousing and items.

Basically it includes:

- A stock of items holds to meet future demands.
- A list of goods and services stored in stock for business.

**Corresponding author*: apandit@kol.amity.edu

Ramakant Bhardwaj, Jyoti Mishra, Satyendra Narayan and Gopalakrishnan Suseendran (eds.) Mathematics in Computational Science and Engineering, (147–166) © 2022 Scrivener Publishing LLC

- A considerable amount of fund is required for it.
- It constitutes significant part of assets.

The flow diagram for Inventory System is as follows:

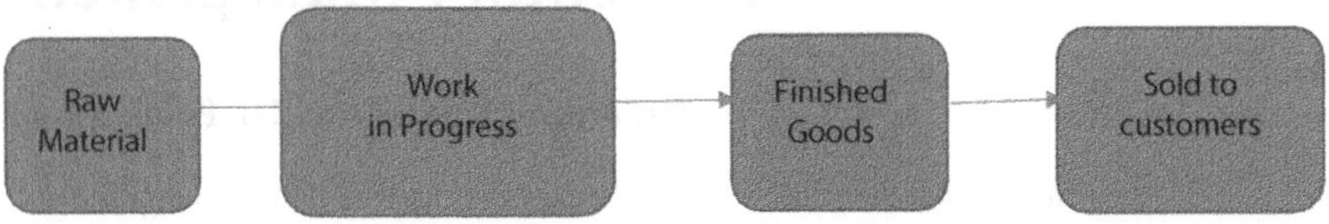

Firstly, the raw materials are used and they are processed to make products or through raw materials production of goods and services are held and then they are send in the markets or sold to customers. In this process there we see two terms:

The main objective of this inventory management system is:

- To minimize the investment in the inventory.
- To maximize the profit.
- To run the production smoothly.

Model: *When we deliberately create a system to find the solution for a problem.*

6.1.2 Simulation

A simulation is an imitation of the operation of the real world process or system [1]. In order to simulate anything there must be a model that will be the main characteristic behavior of the chosen physical or abstract system or process. The model represents the system itself while the simulation system represents overtime performance.

It is experiment testing, before launching it into the market. Simulation (Figure 6.1) involves developing a model of some real phenomena and then performing experiments on the model evolved. So, it is the process of designing a model of a real system for the tenacity of understanding the behaviour of system (decision variable) conducting experiments.

The simulation is used in many contexts such as technology simulation, performance optimization, test safety engineering, training education and various video games you play in your home. Often computer experiments are used to study simulation models. It is used for scientific modeling of natural systems or human systems. It is used when the real system cannot be used because it is not possible, not accessible or dangerously unacceptable to connect the system to real life or it is not designed or does not exist.

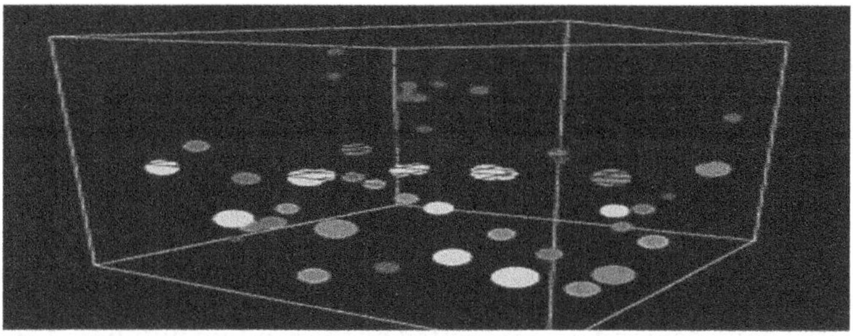

Figure 6.1 Schematic representation of simulation.

The key issues in simulation are acquisition of valid source of information means collection of valid and correct information about the relevant solution of the key characteristics and behaviour.

In a treadmill, before we start to run on the track or a field or athletic tracks we use treadmill in gym or in our house. We do not move from one place to another place, we walk on the same place but we feel like walking or running on a track, so treadmill is kind of a simulator which puts the user in the same or similar kind of simulation which is the biggest role of simulator in a treadmill.

Solution that we get from simulation models and analytical models:

1. **Analytic Model:**
   The solutions we find in the analytical models are accurate, precise and reflect the real state of the system. Moreover, solutions like these are often the appropriate solution, reflecting the ethical situation in the system. However, these solutions include complex calculations that require equal time. Also, it is not easy to find a system solution without small programs.
2. **Simulation Model:**
   In the simulation model, the solutions are inaccurate and certainly incorrect, which is why there is so much uncertainty leading to various mathematical errors. However, these solutions are quick to produce, without the complex calculations involved. Although inaccurate, they still have to look at all the right behaviors of the system.

   For example, NASA uses space shuttle, airplane simulators so that its astronauts can practice shuttle flying. In many video games computer simulators are used to mimic real-life games such as boxing, racing, hockey, cricket and the world-famous PUBG and many other such games.

Now we will see some simulation examples related to inventory system to understand it better. Before that we define random number:

Random Number: *is defined as the numbers which are chosen randomly to solve any problem, we can get these sets of numbers from software like excel, maple, etc. These numbers get changed every time when the software is opened, every time the number gets refreshed.*

1) Suppose we have to perform the simulation of the inventory system given the daily demand is represented by the random numbers 4, 3, 8, 2, 5 and the demand probability is:

Demand	0	1	2
Probability	0.2	0.5	0.3

Let the initial inventory = 4 units

⇨ Now to solve the problem we need to draw two tables one for the data and one for the simulation part.

Data table–

Demand	Probability	Cumulative Probability	Random No. mapping
0	0.2	0.2	1 - 2
1	0.5	0.7	3 - 7
2	0.3	1.0	8 - 9

Simulated table–

Days	Beginning Inventory	Random No. for Demand	Demand	End Inventory	Shortage
1	4	4	1	3	0
2	3	3	1	2	0
3	2	8	2	0	0
4	0	2	0	0	0
5	0	9	2	0	2

Here, we have simulated only for five days because we have five random numbers given in the question. So, in this we can say that the shortage has occurred on the very 5th day.

2) Let us simulate the given data and find the average weekly cost of inventory.

Let the initial inventory = 25 units

Demand per customer	Probability
0	0.1
1	0.3
2	0.25
3	0.2
4	0.15

Demand per customer	Probability
1	0.1
2	0.15
3	0.4
4	0.35

⇨ Firstly we will choose random number as:

RN1= No. of customers and RN2= No. of demand per customers

No. of customers	Probability	Cumulative Probability	Random Mapping
0	0.1	0.1	0 – 9
1	0.3	0.4	10 – 39
2	0.25	0.65	40 – 64
3	0.2	0.85	65 – 84
4	0.15	1.0	85 – 99

Here, since of the probability is 10%, so we have given difference of 10 digits over there

Demand per Customer	Probability	Cumulative Probability	Random Mapping
1	0.1	0.1	0 – 9
2	0.15	0.25	10 – 24
3	0.4	0.65	25 – 64
4	0.35	1.0	65 – 99

Now, we will take the consideration of inventory policy for this question, i.e., we will place the order when inventory is Q < R and there is no outstanding order Q=25, R=10.

We have cost:

Ordering Cost = \$45 per order
Holding Cost = \$ 1 per unit per week
Backorder = \$ 5 per unit per week

Backorder cost: *Suppose if in the case when the demand is not fulfilled and you are out of stock so at that time this comes and it gets added to the costs till everything gets back to the normal condition.*

Here, in the model it is given that it takes 2 weeks to receive the order, which means our lead time is 2 weeks.

Now, we will simulate the no. of customers and the demand per customers. Firstly, we will require some sets of random numbers which we can and further we will get from Excel. So, that means RN_1 and RN_2 are merely assumed. We choose no. of RN_2 as per the no. of customers obtained in RN_1. For example, if we choose No. of customers = 3, then we choose three different random values from our Random number set RN_2.

In recent years, it has become an important topic for both academics and professionals on how to manage customer relationships. Different clients have different economic values in the company, so they should be treated a little differently. Managing customers in different ways is beneficial but at the same time companies need to keep in mind that building multiple relationships is not always the best, most importantly building a strong customer relationship [4].

Companies need to consider customer behavior and profitability in order to effectively manage individual customer relationships. In addition to profit, other factors may be involved, such as how companies view their customers, such as the company's attention, how close they are to the customer, and how much they know about the consumer market [5]. Ulaga and Egert [6] mean that it can be very expensive for companies to get the same service for all their customers. It is better to split the service for more reasons than financial.

By acquiring specialized service for very important customers, one can create a situation where the pay firm can find the status of a major supplier and specific customers, and thus make that customer more dependent on that particular supplier. The controversial situation is that companies may find themselves in a position to be a backup provider. In today's business environment, independently of the business environment, it differs not only in product and price, but also in the structure of consumer value. For example, pricing can be a new way of creating additional services that other companies do not currently accept [6]. Easily solved by other strategies or mathematical analysis.

It is very popular in many industries and there are several advantages of it like it helps to reduce the cost in assessing different situations.

There are different types of simulation which can be classified into three categories:

1. Monte Carlo simulation.
2. Operational gaming.
3. System simulation.

3. Monte Carlo Simulation Model:

It is a method based on probability or opportunity. Uses random numbers by process or calculation. He needs to make decisions under uncertainty. There are a wide range of computational algorithms based on random sampling to obtain these numerical results. The main idea is to use random to solve a problem that can be decisive in the goal. This method is often used for physical or mathematical problems and is very useful when other methods are difficult or impossible to use.

The Monte Carlo method was developed in the 1940s by atomic bomb scientists, who named the city of Monaco famous for its casinos and sports. Its main purpose is to use random samples of parameters or inputs to test system performance and complex process. Scientists were dealing with physics problems such as neutron dispersed models dello, which were very difficult to find a solution for analysis - so they had to be quantitatively tested. They first entered one of the computers - MNIAAC - but their models delo included so many sizes that full statistical testing was slow. The imitation of Monte Carlo has been remarkably effective in finding solutions to these problems. Since then, Monte Carlo methods have been used for a variety of scientific, engineering, and financial purposes - and commercial applications in almost every industry.

The imitation of Monte Carlo is a computerized mathematical formula that allows people to account for risks in price analysis and decision making. This approach is used by experts in a wide range of fields such as finance, project management, energy, manufacturing, engineering, research and development, insurance, oil and gas, transportation and the environment.

The imitation of Monte Carlo provides the decision maker with various possible outcomes and opportunities of any choice. It shows the extreme potential - the consequences of the crash and the most decisive decision - and all the possible consequences of the decisions on the road.

This method was first used by scientists working in the atomic bomb; Named Monte Carlo, the tourist city of Monaco is famous for its casinos. Since its introduction in World War II, Monte Carlo imitations have been used to sample a variety of physical and mental educators.

The Monte Carlo Simulation analyzes risk by creating models for potential outcomes by incorporating multiple values – the distribution of opportunities – of any uncertain aspect. It then calculates the results over and over, each time using a different set of random values from the possible activities. Depending on the amount of uncertainty and the extent to which it is specified, the imitation of Monte Carlo may involve thousands or tens of thousands of re-calculations before it is completed. The imitation of Monte Carlo reveals the distribution of possible result values.

Through the distribution of opportunities, variables can have different possibilities for different outcomes that occur. Opportunity distribution is the most logical way to describe the volatility of risk analysis.

Common distribution of opportunities includes:

- Normal
  The user simply defines the mean or expected value and general deviation to explain the variation in terms of meaning. Moderate average values are more likely. It is similar and describes many natural phenomena such as human heights. The dynamic examples described in the general distribution include the prices and the price of the energy.
- Lognormal
  Prices are definitely distorted, inconsistent with the standard distribution. It is used to represent values that are not less than zero but have unlimited positive energy. Examples of variables described in the distribution of physical activity include commodity prices, stock prices, and oil reserves.
- Uniform
  All values have an equal chance of appearing, and the user only specifies the minimum and maximum. Examples of variables that can be evenly distributed include production costs or future sales costs for a new product.
- Triangular
  User defines a very low value, which may be too high. Round values are more likely to occur. The variables that can be defined by a triangular distribution include past sales history for each unit of time and levels of items.
- PERT
  The user defines low-level, probability and intensity as a triangular distribution. Round values are likely to occur. However the number between potential and extreme is greater than triangle. That is, there is less emphasis on

extremes. An example of a PERT distribution is a description of the length of work in a project management model.

- Discrete
  User defines specific values that are possible with each opportunity. An example would be court results: a 20% chance of a good decision, a 30% chance of a bad decision change, a 40% chance of a solution, and a 10% mistrial chance.

At the time of the Monte Carlo match, prices were randomly sampled from the distribution of input opportunities. Each sample set is called an iteration, and the result from that sample is recorded. The imitation of Monte Carlo does this hundreds or thousands of times, and the result is a wide range of potential side effects. In this way, the imitation of Monte Carlo gives a more complete picture of what is possible. It tells us not only how it is possible but also how often it is possible.

• The imitation of Monte Carlo offers many advantages over deterministic analysis, or "one-point estimates":

- Potential consequences. The results show not only the probability, but also the probability of each individual outcome.
  Clicking Results. Thanks to the data generated by the imitation of Monte Carlo, it is easy to create graphs of different outcomes and possibilities. This is important for the information obtained from other participants.
  Sentiment analysis. In just a few cases, decision-making analysis makes it difficult to determine which variables are most affected. In the Monte Carlo copy, it is easy to see which inputs have the greatest impact on the results of the following lines.
  Condition Analysis: In defined models, it is very difficult to compare different price combinations for different inputs to actually see the effects of different conditions. Using a copy of Monte Carlo, analysts can see clear results of what ideas are put together when certain results come out. This is very important in the search for further analysis.
- Input integration. In the Monte Carlo simulation, it is possible to model the interdependent relationship between input variables. It is important to present precisely that, in fact, when some things go up, some go up or down properly.

Improving the imitation of Monte Carlo is the use of the Latin hypercube sample, which is more accurate in the whole range of distribution functions. Whenever there is a large uncertainty where we need to make an estimate, a prediction or a decision, we are advised to pay attention to the example of Monte Carlo – if we do not do so, our guesses or predictions will be accurate, with negative consequences for our decisions. Sam Simter, a well-known authority on savage, simulation and other methods of measurement, says that "many people, when faced with uncertainty ... fall into the trap of entering the uncertain number specified by a common denominator."

Like the physics problems of the 1940s – most business activities, programs and processes are too complex to find a solution for analysis. But we can create a spreadsheet model that allows us to test by price through our system – we can change numbers, ask 'what if' and see the results. This is especially true if we have one or two limitations that we can consider. But many business situations involve large-scale uncertainty – for example, volatile market demand, unplanned competitor plans, cost uncertainty, and much more – such as 1940s physics problems. If our situation sounds like this, we may find that the way Monte Carlo works wonders for us.

To apply the Monte Carlo simulation, we must be able to create a modeling model for our business, program, or process. One of the easiest and most popular ways to do this is to create a spreadsheet model using Microsoft Excel – and then use the Analytic Solver Simulation of Frontline Systems as a simulation tool. Other methods include coding in a programming language such as Visual Basic, C ++, C # or Java – with Frontline's Solver Platform SDK or using a special simulation language. We will also study (or review) the basics of opportunities and statistics. To eliminate the uncertainty of our model, we will replace certain fixed numbers – for example in spreadsheets – with functions that draw random samples from the distribution of opportunities. And to analyze the effects of simulation, we will use calculations like definitions, standard deviations, percentages and charts and graphs.

This is mainly used for three different problems–

1. Good class performance.
2. Price aggregation.
3. Performance from opportunity distribution.

A shop sells cakes. The past data of demand per day for 50 days is given below. Using the data and following sequence of the random numbers, generate the demand for next 10 days. Also find out the average demand per day.

We will make tables to solve the problem. Here, total number of days = 50.

No. of cakes per day	No. of days
0	2
5	11
10	8
15	21
20	5
25	3

No. of cakes	No. of days	Probability	Cumulative probability	Random no. range
0	2	0.04	0.04	00 to 03
5	11	0.22	0.26	04 to 25
10	8	0.16	0.42	26 to 41
15	21	0.42	0.84	42 to 83
20	5	0.10	0.94	84 to 93
25	3	0.06	1.00	94 to 99

Simulation–

Days	Random Nos.	Range	Demand
1	15	04 to 25	5
2	68	42 to 83	15
3	29	26 to 41	10
4	44	42 to 83	15
5	88	84 to 93	20
6	2	00 to 03	0
7	58	42 to 83	15
8	97	94 to 99	25
9	35	26 to 41	10
10	65	42 to 83	15

So, these are the forecasted or predicted demand for next 10 days. Now, here the total demand for days is 130

Therefore, the Average demand is calculated as Average Demand = Total demands / No. of days

= 130/10

= 13 Therefore, average demand is 13 cakes per day.

Monte Carlo Simulation is a computer simulation. In this simulation the problem is solved by simulating the original data with some random numbers which will be generated in Excel. So, we will use Excel Software for our study.

Now, here are some real past inventory data for 21 consecutive working days excluding Sundays for a confectionary bakery shop which was collected before the lockdown period, from 20/02/20 to 14/03/20. We will perform Monte Carlo Simulation on this data and we will find how we can estimate and predict the inventory for the next 10–15 days or few weeks. At the end we will compare the actual data and the simulated data and will also plot the graph between the actual data and the data simulated.

⇨ First we will give the actual real data which was collected from a confectionary bakery shop.

We have collected the data for 10 different items and their daily demand, and this has continued for 21 days starting from 20/02/20 and concluding to 14/03/20. For each item the cost (in Rs) is given in the table.

Now, we found out the total no. of items sold in 21 days:

Avg. daily sales units = total items sold in 21 days/total no. of days, probability that items will be sold is

Probability = Avg. daily sales for 1 item/Sum of all Avg. daily sales, cost per unit for each item and Sales per day for each items, i.e., *Sales = Cost per unit * Avg. daily sales*

Now, we will do the process of the simulation which involves finding of probabilities for different items, then finding the cumulative probability, random no. coding (according to the probabilities of the items).

Item	Sales/day	Probability	Cumulative probability	R.N. coding
Bun Bread	3.86	0.08	0.08	00 to 07
Pineapple Cake	4.57	0.10	0.18	08 to 17
Vanilla Cake	4.67	0.10	0.27	18 to 26
Cupcakes	4.71	0.10	0.37	27 to 36
Cookies	4.76	0.10	0.47	37 to 46
Biscuit	4.81	0.10	0.57	47 to 56
Pastries	4.90	0.10	0.67	57 to 66
Chocolate Cake	5.10	0.11	0.78	67 to 77
Cream Milkshake	5.24	0.11	0.89	78 to 88
Bread	5.33	0.11	1.00	89 to 99
TOTAL	47.95	1.00		

Now, for each working day the shop remains open for 8 hours, from 11:00 a.m. to 7:00 p.m. Therefore, on an average an order is placed for 48 units per day. So, the average inter-order time = 8/48

= 1/6 hours = 10 minutes

Now, for 10 items and their prices are given, so the simulation is as –

ITEM NO / ITEM / PRICE PER UNIT IN RUPEES/RANDOM NO: (1-BUN BREAD-RS 30-RN: 00 to 07), (2-PINEAPPLE CAKE-RS 130-RN: 08 to 17), (3-VANILLA CAKE-RS 120-RN: 18 to 26), (4-CUPCAKE-RS 40-RN: 27 to 36), (5-COOKIES-RS 45-RN: 37 to 46), (6-BISCUIT-RS 35-RN: 47 to 56), (7-PASTRIES-RS 60-RN: 57 to 66),(8-CHOCOLATE CAKE-RS 140-RN: 67 to 77), (9-CREAM MILKSHAKE-RS 70-RN: 78 to 88), (10-BREAD-RS 30-RN: 89 to 99)

Simulation Part

TIME INTERVAL	MONTE CARLO SIMULATION FOR ITEM THAT CAN GET SOLD DURING VARIOUS TIME INTERVAL IN A WORKING DAY																	
	DAY 1			DAY 2			DAY 3			DAY 4			DAY 5			DAY 6		
	R.N.	ITEM NO	PRICE (Rs)	R.N.	ITEM NO	PRICE (Rs)	R.N.	ITEM NO	PRICE (Rs)	R.N.	ITEM NO	PRICE (Rs)	R.N.	ITEM NO	PRICE (Rs)	R.N.	ITEM NO	PRICE (Rs)
11:00-11:10	11	2	130	19	3	120	92	10	30	77	8	140	7	1	30	7	1	30
11:10-11:20	86	9	70	53	6	35	51	6	35	11	2	130	94	10	30	20	3	120
11:20-11:30	70	8	140	29	4	40	50	6	35	70	8	140	20	3	120	6	1	30
11:30 - 11:40	33	4	40	79	9	70	41	5	45	3	1	30	73	8	140	25	3	120
11:40-11:50	20	3	120	11	2	130	93	10	30	41	5	45	80	9	70	16	2	130
11:50-12:00	23	3	120	26	3	120	51	6	35	33	4	40	4	1	30	30	4	40
12:00-12:10	92	10	30	67	8	140	17	2	130	93	10	30	56	6	35	29	4	40
12:10-12:20	84	9	70	93	10	30	71	8	140	7	1	30	50	6	35	85	9	70
12:20-12:30	97	10	30	27	4	40	52	6	35	49	6	35	42	5	45	36	4	40
12:30-12:40	36	4	40	23	3	120	19	3	120	46	5	45	3	1	30	93	10	30
12:40-12:50	67	8	140	69	8	140	45	5	45	22	3	120	99	10	30	34	4	40
12:50-13:00	10	2	130	35	4	40	77	8	140	38	5	45	3	1	30	67	8	140
13:00-13:10	64	7	60	70	8	140	31	4	40	44	5	45	65	7	60	73	8	140
13:10-13:20	63	7	60	46	5	45	55	6	35	2	1	30	37	5	45	32	4	40
13:20-13:30	60	7	60	38	5	45	67	8	140	71	8	140	90	10	30	19	3	120
13:30 - 13:40	62	7	60	89	10	30	33	4	40	25	3	120	0	1	30	72	8	140
13:40-13:50	75	8	140	16	2	130	81	9	70	58	7	60	61	7	60	5	1	30
13:50-14:00	50	6	35	77	8	140	55	6	35	2	1	30	39	5	45	27	4	40
14:00-14:10	92	10	30	23	3	120	65	7	60	12	2	130	36	4	40	50	6	35
14:10-14:20	14	2	130	23	3	120	24	3	120	64	7	60	56	6	35	86	9	70
14:20-14:30	42	5	45	11	2	130	0	1	30	16	2	130	25	3	120	13	2	130
14:30-14:40	53	6	35	66	7	60	41	5	45	51	6	35	47	6	35	24	3	120
14:40-14:50	40	5	45	98	10	30	25	3	120	55	6	35	93	10	30	12	2	130
14:50-15:00	73	8	140	27	4	40	34	4	40	95	10	30	22	3	120	30	4	40
15:00-15:10	76	8	140	81	9	70	8	2	130	99	10	30	46	5	45	80	9	70
15:10-15:20	36	4	40	83	9	70	92	10	30	29	4	40	7	1	30	17	2	130
15:20-15:30	49	6	35	38	5	45	81	9	70	43	5	45	92	10	30	53	6	35
15:30 - 15:40	5	1	30	17	2	130	17	2	130	93	10	30	40	5	45	68	8	140
15:40-15:50	36	4	40	12	2	130	54	6	35	64	7	60	31	4	40	4	1	30
15:50-16:00	54	6	35	67	8	140	58	7	60	6	1	30	88	9	70	56	6	35
16:00-16:10	69	8	140	52	6	35	44	5	45	31	4	40	8	2	130	58	7	60
16:10-16:20	6	1	30	1	1	30	47	6	35	3	1	30	40	5	45	40	5	45
16:20-16:30	21	3	120	76	8	140	94	10	30	6	1	30	22	3	120	3	1	30
16:30 16:40	14	2	130	72	8	140	61	7	60	76	8	140	75	8	140	53	6	35
16:40-16:50	14	2	130	80	9	70	47	6	35	36	4	40	96	10	30	58	7	60
16:50-17:00	19	3	120	43	5	45	55	6	35	12	2	130	43	5	45	83	9	70

(Continued)

(*Continued*)

17:00-17:10	21	3	120	70	8	140	52	6	35	90	10	30	18	3	120	49	6	35
17:10-17:20	12	2	130	38	5	45	54	6	35	9	2	130	90	10	30	55	6	35
17:20-17:30	42	5	45	94	10	30	91	10	30	74	8	140	54	6	35	37	5	45
17:30 - 17:40	1	1	30	78	9	70	16	2	130	15	2	130	42	5	45	49	6	35
17:40-17:50	55	6	35	68	8	140	23	3	120	11	2	130	1	1	30	0	1	30
17:50-18:00	48	6	35	12	2	130	2	1	30	54	6	35	42	5	45	15	2	130
18:00-18:10	60	7	60	59	7	60	5	1	30	91	10	30	17	2	130	77	8	140
18:10-18:20	20	3	120	44	5	45	68	8	140	2	1	30	47	6	35	61	7	60
18:20-18:30	58	7	60	1	1	30	69	8	140	17	2	130	81	9	70	24	3	120
18:30-18:40	44	5	45	77	8	140	99	10	30	87	9	70	51	6	35	75	8	140
18:40-18:50	39	5	45	18	3	120	20	3	120	84	9	70	93	10	30	79	9	70
18:50-19:00	40	5	45	55	6	35	69	8	140	85	9	70	27	4	40	22	3	120

Now, here since our inter-order time is 10 minutes, so we have divided each hour into 6 parts of equal time interval and this has continued for all 8 hours from 11:00 a.m. till 07:00 p.m. in the evening Hence we can say on average that 1 item is ordered in every 10 minutes.

Now, we have taken random numbers from excel and used to simulate for the data. So, according to the random numbers **(RN)**, the **Item No.** will get assigned and accordingly the cost per unit of each item, i.e. **The Price** of each will also be updated on the display. So, if every time as the random number changes, all the values will also get changed accordingly.

So, we will continue this and calculate the total sales for the day. This will give us the sales for the first day.

Now, we will do this calculation for 2nd day and will find out the total sales of that day. And we will continue with the same process for up to 6 days.

Then in the next step we will write all the 10 items serially as per their average daily sales. Then we will calculate the total sales for each day (Day 1, Day 2 ...Day 6) and will calculate the total numbers sold for each items during these 6 days.

Then we will calculate the cost of total sales during these 6 days for each items i.e.

Total Cost = Price* Total no. of items sold during 6 days

Now, if we do the summation from ***column*** or summation from ***row*** we will get the same value of the **Total cost** of sales.

Now, we see cost of Average sales per day (in Rs) which will be again the same if we do ***column sum*** or ***row sum***. It is calculated as:

Average Sales per day = Total Sales Cost / No. of Days (i.e. 6 days)

Now, we have to keep this thing in our mind that every time the whenever random number will change, all the values will get changed accordingly.

Item No	Item Type	Day 1	Day 2	Day 3	Day 4	Day 5	Day 6	Total of Item for 6 days	Price (Rs)/unit	Total Sales(Rs) of Item for 6 days	Average Sales(Rs)/day
1	Bun Bread	3	2	3	8	7	6	29	30	870	145.00
2	Pineapple Cake	6	6	4	8	2	5	31	130	4030	671.67
3	Vanila Cake	6	6	5	2	5	6	30	120	3600	600.00
4	Cup Cake	4	4	3	4	3	7	25	40	1000	166.67
5	Cookies	6	6	4	5	9	2	32	45	1440	240.00
6	Biscuit	6	3	12	4	7	7	39	35	1365	227.50
7	Pastries	6	2	3	3	2	3	19	60	1140	190.00
8	Chocolate Cake	6	10	6	5	2	6	35	140	4900	816.67
9	Cream Milkshake	2	5	2	3	3	5	20	70	1400	233.33
10	Bread	3	4	6	6	8	1	28	30	840	140.00
Total no of items		48	48	48	48	48	48	288			
Total Sales (Rs)		**3660**	**4085**	**3270**	**3315**	**2690**	**3565**			20585	**3431**

Now, we have the original data in which we have the Avg. sales, Probabilities that item will be sold, total items sold during 21 days, their price per unit (in Rs) and Total Sales (in Rs) per day for each of the 10 items.

ACTUAL SALES DATA

Items	Vanilla Cake	Pastries	Cupcakes	Bread	Biscuit	Cookies	Chocolate Cake	Bun Bread	Pineapple Cake	Cream Milkshake	GRAND TOTAL FOR 10 ITEMS
Total Number sold in 21 days	98	103	99	112	101	100	107	81	96	110	1007
Average daily sales (units)	4.67	4.90	4.71	5.33	4.81	4.76	5.10	3.86	4.57	5.24	47.95
Probability that item will be sold	0.097	0.102	0.098	0.111	0.100	0.099	0.106	0.080	0.095	0.109	1.000
Price /Unit (in Rs)	120	60	40	30	35	45	140	30	130	70	
Sales (Rs)/day	560.00	294.29	188.57	160.00	168.33	214.29	713.33	115.71	594.29	366.67	3375

And also we have the simulated data in which we found that the value we get from simulation always depends on the random numbers which get changed every time as we refresh or get changed always as we open it in Excel.

SIMULATED SALES DATA

Items	Vanilla Cake	Pastries	Cupcakes	Bread	Biscuit	Cookies	Chocolate Cake	Bun Bread	Pineapple Cake	Cream Milkshake	GRAND TOTAL FOR 10 ITEMS
Total Number sold in 6 days	20	28	23	36	33	27	43	14	44	20	288
Average daily sales (units)	3.33	4.67	3.83	6.00	5.50	4.50	7.17	2.33	7.33	3.33	48.00
Price/Unit (in Rs)	120	60	40	30	35	45	140	30	130	70	
Sales (Rs)/day	400.00	280.00	153.33	180.00	192.50	202.50	1003.80	70.00	953.33	233.33	3668

Now, after comparing the original data and the simulated data in this case we find that the Average Sales (in Rs) is more in the simulated data than the actual data. So, we can say for coming few days the sales value will rise a bit. Now, we will plot a graph for above following data. The graph will be between the original data and the simulated data and will be linked to the values so if the values in the data set will change, then it will directly affect the Graph.

Items	Vanilla Cake	Pastries	Cupcakes	Bread	Biscuit	Cookies	Chocolate Cake	Bun Bread	Pineapple Cake	Cream Milkshake	Total (Rs)/ day
Actual Sales (Rs)/day	560.00	294.29	188.57	160.00	168.33	214.29	713.33	115.71	594.29	366.67	3375
Simulated Sales (Rs)/day	400.00	280.00	153.33	180.00	192.50	202.50	1003.80	70.00	953.33	233.33	3668

Graph –

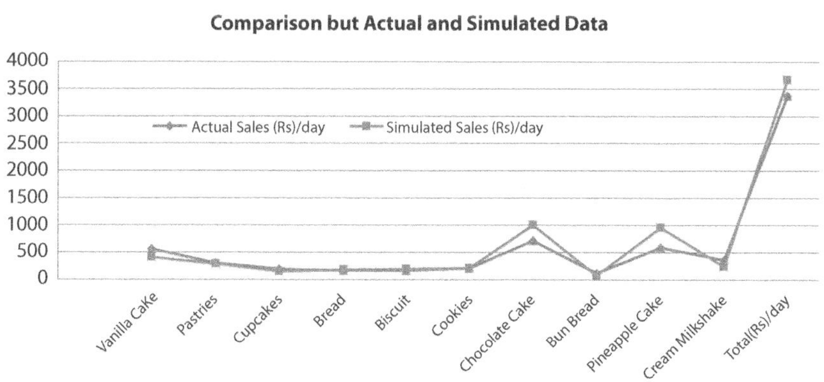

Now, here in the graph we can see that the ***blue line*** represents the Actual Sales per day (in Rs) and the ***red line*** represents the simulated Sales per day (in Rs). So, we can predict the Sales for the future for the next 10 days or we can calculate in the same way to predict for like few weeks as well.

6.2 Conclusion

Bona (2014) states that alternative demand planning and forecasting future demand is an important phase in resource planning because it supports other planning activities such as production planning and material planning. As markets today change rapidly, supply chains need to be flexible and efficient. This can be attributed to the uncertainty of customer demand but also to increased customer expectations.

Covering and planning the necessary uncertainty is important for an organization so that it does not lose customers [3]. A variety of uncertainties are required in the long run as well as in the long run [3]. Short-term uncertainties may include daily variations in operations, cancellation or order speed, equipment failure, etc. Uncertainty over time may be attributed to changes in raw material/final product prices, annual demand changes, or production ratios.

Underestimating the impact of uncertainty can lead to situations where plan decisions do not protect the company from threats by taking advantage of the opportunities offered by high levels of uncertainty. An example of this would be the uncertainty about product demand for the product distribution system. Ignoring demand volatility can lead to more expensive storage costs or consumers are dissatisfied with the risk of market share losses and backlogs. Both of these conditions are less favorable in the current market settings as profit margins are stronger. Mathematical prediction is the first and most important step in service planning, which is a complex process. The effectiveness of traditional methods largely depends on the diversity of different organizations and their areas of conservation (see Bóna and Lénárt [2]). Given the uncertainty in predicting future needs, the lack of a planning system can lead to negative planning decisions compared to the model that causes uncertainty (see Gupta and Costas [3]).

From this project we saw how simulation works, simulation models, their merits and demerits and finally Monte Carlo simulation for a real inventory system. Now, we can say that the simulation always depends upon the random numbers so every time if random number gets refreshed the values of the simulation will also get changed and the value of sales will also change accordingly. In the above solution we found that the average sales for coming days will rise up but as we know it can go a bit down as it all depends on the random numbers. So, now we can predict the sales not only for few days but also for as many days we want to, provided if the inventory system moves as normal all time it does and there is no any case of insufficient stock or any uncertain crisis kind of situation.

References

1. A Case Study of HOMES 71 LTD, Bangladesh, *American Journal of Operational Research,* Vol. 5 No. 3, 2015, pp. 64-73. Doi: 10.5923/j.ajor.20150503.03.
2. Bóna, K., Lénárt, B. (2014), "Supporting demand planning process with Walsh-Fourier based techniques." *Periodica Polytechnica. Transportation Engineering* 42.2 p. 97.
3. Gupta, A., Costas D. M (2003). "Managing demand uncertainty in supply chain planning." *Computers & Chemical Engineering* 27.8 pp. 1219-1227.
4. Reinartz, W., Krafft M., Hoyer, W., D. (2004). "The customer relationship management process: Its measurement and impact on performance." *Journal of Marketing Research* 41.3 293-305.
5. Gordon, I (2003). "Measuring customer relationships: What gets measured really does get managed."
6. Ulaga, W., Eggert A (2006). "Value-based differentiation in business relationships: Gaining and sustaining key supplier status." *Journal of Marketing,* Vol. 70, No. 1, pp. 119-136.

7

Workability of Cement Mortar Using Nano Materials and PVA

Dr. Mohan Kantharia[1] and Dr. Pankaj Mishra[2*]

[1]Department of Civil Engineering, Amity School of Egineering and Technology, Amity University, Gwalior, Madhya Pradesh, India
[2]Department of Physics, Amity School of Engineering and Technology, Amity University, Gwalior, Madhya Pradesh, India

Abstract

Workability of fresh cement mortar is a very important property in deciding its final strength. More water content increases flow value but decreases final strength. Similarly, less water content decreases flow value but increases strength but very low water content makes the mortar mix very stiff, hence requires more energy in compaction and vibration, etc. Therefore some admixtures are tested with cement mortar to understand their effect on flow value. The admixtures were taken Nano Silica, Nano Alumina, and Nano Zinc Oxide. PVA as polymer was also tested for flow value determination. It was found that initially mixing of Nano Materials and PVA the flow value decreases but on increasing the content of these admixtures the flow value increases. The PVA more than 2% content of weight of cement increases flow value more than control mix.

***Keywords*:** Workability, admixtures, nano materials and PVA

7.1 Introduction

Mortar is a composite of cement and sand. Sand is generally explored from a river. Over exploitation of natural minerals like limestone for cement, sand and aggregate is a threat to sustainability. Hence researchers are trying to get substitutes for cement and sand so that our natural reserves can be maintained

**Corresponding author*: pmishra@gwa.amity.edu

Ramakant Bhardwaj, Jyoti Mishra, Satyendra Narayan and Gopalakrishnan Suseendran (eds.) Mathematics in Computational Science and Engineering, (167–180) © 2022 Scrivener Publishing LLC

for the long term. In this attempt, various experimentations are being done to reduce the quantity of the cement and sand. Various industry by-products like fly ash, slag, rice husk ash, and manufactured sand have been added to cement mortar and concrete. Admixtures are added to increase the strength and workability. Good workability is essential for good results in strength. There are various methods to find the workability of cement mortar and concrete. Workability is not dependent on a single parameter; it is a function of many factors like the shape of the particle, size of aggregate, water content, admixture, temperature, etc. In this work Flow value of cement mortar is determined with various nano admixtures and PVA.

7.2 Literature Survey

[1] investigated the effect of lime in cement mortar, and enhanced workability, flexural strength, bond strength and compressive strength. ASTM standard C-270 specifies various types of cement and lime. It specifies both property specifications and proportion specifications. Generally, in Brick masonry construction practices cement mortar is widely used. Evidences of existence of lime mortar, date back to 500 BC. By mixing lime with cement mortar, much variation in properties is observed. Cement lime mortar is useful in restoration of masonry in old monuments. Lime mortar, having high water retentive property, optimizes cement hydration. Mixing lime can help in autogenous healing of hair cracks, as the hydrated lime and CO_2 combines to form limestone deposits. Lime mortar has elasticity property.

[2] reported that use of polymer (acrylic latex) enhances water resistance, penetration of salt solution, workability and flexural strength of cement mortar. Air entrainment also increases giving adverse effect on compressive strength but freeze and thaw resistance increases. The paper reports on new latex reduces wet density. For constant flow value polymer/cement ratio has been worked out. Chloride penetration and water absorption is reported to be reduced.

[3] studied thin coating of SBR, PAE, St/BA emulsions for self-leveling polymer mortar. Generally, Industrial floors are finished with top 5-15 mm coat of self-leveling polymer mortar. The properties depend upon the types of polymer used in mortar. Suitable super plasticizers and thickeners have been used for workability, to reduce drying shrinkage and bleeding. Ratio of polymer to cement kept 50% and 75%. Consistency change, adhesion in tension, density and flow were studied. The conclusions drawn from results are as follows. Higher density of self-leveling mortar was found at

50% of polymer cement ratio, irrespective of type of polymer and cement sand ratio. The urethane and epoxy variants had difficulty in consistency change; however, SBR and PAE modified mortar satisfy KS requirements. Adhesion in tension was also better in polymer modified mortars. PAE modified self-leveling mortar at cement sand ratio 1:1 furnished the best result.

[4] studied various concrete mixes with polymers. Concrete cubes of M_{20}, M_{30}, M_{40} were prepared as per IS 10262: 2009 standard and two polymers Glenium B-233 and Glenium ACE-30 were used for admixing with 0.2, 0.4, 0.6%. Compressive strength for hardened state, as well as workability has been determined in fresh concrete. Comparison is made between normal concrete and polymer concrete. It was found that strength reduces with increment in porosity. The maximum benefit of strength was found in M_{20} mix concrete for all variations with polymers. Addition of polymers increased the strength as well as workability. For maintaining the constant slump value water cement ratio slowed with increase of polymer. Compressive strength was less for unmodified concrete.

[5] discussed rheological properties of concrete by adding super plasticizer and nano particles. Addition of nano ZnO and Al-ZnO, particles affects the workability of cement mortar. Cement paste containing nano particle up to 0.4% showed excellent workability. Use of nano particles facilitates higher absorption of the superplasticizer because of high specific surface area (SSA), Yield stress, and viscosity was considerably enhanced.

[6] studied workability and compressive strength of mortar determined by partial addition of Titanium dioxide nano particles by 0.5, 1.0, 1.5, 2.0%. It was investigated that workability of mortar decreases, and compressive strength increases up to 2% use of TiO_2 and maximum improvement is found at 1.0% addition of TiO_2.

[7] carried out investigations for compressive strength, permeability, split tensile strength, and modulus of elasticity. Polymer Rheomix141 (a copolymer of styrene butadiene rubber) content 5% to 10% was used in different cement concrete mixes of M_{20} to M_{60}. The study revealed that 10% polymer use gives appreciable results with increase in workability and reduction in coefficient of permeability with relative increase of polymer content. Polymer content of 10% exhibit promising results which includes increase in compressive strength for various mixes increases from 16.25% to 33.4%, split tensile strength increases 5.1% to 19.8%, and flexure strength 13.2% to 18.4%.

[8] investigated the effect of nano SiO_2 on partially or fully recycled aggregate concrete, the use of 0.4, 0.8, 1.2% of nano silica was observed, and it was found that workability reduces with use of nano particles, and water absorption also reduces. The increment in compressive strength is up to 10-22% and tensile strength also increases slightly.

[9] experimented on nano alumina of average diameter 15 nm for workability and compressive strength. Four samples of admixtures 0.5%, 1.0% 1.5% and 2% were investigated and it was concluded that at 1% alumina best results were furnished for strength. Workability decreases at higher percentage of nano alumina giving low slump value. Therefore, it suggests use of super plasticizer is necessary to work with nano particles.

[10] discussed the polyvinyl alcohol with carbamide in cement paste. The ratio of PVA to carbamide 20:80, 40:60, 80:20, were used to test mechanical properties, consistency, initial and final time, etc., compressive strength was reported to be increased, consistency found to be decreased and setting times were elongated. With increased percentage of polymer, the workability improved, also confirmed by comparison of SEM images which showed that in conventional cement mortar needles ($C_3A.3CaSO_4.32H_2O$), formed C_3A with gypsum reaction, this factor shows a decrease in polymer mortar, SEM images, the polymer mixed cement mortar showed that hydration products surrounded by polymer films therefore hydration was delayed.

[11] studied the cement mortar with cement and sand ratio 1:1 and 1:2 by weight. Ratio of water to cement 0.35 were prepared and curing 7 and 28 days strength were determined with 3% and 5% nano Zinc oxide as admixture. To study microstructure of cement SEM analysis was done and it was found that addition of nano ZnO durability and strength of mortar up to 3% of nano particles. In mix 1:1 compressive strength has been reported increased by 23.88% and in 1:2 ratio, 23.05%. Split tensile strength 39.93% and 61.35% increase. SEM observation also showed that nano particles enhances hydration reaction rate. However, nano particles reduce the workability; therefore use of super plasticizer is necessary.

[12] in their paper analyzed nano silica as admixture in cement mortar and the effect on various properties was discussed. Nano particle size was 0.2 to 0.3 micron, and their 1, 2, 3, 4, 5, 6% of cement fraction is used, water to cement ratio was taken 0.32 and 0.37, the increase in strength is found up to 5% of nano silica addition. Workability decreased and setting time decreased with increase of nano silica.

7.3 Materials and Methods

In this work the details of the materials used are as follows.

Cement 43 grades OPC, sand 1.18mm sieve passed, natural river sand, water cement ratio 0.5, cement sand ratio 1:3 by weight. All nano particle size was 30-50nm and purity 99.5% was used. PVA $(C_2H_4O)_n$ of molecular weight 1,15,000, and pH value 5-7 were used.

Procedure for determining Workability of fresh mortar by flow value.

The mortar is prepared with admixture as specified

a) Flow table is cleaned
b) Conical frustum is kept at the centre of the table
c) Mortar is filled in two layers, after filling half temping is done
d) Again, filled full and temping is done
e) Now Frustum is removed
f) Mortar is allowed to flow
g) At given mechanical arrangement the plate is lifted 25 times in 15 seconds
h) Diameter (D_2) of spread mortar is taken at four equal spaces
i) Initial Diameter = D_1 mm
j) Final Diameter = D_2 mm
k) Now % Flow value is determined by formula

$$\text{Flow Value (100\%)} = \frac{D2 - D1}{D1} \times 100$$

l) Average of the three readings is taken as flow value of control mix mortar (refer to Table 7.2)

Table showing flow value of various mortar mixes with Nano material combinations. Tables 7.1 to 7.9, and Figures 7.1–7.8 show summaries of all results.

7.4 Results and Discussion

Mortar without any admixture has flow value 41.1%. When various nano or PVA admixtures are added then initially flow value decreases as shown in Tables 7.2–7.9 and Figures 7.1–7.8. But increasing the percentage of the

Table 7.1 Flow value of control mortar.

Sr. no.	Initial diameter D_1	Final diameter after tapping D_2	Flow value = $\frac{D2 - D1}{D1} \times 100$	Average flow value
1	100 mm	140.0	41.0	(41 + 41.1 + 41.2) / 3 = 41.1%
2	100 mm	141.1	41.1	
3	100 mm	141.2	41.2	

Table 7.2 Flow value of mortar with nano silica.

% of nano SiO_2	Flow value %
0.50	27.00
1.00	28.80
1.50	32.20
2.00	38.20

Table 7.3 Flow value of admixture with nano alumina.

Nano Al_2O_3 %	Flow value %
0.5	22
1	23.9
1.5	24.8
2	28

Table 7.4 Flow value of cement mortar with nano Zinc oxide.

% of nano ZnO	Flow value %
0.50	21.60
1.00	23.30
1.50	27.70
2.00	28.20

Table 7.5 Flow value of admixture with PVA.

% of PVA	Flow value %
0.50	25.00
1.00	36.60
1.50	40.00
2.00	43.33

Table 7.6 Flow value of admixture with nano silica + PVA.

% of PVA + nano SiO_2	Flow value %
0.5+0.5	30.16
1.0+1.0	38.50
1.5+1.5	44.30
2.0+2.0	49.30

Table 7.7 Flow value of admixture with nano alumina + PVA.

% of PVA + nano Al_2O_3	Flow value %
0.5+0.5	29.80
1.0+1.0	31.56
1.5+1.5	41.30
2.0+2.0	44.60

Table 7.8 Flow value of admixture with nano ZnO + PVA.

% of PVA + nano ZnO	Flow value %
0.5+0.5	34.50
1.0+1.0	36.23
1.5+1.5	40.56
2.0+2.0	42.36

Table 7.9 Flow value of admixture with nano alumina + nano silica + PVA.

% of PVA + nano SiO_2 + nano Al_2O_3	Flow value %
0.5+0.5+0.5	30.00
1.0+1.0+1.0	34.50
1.5+1.5+1.5	38.13
2.0+2.0+2.0	42.60

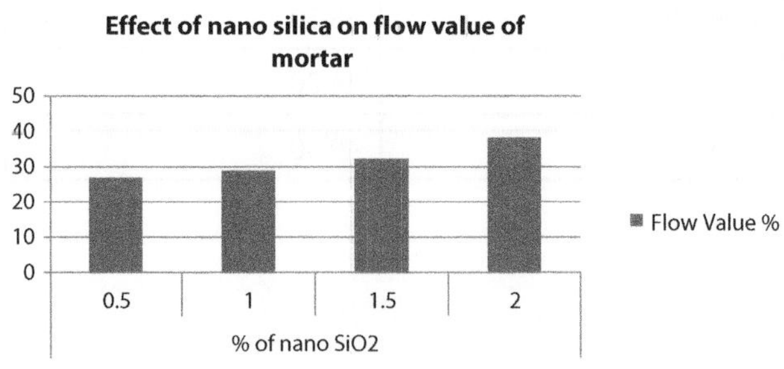

Figure 7.1 Graph for flow value of cement mortar with nano silica.

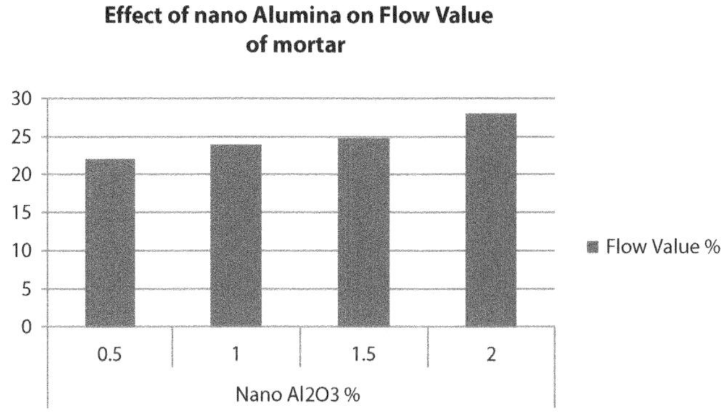

Figure 7.2 Graph for flow value of cement mortar with nano Alumina.

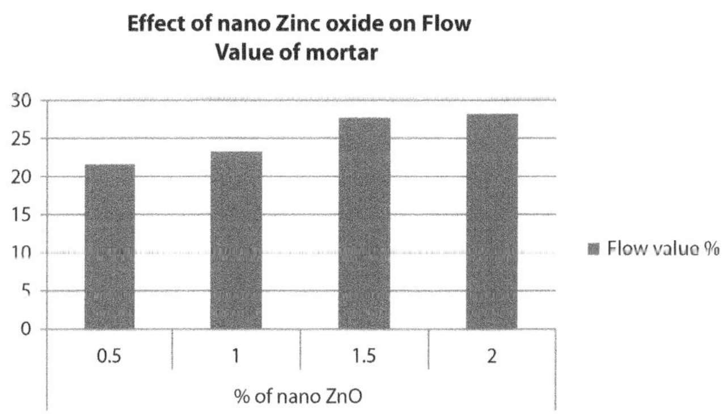

Figure 7.3 Graph for flow value of cement mortar with nano zinc oxide.

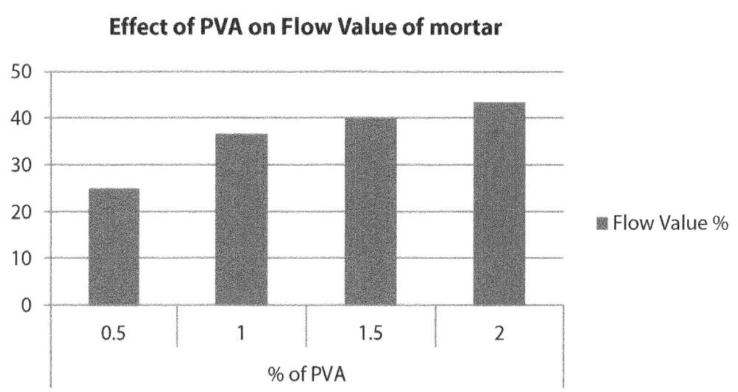

Figure 7.4 Graph for flow value of cement mortar with PVA.

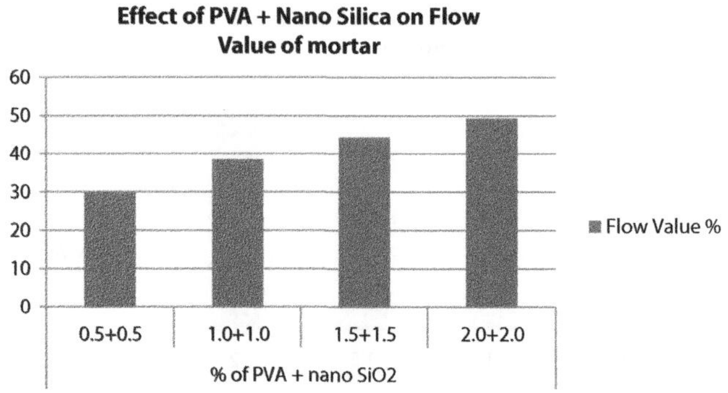

Figure 7.5 Graph for flow value of cement mortar with nano silica + PVA.

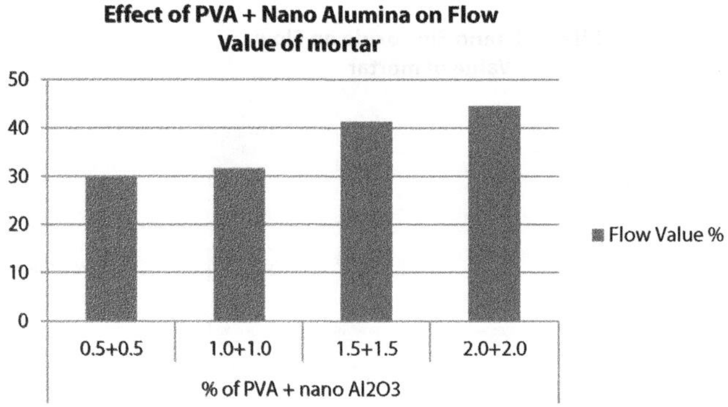

Figure 7.6 Graph for flow value of cement mortar with nano Alumina + PVA.

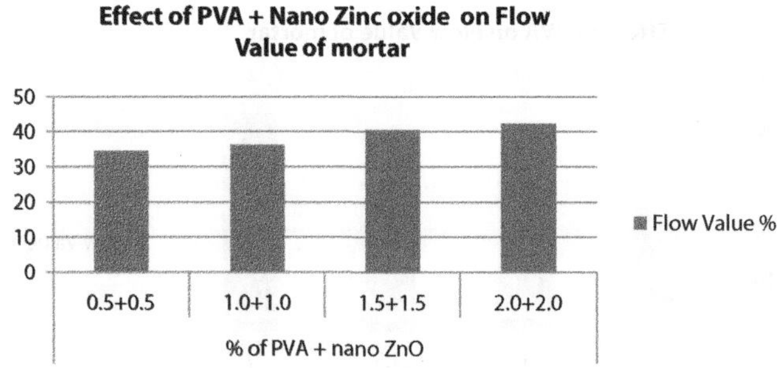

Figure 7.7 Graph for flow value of cement mortar with nano Zinc oxide + PVA.

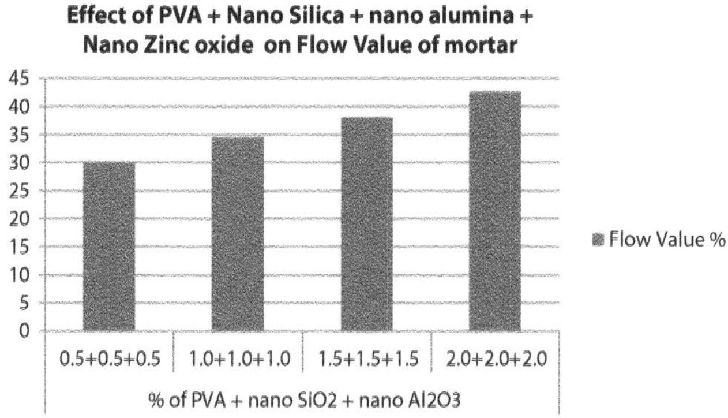

Figure 7.8 Graph for flow value of cement mortar with nano silica + nano alumina + nano Zinc oxide + PVA.

admixtures it was found that workability increases with increase of admixtures but this value remains less than the control mix. When PVA added 2% then workability value become more than the control mix. Mortar mixes have been experimented with binary mix of nano and PVA also and found that less than 2% content the workability is less than the control mix. The reason may be due to more specific area of the nano particles the more liquid required for flow ability. At lower percentage of PVA thin film may decrease the workability but when the surface film become thick it causes the increase in workability.

7.5 Conclusion

Workability is an important factor in achieving the final strength of the concrete. Hence various admixtures have been experimented with to achieve better workability. It is found that workability reduces while nano silica, nano alumina, nano zinc oxide added 0.5 to 2.0%. and flow value is also less in case of PVA 0.5 to 1.5%. In these cases flow value is between 27-38.2% in case of nano silica addition, 22-28% with nano alumina, 21.6-28.2% with nano Zinc oxide, and 25-43% with PVA. PVA at 2% gives the flow value more than control mix. When nano alumina and PVA added 0.5% to 1% the flow value increases but less than the control mix; however, at 1.5 and 2.0% flow value is more than the control mix; these values are 41.3% and 44.6%. Similarly, with alumina and

PVA combination at 2% mix we get flow value 44.6%. Hence it may be concluded that PVA + Nano particles increases flow value at 2% both content.

References

1. M. Tate, "The Most Important Property of Cement-Lime Mortar in Masonry Construction Is.", *International Building Lime Symposium, March 9-11, Orlando, Florida*, pp. 1-13, 2005.
2. L. M. Saija, "Waterproofing of Portland Cement Allortars with a Specially Designed Polyacrylic Latex", *Cement and Concrete Research*, Vol. 25, No. 3, pp. 503-509, 1995.
3. J. Do, Y. Soh, "Performance of Polymer-Modified Self-Levelling Mortars with High Polymer–Cement Ratio for Floor Finishing", *Cement and Concrete Research* 33, pp. 1497-1505, 2003.
4. S. Sahu, S. P. Mishra, "Effect of Polymers in Fresh and Hardened State of Cement Concrete", *International Journal of Innovative Research in Science, Engineering and Technology*, Vol. 3, Issue 8, pp. 15241-15247, 2014.
5. E. Ghafari, S.A. Ghahari, Y. Feng, F. Severgnini, N. Lu," Effect of Zinc oxide and Al-Zinc oxide nanoparticles on the rheological properties of cement paste", *Composites Part B* 105, pp. 160-166, 2016.
6. A.Nazari, S. Riahi, S. Riahi, S. F. Shamekhi, A. Khademno, "Assessment of the effects of the cement paste composite in presence TiO_2 nanoparticles", *Journal of American Science* 6(4), nanoparticles in Concrete, pp. 43-46, 2010.
7. V. Bhikshma, K. Jagannadha Raob, B. Balajia, "An Experimental Study on Behavior of Polymer Cement Concrete", *Asian Journal of Civil Engineering (Building and Housing)* Vol. 11, No. 5, pp. 563-573, 2010.
8. K. H. Younis, S. M. Mustafa "Feasibility of Using Nanoparticles of SiO_2 to Improve the Performance of Recycled Aggregate Concrete", *Advances in Materials Science and Engineering*, Vol. pp. 1-11, 2018.
9. A.Nazari, S. Riahi, S. Riahi, S. F. Shamekhi and A. Khademno, "Influence of Al2O3 Nanoparticles on the Compressive Strength and Workability of Blended Concrete", *Journal of American Science* 6(5), nanoparticles in Concrete, pp. 6-9, 2010.
10. E.S.M. Negim, G.Z. Yeligbayeva, R. Niyazbekova, R. Rakhmetullayeva, A.A.Mamutova, R.Iskakov, M. Sakhy, G.A.Mun, "Studying Physico-Mechanical Properties of Cement Pastes in Presences of Blend Polymer as Chemical Admixtures", *International Journal of Basic and Applied Sciences*, 4 (3), pp. 297-302, 2015.

11. D. Nivethitha, S Dharmar, "Influence of Zinc Oxide Nanoparticle on Strength and Durability of Cement Mortar", *International Journal of Earth Sciences and Engineering*, Vol. 9, No. 3, pp. 175-181, 2016.
12. A. Kumar, G. Singh, Effect of Nano Silica on the Fresh and Hardened Properties of Cement Mortar, *International Journal of applied Engineering Research* Vol. 13, pp. 11183-11188, 2018.

8

Distinctive Features of Semiconducting and Brittle Half-Heusler Alloys; LiXP (X=Zn, Cd)

Madhu Sarwan[1]*, Abdul Shukoor V.[2] and Sadhna Singh[2]

[1]Department of Physics, Govt. College Harrai, Chindwara (M.P.), India
[2]High Pressure Lab, Department of Physics, Barkatullah University, Bhopal, India

Abstract

To explore various properties of half-Heusler alloys, LiZnP and LiCdP, the well-established density functional theory has been employed. These calculations are accomplished using the pseudopotential plane-wave method as implemented in the Quantum espresso package. The generalized gradient approximation is used for exchange-correlation energy. The structural quantities (lattice constant, bulk modulus, equilibrium volume and second-order elastic constants) along with electronic band structure, the corresponding density of states and charge density plot are reported in this chapter. We have found that the half-Heusler alloys LiZnP and LiCdP are direct bandgap semiconductors and are brittle in nature. A strong covalent bonding is observed between X (Zn/Cd) and P atom and ionic bonding is found between Li and P atom. The thermodynamic properties of these compounds are fascinating and have many technological applications. No studies have been carried out to explore the thermodynamic features of LiXP yet. In addition, the thermodynamic nature of LiXP with pressure and temperature is described in this chapter.

Keywords: Half-Heusler alloys, density functional theory, thermodynamic properties, elastic properties

**Corresponding author*: madhusarwan@gmail.com

Ramakant Bhardwaj, Jyoti Mishra, Satyendra Narayan and Gopalakrishnan Suseendran (eds.) Mathematics in Computational Science and Engineering, (181–198) © 2022 Scrivener Publishing LLC

8.1 Introduction

Half-Heusler alloys with XYZ structure have become a dominant class of solids and are capable entrants of novel advances in technology. The Heusler alloys are considered vital materials because of their broad bandgap, vast polarization and greater Curie temperature. The semiconductor half-Heusler compounds are the novel version of the Heusler family, which form in the C1b structure with F-43m symmetry. Because of these superior characteristics, they have been in the limelight of consideration by many researchers recently [1]. Among the semiconducting Heusler compounds, the transition metal-based Li-half-Heusler alloys are significant owing to their promising utilization in the domain of optics and photonics. This is mainly caused by their large absorption coefficient and adjustable bandgap energies [2]. Besides, Li-based alloys keep on being examined in solar cell and lithium battery applications [3–5].

These alloys belong to the A^{I} B^{II} C^{V} group and are also called "filled zinc blend structures" or "filled tetrahedral compounds". Some of these compounds were characterized by Nowtony *et al.* [6] and Juza *et al.* [7] way back in the mid-20th century. The awareness in this family has been revitalized after the prediction of their semiconducting behavior with a direct bandgap by Wood *et al.* [8] and Carlson *et al.* [9]. Among these motivating classes of materials, a few compounds like LiCdP, LiCdAs, and AgMgAs were produced, although only a couple of them have been explored. Beleanu *et al.* [10] have studied the structural and electronic properties of LiMgZ (Z = P, As and Sb) experimentally. Their estimations established the semiconducting behavior with a direct bandgap of 1.0 – 2.3 eV in all alloys. Murtaza *et al.* [11] explored the structural, elastic, electronic, and optical properties of XYZ (X=Li, Na, K; Y=Mg, Zn; Z=N, P, As, Sb, Bi) alloys by employing density functional theory (DFT). These alloys are guaranteed to be a promising candidate for optoelectronic and anode materials in lithium batteries. For another alloy LiZnP, optical bandgap and electrical resistivity had been examined [12, 13]. They concluded that LiZnP is a semiconductor with a large bandgap of 2.04 eV. For full-Heusler alloys (A_2BC), the unit cell comprises four face-centered cubic sublattices merging together with atom A at (0, 0, 0) and (1/2, 1/2, 1/2), B at (1/4, 1/4, 1/4), and C at (3/4, 3/4, 3/4) coordinates. In half-Heusler alloys, A is at (0, 0, 0), and B is at (1/4, 1/4, 1/4) and C atom is either at (1/2, 1/2, 1/2) or at (3/4, 3/4, 3/4) [14]. In this chapter, we are exploring various information on the structural, elastic, electronic and thermodynamic behavior of LiXP (X=Zn, Cd) where atom-P is at (0, 0, 0), X is at (0.25, 0.25, 0.25) and Li is at (0.5, 0.5, 0.5). Here, the most electro-negative components are at (0.5, 0.5, 0.5) and at (0, 0, 0) that form a rock salt lattice, while the component at (0.25, 0.25, 0.25) fills half of the interstitial sites and it can be identified as a stuffed AC or BC zinc blend structure [15]. Clearly, the data about the elastic properties of LiXP (X= Zn and Cd) is undersupplied in the literature. To the best of our understanding, no data is available for elastic constants of LiXP (X= Zn and Cd). The outcomes of this work will be beneficial for experimentalists and will lead to the effective exploitation of these materials for advanced technical and industrial purposes.

8.2 Computation Method

Quantum espresso open-source code is used to study the electronic-structural properties. The computation has been performed within the framework of density functional theory (DFT) as accomplished in the Quantum espresso suite [16, 17]. The GGA approximation of Perdew Burke Ernzerhof (PBE) has been employed for exchange-correlation. For wave function, an energy cut-off of 45 Ry is used while 300 Ry is applied for density. The Brillouin Zone (BZ) was sample using 13*13*13 k-mesh (Monkhorst pack) with Marzari-Vanderbilt cold Gaussian [18] smearing 0.005 Ry. The elastic properties have been investigated using ElaStic [19] code with input files from Quantum espresso and the calculations have been carried out at the equilibrium structure with relaxed atomic positions. The bulk modulus, shear modulus, and Young's modulus have determined with the help of three forms; Voigt, Reuss, and Hill elastic moduli [20–22].

8.3 Result and Discussion

8.3.1 Structural Properties

The half-Heusler compounds, LiZnP and LiCdP, crystallize in cubic MgAgAs-type structure (space group F-43m) in which Li atoms occupy at (0.5, 0.5, 0.5), Zn/Cd at (0.25, 0.25, 0.25), and P at (0, 0, 0) symmetry. The Crystal structure of LiXP half-Heusler is presented in Figure 8.1. The equilibrium cell volume and lattice parameter have been determined using the popular Birch-Murnaghan equation of state by fitting total volume with respect to energy.

In Figure 8.2(a-b) we plotted energy as a function of volume and the pattern of the curve indicates that these compounds are stable at equilibrium

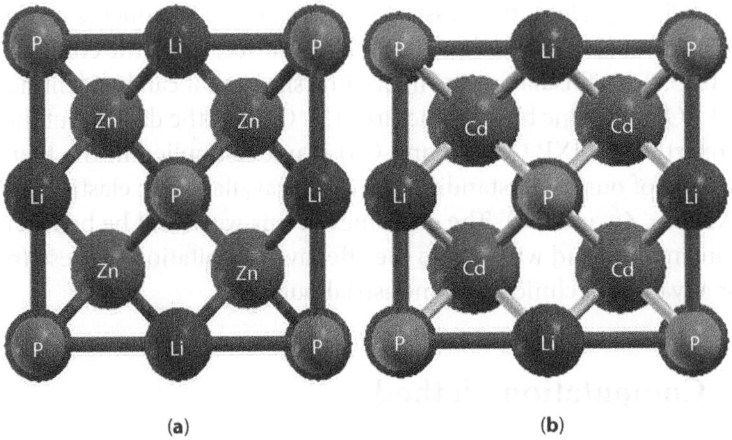

Figure 8.1 Crystal structure of half-Heusler alloys; (a) LiZnP and (b) LiCdP.

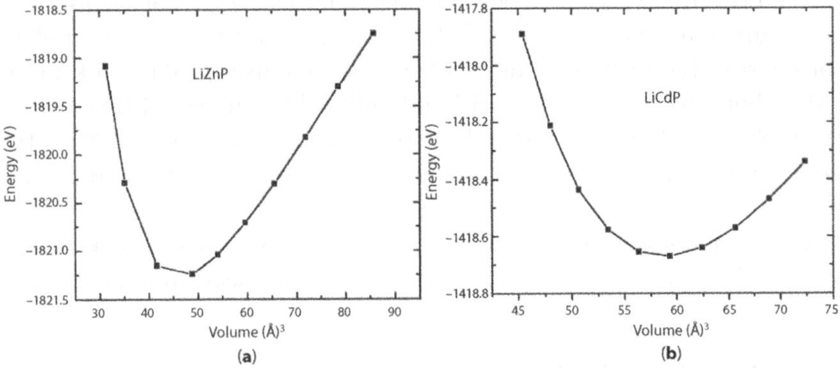

Figure 8.2 Total energy vs. volume for half-Heusler alloys; (a) LiZnP and (b) LiCdP.

lattice constant. The obtained results for ground-state properties of LiZnP and LiCdP are tabulated in Table 8.1 and are compared with available experimental/theoretical results. Moreover, we have made a comparison with other Li-based half-Heusler compounds to understand their common trend [15, 23–34]. The calculated lattice constant of LiZnP (5.713 Å) and that of LiCdP (6.153 Å) are in very good agreement with reported experimental results which are 5.779 Å and 6.087 Å for LiZnP and LiCdP, respectively. The lattice constants of other Li-based half-Heusler alloys also lie between 5 to 7 Å.

Table 8.1 Ground state structural properties of transition metal-based Li half-Heusler alloys.

Compound	a_{cal} (Å)	a_{exp} (Å)	V_0 (Å)^3	B_0 (GPa)	B_0'
LiZnP	5.713 [Present] 5.707 [15] 5.77 [1] 5.576 [25] 5.61 [26]	5.779 [15] 5.76 [23] 5.755 [24] - -	46.269 - - - -	52.71 - - - -	4.46 - - - -
LiCdP	6.153 [Present] 6.118 [15] 5.98 [27] 6.14 [1] 5.962 [29]	6.087 [15] 6.096 [24] 6.10 [24] -	58.978 - - -	60.47 [Present] 52.80 [15] 66.22 [27] 60.158 [28]	4.14 - 4.44 [27] 4.07 [28]
LiScSi	6.26 [30]	-	61.38 [30]	54.71 [30]	-
LiScPb	6.78 [31]	-	-	37.39 [31]	-
LiScC	5.321 [32]	-	-	-	-
LiScGe	6.302 [2] 6.306 [32]	-	-	33.09 [2]	-
LiMnSi	5.778 [33]	-	-	-	-
LiZnSb	6.41 [34]	-	-	41.8 [34]	4.01 [34]

8.3.2 Elastic Properties

The elastic properties of materials are very imperative as they are related to many solid-state properties [35]; particularly they can give an idea about structural stability, bonding, and anisotropic nature in the formation of compounds [36]. To study the mechanical stability of LiZnP and LiCdP, we have computed the second-order elastic constants at equilibrium lattice constants and the results are tabulated in Table 8.2. It is found that C_{11}, C_{12}, and C_{44} are 102.61 GPa, 46.88 GPa, and 89.19 GPa for LiZnP and 78.02 GPa, 51.69 GPa, and 58.48 GPa for LiCdP. It is noticeable that value of C_{11} is larger than C_{44} for LiZnP and LiCdP. It indicates that these compounds exhibit stronger unidirectional strength against the deformation force compared to shear twisting. The values of second-order elastic constants are compared with other reported theoretical data along with some additional Li-based half-Heusler compounds (in Table 8.2). These values of elastic constants are found positive and hence obey Born-Huang stability criteria [37]; C_{11}, C_{44},

Table 8.2 Elastic properties of transition metal-based Li-half-Heusler alloys.

	LiZnP	**LiCdP**	**LiScGe**	**LiScSi**
C_{11} (GPa)	102.61 [Present] 145 [25] 149 [26]	78.02 [Present] 115 [29] 92.63 [42]	70.429 [2]	106.66 [2] 95.596 [30]
C_{12} (GPa)	46.88 [Present] 41 [25] 47 [26]	51.69 [Present] 41 [29] 53.04 [42]	35.507 [2]	27.849 [2] 34.276 [30]
C_{44} (GPa)	89.19 [Present] 75 [25] 76 [26]	58.48 [Present] 48 [29] 42.15 [42]	14.063 [2]	25.045 [2] 19.032 [30]
B_V=B_R=B_H (GPa)	65.46 [Present] 76 [26]	60.47 [Present] 62.24 [42]	47.47 [2]	50.468 [2] 54.716 [30]
G_V	64.66 [Present]	40.35 [Present]		33.483 [2] 23.054 [30]
G_R	47.43 [Present]	24.60 [Present]		
G_H	56.04 [Present] -	32.48 [Present] 43 [29]	25.704 [2]	
E (GPa)	130.81 [Present] 127 [25]	82.64 [Present] 105 [29]		
B/G	1.167 [Present]	1.861 [Present]		
ν	0.166 [Present] 0.21 [25]	0.272 [Present] 0.23 [29]		
A	3.200 [Present] -	4.44 [Present] 0.174 [29]		

and $C_{11}+2C_{12}$ are greater than zero and bulk modulus lies in between C_{11} and C_{12}. It confirms that LiZnP and LiCdP are mechanically stable. From elastic constants, we have determined shear modulus (G), Young's modulus (E), Pugh ratio (B/G), Poisson's ratio (ν), and Zener anisotropy (A). From these mechanical properties, one can estimate the strength of the material. The LiZnP and LiCdP carry no large Young's and shear moduli as shown in Table 8.2, directing their less hardness behavior. The ductility and brittleness of these materials are predicted using Pugh's ratio (B/G) [38]. According to this, if the B/G ratio is smaller than 1.75 it is brittle, otherwise, it is ductile in nature. Our calculated value of the B/G ratio is 1.167

for LiZnP and 1.861 for LiCdP indicating that LiZnP is brittle and a slight ductile nature for LiCdP. Besides, the brittle/ductile feature can be further justified by the Poisson's ratio (ν) also with a critical value of above 0.33 for ductility. From Table 8.2, the calculated values of Poisson's ratio are 0.166 for LiZnP and 0.272 for LiCdP indicating the dominancy of brittle behavior. Moreover, the Poisson's ratio fixes the bonding nature of the materials and these assigned values are as follows: covalent (0.1), ionic (0.25), and metallic (0.33) materials [39]. The present study illustrates that covalent and ionic bondings are dominant in LiZnP and LiCdP, which further corroborates their brittle behavior. Another key parameter, Zener Anisotropy factor/ratio (A) reveals elastic anisotropy in materials. Any value of 'A' that is differed from one (smaller or greater than one) indicates level of elastic anisotropic behavior of the materials [40]. The calculated Zener ratios are 3.20 and 4.44 for LiZnP and LiCdP respectively, which indicate that both compounds are elastically anisotropic material. The high Poisson's ratio (> 0.25) also confirms the high elastic anisotropy [41]. As LiZnP and LiCdP are less explored materials, very few data are available for comparison for these compounds. Therefore, we have presented the results of available data on LiScGe and LiScSi for predicting the common trend in the results of these transition metal-based Li-half-Heusler alloys [32–33, 36–37, 42].

8.3.3 Electronic Properties

The understanding of electronic properties is very essential for experimental modeling and fabricating energy devices [2]. To understand the electronic properties, the band structure, total density of states (DOS), partial density of states (PDOS), and charge density plots of LiZnP and LiCdP have been analyzed. The non-spin-polarized band structures along the direction of the Brillouin zone are illustrated in Figure 8.3 (a–b). In both the compounds, bandgaps are observed at the Fermi level that authorizes the semiconductor nature of these compounds. The Figure 8.3 (a–b) reveals that LiZnP and LiCdP are direct bandgap semiconductors as the top of the valence band and bottom of the conduction band are located at Γ point. The estimated bandgap has been found to be 2.03 eV for LiZnP and 0.78 eV for LiCdP and is in fair compromise with experimental results. We have also compared our results with other Li-based half-Heusler alloys [23–24, 27–29, 31–34, 43–45] and they are listed in Table 8.3. From this table, it can be observed that the band gap values lie between 0.5 eV to 2.5 eV.

These half-Heusler alloys do not exhibit metallic or half-metallic behavior as there is no p or d-states in lithium as well as due to the absence of p-d and d-d interactions near the Fermi level. The semiconducting behavior of

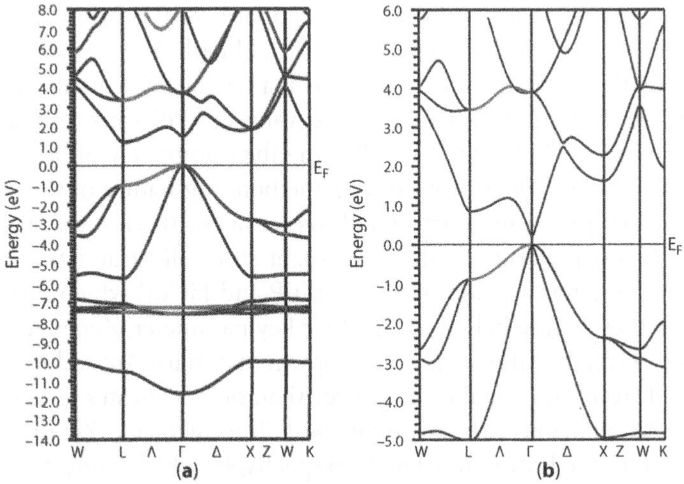

Figure 8.3 (a-b) Band structure of half-Heusler alloys; (a) LiZnP and (b) LiCdP.

Table 8.3 Energy band gap (eV) of transition metal-based Li half-Heusler compounds.

Compound	Present	Experiment	Others
LiZnP	2.03	2.04 [23]	2.05[12], 2.1[43], 2.1[9], 1.353[32], 1.15[44]
LiCdP	0.78	0.850[15] 1.30 [24]	1.51[12], 0.555[32], 0.635[27] 0.76[44], 1.3[43], 0.61[28], 0.55[45], 0.91[29]
LiZnAs	-	-	1.12 [44]
LiInSi			0.70 [44]
LiMgP			1.51 [44], 2.43 [43]
LiMgAs			1.55 [44]
LiMgBi			0.62 [44]
LiZnN			1.21 [44]
LiZnSb			0.57 [34]
LiScPb			0.702 [31]
LiScC			0.980 [32]
LiScGe			0.805 [32]
LiMnSi			0.915 [33]

these compounds primarily depends on transition metal Zn/Cd and main group element; phosphorous (P), as there is strong interaction in s-states of Zn/Cd and p-states of phosphorous near the Fermi level. The density of states (DOS) at equilibrium lattice constants are displayed in Figure 8.4 (a–d). These figures suggest that the lowest-lying bands are appeared mainly due to the s-state of phosphorous in LiZnP and LiCdP which results in a peak in the DOS at around -10.0 eV. The highest peak, for LiZnP in valance band around -3.0 eV and LiCdP in -2.7 eV, is due to the d-state of Zn and Cd. The major cause for the appearance of the conduction band, above the Fermi level, is the *p*-states of the phosphorous atom. Moreover, it can be perceived from Figure 8.4, the *d*-states of Zn and Cd atom contribute mostly to the formation of the edged peak in the present alloys.

Furthermore, to elucidate the type of bonding and charge transmission among these LiXP alloys, the charge density plots have been obtained along the (100) plane and are demonstrated in Figure 8.5 (a–b). It can be observed that for Li and P atoms, the charge distributions are spherical in shape and isolated which characterizes ionic bonding between them.

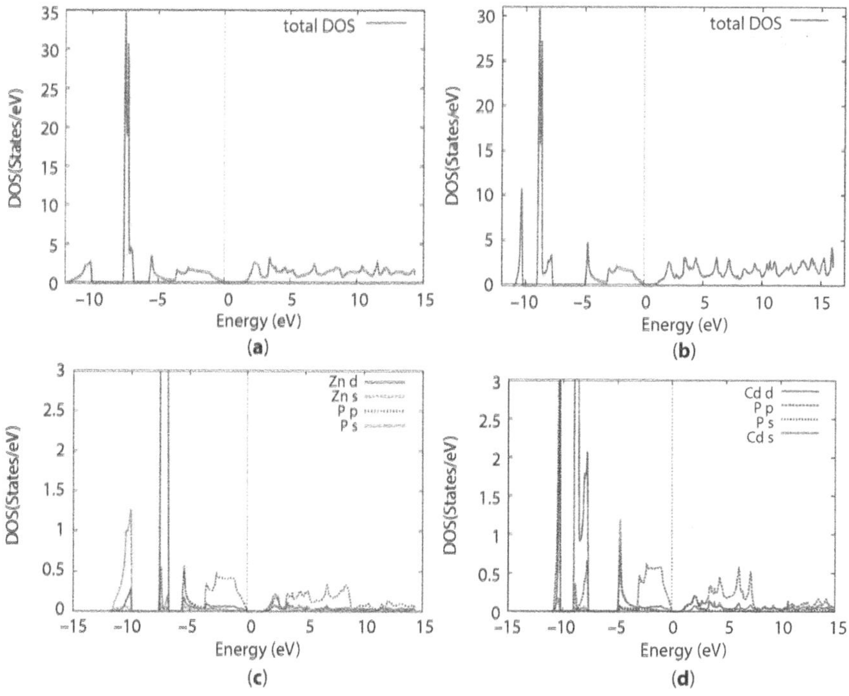

Figure 8.4 Total density of states (a–b) and Partial density of states (c–d) of half-Heusler alloys; LiZnP and LiCdP.

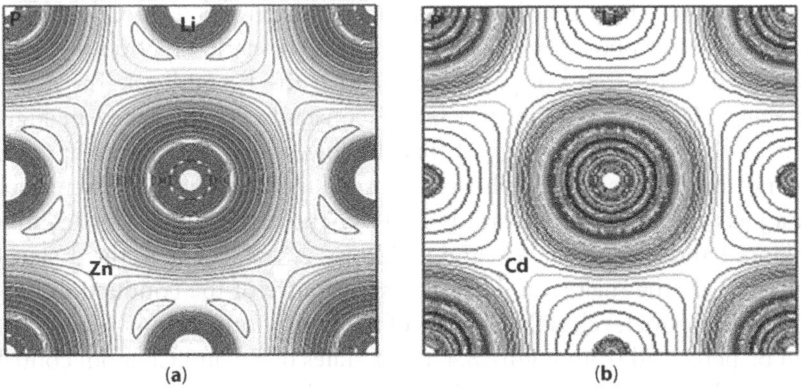

Figure 8.5 Charge density plots for half-Heusler alloys; (a) LiZnP and (b) LiCdP.

The electronegativity of the P atom is 2.19 and the Li atom is 0.98, providing the information of charge transfer from Li to P. Therefore, the charge accumulation takes place at P and depletion occurs at Li (see Figure 8.5(a–b)). On the flipside, for X and P atoms, the charge density counter is mixed or overlapped that shows covalent bonding between the elements. Consequently, one can say that LiXP compounds exhibit ionic and covalent bonding simultaneously.

8.3.4 Thermodynamic Properties

In materials science, the thermodynamic properties play a crucial role as they offer remarkable information about the materials such as stability, atomic interactions, etc. under extreme pressures and temperatures. The information of these peculiar thermodynamic characteristics of materials is much significant in advanced technological applications. The properties, *viz.*, heat capacity, thermal expansion coefficient, Grüneisen parameter, entropy, and Debye temperature, have been investigated to expose the thermodynamic behavior of LiZnP and LiCdP alloys, using quasi-harmonic Debye model as implemented in Gibbs program [46]. These thermodynamic properties of LiZnP and LiCdP are calculated at zero pressure and room temperature and are arranged in Table 8.4. Figure 8.6 (a–b) discloses the consequence of Debye temperature θ_D during high pressures and temperatures. It is obvious from this figure at low pressures, chiefly at 0 GPa, the value of θ_D reduces gradually as the temperature varies from 0 to 1000 K. In general, the values of θ_D of both compounds decrease gradually with temperature while increase with pressure. The computed values of Debye

Table 8.4 Thermodynamic properties of transition metal-based Li half-Heusler alloys at T=300K and P=0 GPa.

Compounds	θ_D (K)	γ	C_V (J/molK)	α (10^{-5} /K)	S (J/molK)
LiZnP	392.65	2.65	68.8	13.0	82.7
LiCdP	314.11	2.15	70.9	9.20	98.3

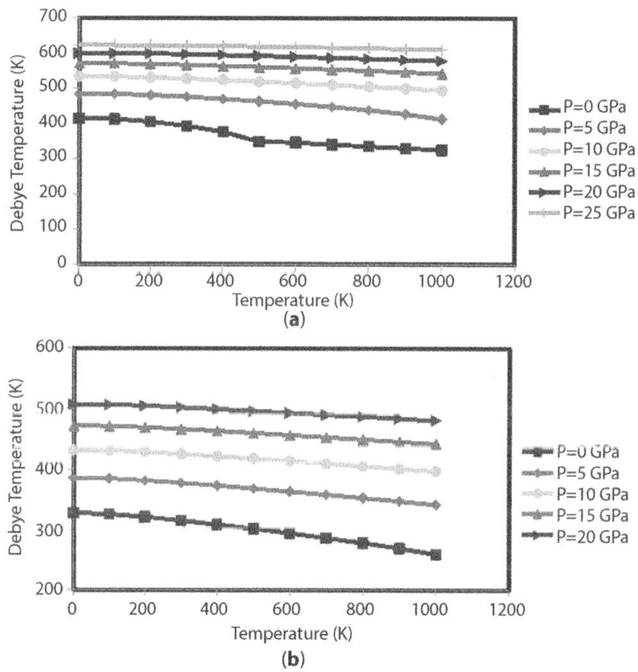

Figure 8.6 Variation of Debye temperature with temperature for (a) LiZnP and (b) LiCdP.

temperatures (at T = 300K and P = 0 GPa) for LiZnP is 392.65 K whereas a lower value of 314.11 K is reported for LiCdP.

The pressure and temperature dependency of the Grüneisen parameter, γ is depicted in Figure 8.7(a–b). It can be viewed that this parameter improves slightly with rising temperature, whereas it reduces with pressure for a given pressure range of 0 GPa to 25 GPa, Figure 8.8 (a–b) depicts the bulk modulus as a function of temperature from 0 to 1000 K and pressure 0 to 25 GPa. We can say that the values of B are nearly constant up to 100 K.

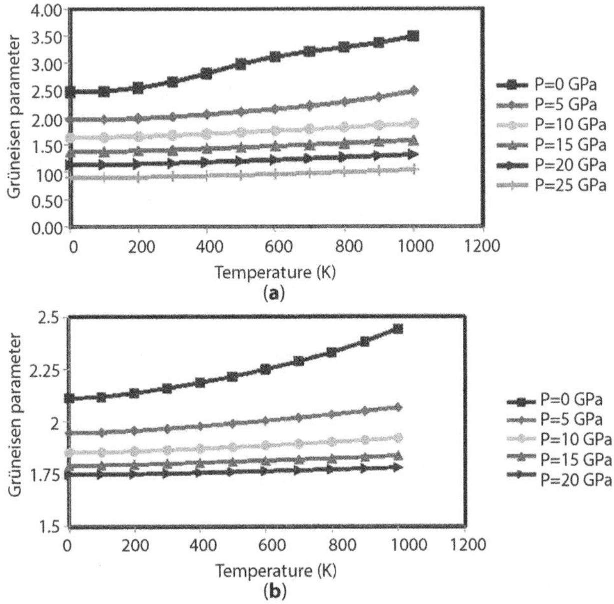

Figure 8.7 Variation of Gruneisen parameter with temperature for (a) LiZnP and (b) LiCdP.

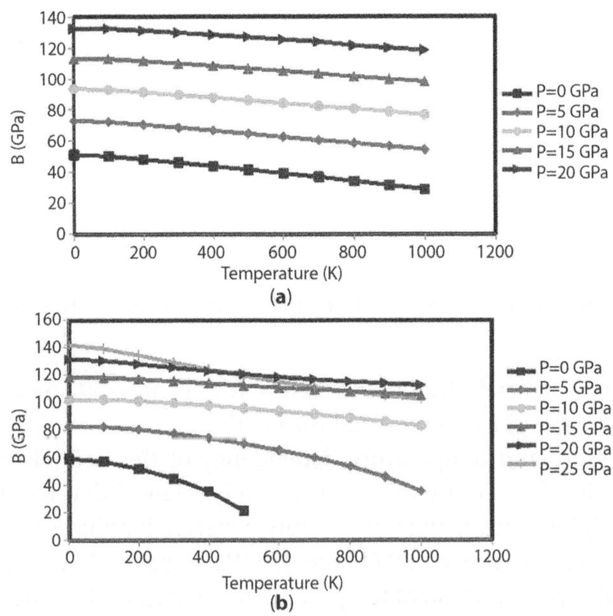

Figure 8.8 Variation of bulk modulus with temperature for (a) LiZnP and (b) LiCdP.

After 100 K bulk modulus varies inversely with temperature for a given pressure, however, it grows with pressure for a given temperature. This can be inferred as, these alloys respond towards pressure contrary to that of temperature, which means that an upsurge in the temperature of material fallouts in the weakening of its hardness.

The specific heat capacity is an essential quantity to enlighten the knowledge of lattice vibration in materials. It is well known that heat capacity describes how much heat a material can store, and its temperature raises on heating. This parameter helps us to understand the vibration properties. Hence the specific heat capacity at constant volume C_V has been computed by varying temperature and pressure as revealed in Figure 8.9(a–b). It is observed that C_V rises steeply till T = 300 K and beyond that, it increases gradually. At high-temperature C_V reaches to Dulong-Petit limit [31], further than up to about 600K Cv remains constant. For both compounds, when the temperature is less than 300 K, specific heat capacity significantly depends on both temperature and pressure (C_V proportional to T^3). It can be also concluded from Figure 8.9(a–b) that the effect of temperature on

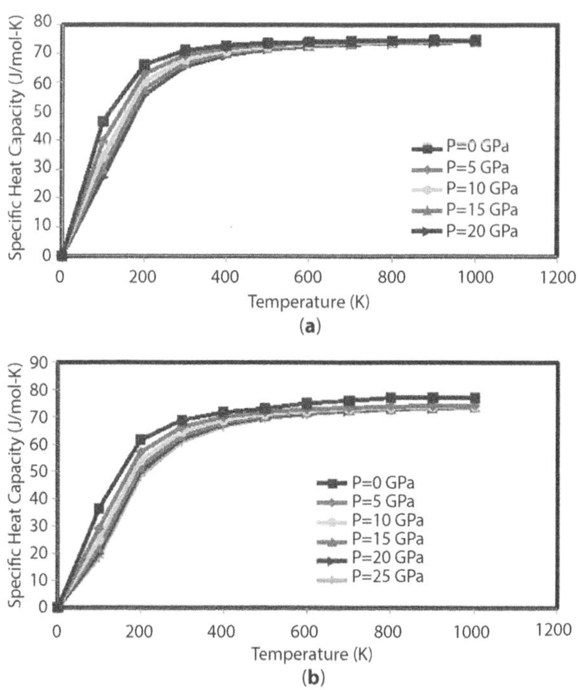

Figure 8.9 Variation of specific heat capacity with temperature for (a) LiZnP and (b) LiCdP.

C_V is substantially greater than the pressure in these half-Heusler alloys. At room temperature and zero pressure, the values of C_V are 68.8 J $mol^{-1}K^{-1}$ for LiZnP and 70.9 J $mol^{-1}K^{-1}$ for LiCdP.

The variation of thermal expansion coefficient (α) with temperature and pressure is presented in Figure 8.10 (a-b). It is understandable from this figure (Figure 8.10 (a–b)) that, at low pressures (i.e., P=0 and 5 GPa), α increases proportionally to an extent till temperature reaches 1000 K. However, after pressure 5 GPa, (P=10, 15, 20, 25 GPa) its value increases progressively with temperature. In conclusion, the thermal expansion coefficient lowers with pressure while it rises with temperatures. This is probably arising from the inadequacies of quasi-harmonic approximation while employing at high temperature and low pressure. Further, the effects of temperature and pressure on entropy (S) of LiZnP and LiCdP are presented in Figure 8.11 (a–b). It provides the knowledge that the entropy is directly proportional with temperature whilst with pressure it demonstrates an inverse proportionality. The obtained values of S for LiZnP and

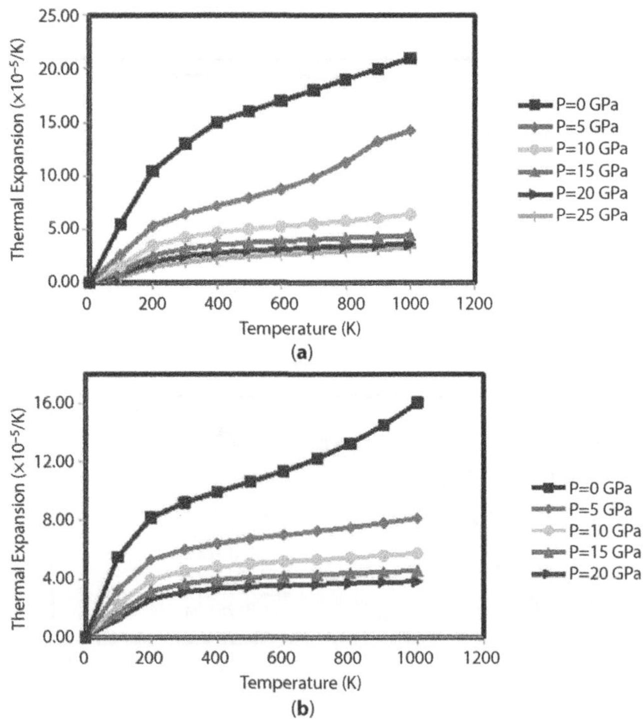

Figure 8.10 Variation of thermal expansion coefficient with temperature for (a) LiZnP and (b) LiCdP.

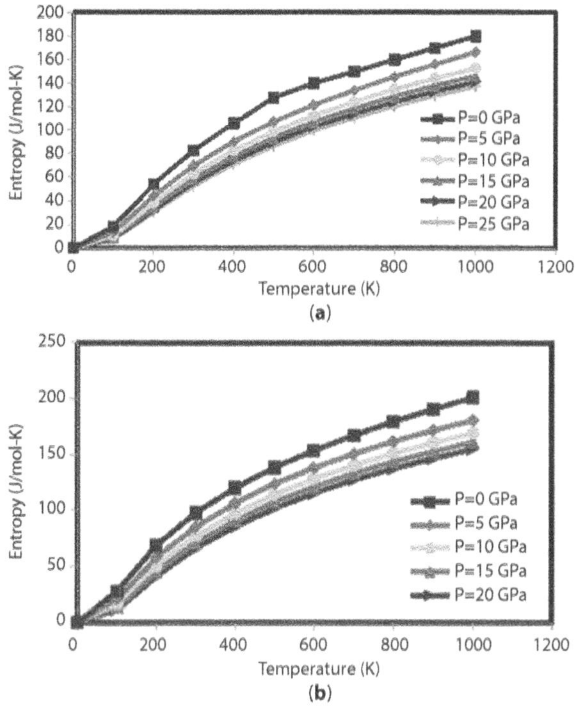

Figure 8.11 Variation of entropy with temperature for (a) LiZnP and (b) LiCdP.

LiCdP are 82.7 $Jmol^{-1}K^{-1}$ and 98.3 $Jmol^{-1}K^{-1}$. These values are also charted in Table 8.4. To our understanding, there is no other reported study of thermodynamic properties available for the comparison.

8.4 Conclusions

At ambient conditions LiZnP and LiCdP crystallize in FCC lattice having C1b structure with face group F-43m (216). To the best of our knowledge very few studies have been performed to describe the elastic constants of these compounds and hence our computed results will add valuable information to the current literature. These materials have moderate hardness as the main contributions to the interatomic boding are due to covalent and ionic bonding along with the brittle nature. Zener anisotropy factor confirms that these compounds are elastically anisotropic. The band structure displays a direct band gap in these LiXP half-Heusler alloys validating their semiconductor behavior. The covalent bonding between X and P is obtained while ionic bonding is also present between Li and P atom due to transfer of charges from Li to P. The thermodynamic properties have been studied for the first time. Furthermore, the authors are not aware of any other published data of thermodynamic properties of LiXP yet. This work can probably be used to overcome the scarcity of data on the above properties of these half-Heusler alloys.

Acknowledgement

The authors are thankful to UGC, New Delhi, for financial support under SAP-DSA-I. Author 1 (MS) is grateful to UGC for financial support from UGC under Grant No.F./PDFSS-2015-17-MAD-11704.

References

1. M. K. Yadav and B. Sanyal, *J Alloys Compd.* 622 (2015) 388–393.
2. A. Amudhavalli, R. Rajeswarapalanichamy, K. Iyakutti, *Comp. Cond. Matter.* 14 (2018) 55–66.
3. M. P. Bichat, J. L. Pascal, F. Gillot, F. Favier, Phys. *Chem. Solids*, 67 (2006) 1233.
4. M. P. Bichat, L. Monconduit, J. L. Pascal, F. Favier, *Ionics*, 11 (2005) 66.
5. D. Kieven, A. Grimm, A. Beleanu, C.G.F. Blum, J. Schmidt, T. Rissom, I. Lauermann, T. Gruhn, C. Felser, R. Klenk, *Thin Solid Films*, 519 (2011) 1866.
6. H. Nowotny, K. Bachmayer, Monatsh. *Chem.* 81 (1950) 488.
7. R. Juza, F. Hund, Z. *Anorg. Chem.* 275 (1948) 1.
8. D.M. Wood, A. Zunger, R. de Groot, *Phys. Rev. B* 31 (1985) R2570.
9. A.E. Carlsson, A. Zunger, D.M. Wood, *Phys. Rev.* B 32 (1985) R1386.
10. A. Beleanu, M. Mondeski, Q. Juan, F. Casper, C. Felser, F. porcher, *J. Phys. D: Appl. Phys.* 44 (2011) 475302.
11. M. Arif, G. Murtaza, R. Ali, R. Khenata, Y. Takagiwa, M. Muzammil, S. B. Omran *Indian J. Phys.* 90 (2016) 639.
12. T. Katoh, K. Kuriyama, *Phys. Rev. B.* 37 (1988) 7140–7142.
13. K. Kuriyama, T. Kato, K. Kawada. *Phys Rev B.* 49 (1994) 11452–11455.
14. Galanakis and P.H. Dederichs: *Lect. Notes Phys.* 676 (2005) 1-39.
15. H. C. Kandpal, C. Felser, and R. Seshadri, *J. Phys. D.* 39 (2006) 776.
16. P. Giannozzi, S. Baroni, N. Bonini, M. Calandra, R. Car, C. Cavazzoni, D. Ceresoli, G. L. Chiarotti, M. Cococcioni, I. Dabo, A. D. Corso, S. de Gironcoli, *J. Phys. Condens. Matter* 21 (2009) 395502.
17. F. Giustino, Rev. *Mod. Phys.* 89 (2017) 015003.
18. N. Marzari, D. Vanderbilt, A. D. Vita, M. C. Payne, *Phys. Rev. Lett.* 82 (1999) 3296-3299.

19. Rostam Goles, *Comp. Phys. Commun.* 184 (2013) 1861–1873.
20. W. Voigt, *Ann. Phy.* 274 (1889) 573–587.
21. A. Reuss, Z. *Ang. Math. Mech.* 9 (1929) 49–58.
22. R. Hill, Proc. *Phy. Soc. Lond.* 65 (1953) 909.
23. K. Kuriyama, T. Katoh, *Phys. Rev. B* 37 (1988) 7140.
24. R. Bacewicz, T.F. Ciszek, *Appl. Phys. Lett.* 52 (1988) 1150.
25. A. Bouhemadou, R. Khenata and F. Zerarga, *Solid State Communications* 141 (2007) 288–291.
26. M. B. Kanoun, A. E. Merad, G. Merad, J. Cibert, H. Aourag, *Solid-State Electron.* 48 (2004) 1601.
27. F. Kalarasse, B. Bennecer, A. Mellouki, L. Kalarasse, *Comp. Mat. Sci.* 43 (2008) 791.
28. A. Bouhemadoua, S. Bin-Omranb, D. Allalia, S.M. Al-Otaibib, R. Khenatad, Y. Al-Dourie, M. Chegaarf, A. H. Reshakg, *Mater. Res. Bulletin* 54 (2015) 337.
29. A. Bouhemadou, R. Khenata, *Semicond. Sci. Technol.* 23 (2008) 105024.
30. Y. Mogulkoc, Y. Ciftci, M. Evecen. *J. Phase Trans.*: 10.1080/01411594 (2018) 1515433.
31. Y. Mogulkoc, Y. Ciftci, *SAÜ Fen Bilimleri Enstitüsü Dergisi.* 21(1) (2017) 1278–1285.
32. S. Kacimi, H. Mehnane, A. Zaoui, *J. Alloys Compd.* 587 (2014) 451–458.
33. L. Damewood, B. Busemeyer, M. Shaughnessy, *Phys. Rev. B.* 91 (2015) 064409.
34. D. Guendouz, Z. Charifi, H. Baaziz, F. Soyalp, G. Uğur and Ş. Uğur, *Physica B: Phys. Cond. Matter*, 519 (2017) 39-52.
35. H.Z. Fu, D. H. Li, F. Peng, *Comput. Mater. Sci* 44 (2008) 774–778.
36. L. Liu, G. Xu, A. Wang, *J. Phys. Chem. Solids.*104 (2017) 243–251.
37. M. Born, K. Huang, *Dynamical theory and experiment* I. Berlin: Springer Verlag; 1982.
38. Pugh SFXCII. *Philos Mag.* 45 (1954) 823–843.
39. B. Fatima, S. S. Chouhan, N. Acharya, *Intermetallics.* 53 (2014) 129–139.
40. M. Evecen, Y. O. Ciftci. *J Nanoelectron Optoelectron.* 12 (2017) 100–108.
41. S. Aydin, M. Simsek, *J. Alloys Compd.* 509 (2011) 5219–5229.
42. A. Mellouki, L. Kalarasse, B. Bennecer and F. Kalarasse, *Comput. Mater. Sci.* 42 (2008) 579.
43. K. Kuriyama, K. Kushida, R. Taguichi. *Solid State Commun.* 108 (1998) 429-32.
44. A. Roy, J. W. Bennett, K. M. Rabe, D.Vanderbilt, *Phys. Rev. Lett.* 109 (2012) 037602.
45. H. Mehnane, B. Bekkouche, S. Kacimi, A. Hallouche, M. Djermouni, A. Zaoui, *Superlattices Microstruct.* 51 (2012) 772.
46. M. A. Blanco, E. Francisco, V. Luaña, *Comp. Phys Commun.* 158 (2004) 57.

9

Fixed Point Results with Fuzzy Sets

Qazi Aftab Kabir[1]*, Sanath Kumar H.G.[2] and Ramakant Bhardwaj[3]

[1]*Department of Mathematics, Govt. Gandhi Memorial Science College Jammu, Jammu & Kashmir, India*
[2]*Shanghai United International School, Hongsong East Road Minhang District, Shanghai, China*
[3]*Department of Mathematics, Amity University, Kolkata, West Bengal, India*

Abstract
Topology is the investigation of geometric properties that doesn't rely just upon the specific state of the objects, rather it acts on how the focuses are associated with one another. In fact, topology manages that stay invariant under the continuous transformation of a map. Digital topology has pulled in the consideration of numerous scientists inferable from its potential applications in certain regions, for example, software engineering, picture handling, topology and fixed point theory.

***Keywords*:** Digital metric space, digital topology (DGy), Banach-contraction mapping, $(\tau - \varphi)$ –contractive mappings, Cauchy sequence (CS), Topology (Ty), geometry (GP), fuzzy metric space (FMS), CFP (Common Fixed Point)

9.1 Introduction

Topology (Ty) is the investigation of GP properties that doesn't rely just upon the specific state of the items, rather it acts on how the focuses are associated with one another. In fact, topology manages those properties that stay invariant under the nonstop change of a guide. Author [12] presented an idea of DGy. Computerized Ty is worried about GP and and Ty possessions of computerized picture. The computerized pictures utilized for picture preparing & PC designs. Advanced Ty likewise gives scientific

**Corresponding author*: aftabqazi168@gmail.com

Ramakant Bhardwaj, Jyoti Mishra, Satyendra Narayan and Gopalakrishnan Suseendran (eds.) Mathematics in Computational Science and Engineering, (199–210) © 2022 Scrivener Publishing LLC

premise to picture preparing activity for Two-D and Three-D computerized pictures. The reader may study [1–3, 6, 7, 10] for more details.

Authors [8, 9], defined the idea of commutative and compatible maps and proved the common fixed point results for such maps. Authors [13] gave the idea of contractive mapping, and proved results in digital metric spaces. In this chapter a new common fixed point theorem is proved on satisfying the digital topology with $(\tau - \varphi)$ –Contractive mappings in fuzzy sets.

9.2 Definitions and Preliminaries

Definition 2.1. [12] If $(X, M,*)$ is a FMS

1. $\{x_n\}$ in X is called a Cs iff $\lim_{n\to\infty} M(x_{n+p}, x_n, t) = 1$ for each $p > 0, t > 0$.
2. A sequence $\{x_n\}$ in X is converging to x in X if and only if $\lim_{n\to\infty} M(x_n, x, t) = 1$. A FMS $(X, M,*)$ is said to be complete iff every CS in X is conv. in X.
3. A fuzzy metric space in which every sequence has a convergent subsequence is said to be compact.

Definition 2.2. ([6]) In FMS (X,d,ρ), if a seq $\{x_n\} \subset X \subset \mathbb{Z}^n$ is a CS, $\exists M \in \mathbb{N}$: $\forall n,m > M$, *we have* $x_n = x_m$.
Definition 2.3. ([6]) Sq. $\{x_n\}$ of points of a FMS $(X,d,\rho) \dashrightarrow L \in X$ *if* $\forall \in > 0$, there is $M \in \mathbb{N} : d(x_n, L) < \epsilon$ *for all* $n > M$.
Definition 2.4. ([6]) If $\{x_n\}$ is a seq. of points of a DGM $(X, d, \rho) \dashrightarrow L \in X \, \forall \in > 0, \exists M \in \mathbb{N} : x_n = L$ *for all* $n > M$.*i.e.*, $x_n = x_{n+1} = x_{n+2} = \cdots = L$
Definition 2.5. ([4]) A DGMS (X, d, ρ) is complete if any CS $\{x_n\} \dashrightarrow L$ of (X,d,ρ).
Definition 2.6. ([6]) A FMS (X,d,ρ) is complete.
Definition 2.7. ([8]) If (X,d,ρ) is a DGMS and T: $(X, d, \rho) \to (X, d, \rho)$ is a self-map. If $\exists \lambda \in [0,1)$:

$$d(Tx, Ty) \leq \lambda d\,(x, y) \text{ for all } x, y \in X,$$

Definition 2.8. ([5]) If $X \subseteq Z^n$, (X, d, ρ) is a DGMS. $\nexists$ a seq.$\{x_n\}$ of distinct elements in X:

$$d(x_{m+1}, x_m) < d(x_m, x_{m-1}) \text{ for } m = 1,--$$

Proposition 2.9. ([12]) Every digital contraction map T: $(X, d, \rho) \rightarrow (X, d, \rho)$ is digitally continuous.
Proposition 2.10. ([7]) Seq $\{x_n\}$ of points of a DGMS $(X, d, \rho) \dashrightarrow l \in X \ \exists\ \alpha \in N : n \geqq \alpha, \Rightarrow x_n = l$.
Proposition 2.11. ([11]) Let f and g be digitally compatible mappings of type (R) of a DGMS $(X, d, \rho) \dashrightarrow (X, d, \rho)$. If $\lim_n f\ x_n = \lim_n g\ x_n = t$ for some $t = \in X$.then

a) $\lim_n gf\ x_n = ft$ if f is fuzzy continuous at t.

b) $\lim_n fg\ x_n = gt$ if g is fuzzy continuous at t.

c) $fgt = gft$ and $ft = gt$ if f and g both are fuzzy continuous at t.

Lemma 2.12. If $X > \subseteq \mathbb{Z}^n$, (X, d, ρ) be a DFMS space. $\nexists\ \{x_m\}$ of distinct element in X, :

$$d(x_{m+1}, x_m) < d(x_m, x_{m-1})\ for\ m = one,\ two,\ --$$

Definition 2.13. ([5]) Suppose that (X, d, ρ) is a DGMS and P, Q:$X \rightarrow X$, on X. $\Rightarrow$ *P and* Q are compatible if $d(PQx, Qpx) \leq d(Px, Qx)$ *for all* $x \in X$.

9.3 Main Results

The common fixed point theorem is proved for $(\tau - \varphi)$ –weakly contractive mapping in FMS is now proved.
Definition 3.1. Let (X, d, ρ) is a $(\tau - \varphi)$ –DGMS and P,Q,R,S:$X \rightarrow X$ defined on X.

$$P,\ Q,\ R \text{ and } S \text{ are } (\tau - \varphi) \text{ –weakly Compatible}$$

If $d(PQx, RSx) = d(Px, Rx)\ \forall\ x \in X$.
Definition 3.2. Let (X,d,ρ) is a complete FMS and S *is a mapping from*: X *to itself*. S will be a digital soft $\alpha_1 - \psi_1$ –contractive mapping, $\exists$ functions α: $X \times X \rightarrow [0, \infty)$ and $\psi \in \Psi : x, y \in X$, we have $\psi(d(Sx, Sy)) \geq \alpha(x, y)d(x, y)$.
Theorem 3.3. Let (X, d, ρ) is a complete $(\tau - \varphi)$ –fuzzy metric space, let N be a nonempty closed subset of X. Let P, $Q : N \rightarrow N$ and G, $H : N \rightarrow X$ be a bijective, Fuzzy $(\tau - \varphi)$ –contractive mappings satisfying $Q(N) \subset H(N)$ and for every $x, y \in X$,

i. P^{-1}, Q^{-1} is α – admissible; there exist

$$\alpha(d(P^{-1}x,Q^{-1}y)) \leq (\tau-\varphi)-\left[\left(d_{G,H}(x,y)\right)-\frac{1}{2}(\tau-\varphi)-\left(d_{G,H}(x,y)\right)\right] \tag{9.1}$$

ii. *G,H* is digitally continuous

$$(\tau - \varphi) - (d_{G,H}(x, y)) = (\tau - \varphi) - [\max\{(Gx, Hy), (Gx, Px), (Hy, Qy)\}] \tag{9.2}$$

Then P^{-1}, Q^{-1}, *G* and *H* have a unique common fixed point in *X*.

Theorem 3.4. If (X, d, ρ) is a complete $(\tau - \varphi)$ –FMS, let $N \neq \varnothing \subset$. P^{-1}, Q^{-1} $: N \rightarrow N$ and $G,H : N \rightarrow X$ be a bijective, digital

$(\tau - \varphi)$ –Contractive mappings satisfying $Q(N) \subset H(N)$ and for every $x, y \in X$,

i. P^{-1}, Q^{-1} is α –admissible; there exist

$$\alpha(d(P^{-1}x,Q^{-1}y)) \leq (\tau-\varphi) -\left[\left(d_{G,H}(x,y)\right)-\frac{1}{2}\left(d_{G,H}(x,y)\right)-\left(d_{G,H}(x,y)\right)\right] \tag{9.3}$$

ψ:$[0,\infty) \rightarrow [0,\infty)$ is a $(\tau - \varphi)$ –fuzzy continuous function:
(i) $\psi(\rho) = 0$.

ii. *G, H* is digitally continuous

$$\alpha\left(d_{G,H}(x,y)\right) = (\alpha-\Psi)\max\left\{\begin{matrix}(Gx,Hy),(Gx,P^{-1}x),(Hy,Q^{-1}y),\\ \frac{1}{2}\left((Gx,Q^{-1}y)+(Hy,P^{-1}x)\right)\end{matrix}\right\} \tag{9.4}$$

then P^{-1}, Q^{-1}, *G* & *H* have a unique common FP in *X*.

Solution x_0 is an arbitrary point in X.
Since $Q^{-1}(N) \subset G(N)$, $P^{-1}(N) \subset H(N)$,
The sequences $\{x_n\}$ and $\{y_n\}$ can be defined on X:

$$y_{2n-1} = P^{-1}_{x_{2n-2}} = H_{x_{2n-1}}, y_{2n} = Q^{-1}_{x_{2n-1}} = G_{x_{2n}}, \quad n = 1,2,3,4,\ldots\ldots$$

Suppose that $y_{n_0} = y_{n_{0+1}}$ only f or some n_0. $\Rightarrow$ $\{y_n\}$ is constant if $n \geq n_0$. $\Rightarrow$ let $n_0 = 2k$. Then $y_{2k} = y_{2k+1}$ using (9.3):

$$\begin{aligned}\alpha\left(\left(y_{2k+1}, y_{2k+2}\right)\right) &= (\tau-\varphi)-\left(P^{-1}x_{2k}, Q^{-1}x_{2k+1},\right) \\ &\leq (\tau-\varphi)-\left[\left(d_{G,H}(x_{2k}, x_{2k+1})\right)-\left(d_{G,H}(x_{2k}, x_{2k+1})\right)\right]\end{aligned} \tag{9.5}$$

Where

$$\begin{aligned}&\alpha(d_{G,H}(x_{2k}, x_{2k+1})) \\ &= (\tau-\varphi)-max\begin{Bmatrix}(y_{2k}, y_{2k+1}), (y_{2k}, P^{-1}x_{2k},), (y_{2k+1}, Q^{-1}x_{2k+1}), \\ \frac{1}{2}\alpha-\psi-\left((y_{2k}, Q^{-1}x_{2k+1})+(y_{2k+1}, P^{-1}x_{2k})\right)\end{Bmatrix} \\ &= (\tau-\varphi)-max\begin{Bmatrix}0, 0, (y_{2k+1}, y_{2k+2}), \\ \frac{1}{2}\alpha-\psi-\left((y_{2k}, y_{2k+2})+0\right)\end{Bmatrix} \\ &= (\tau-\varphi)-max\left\{(y_{2k+1}, y_{2k+2}), \frac{1}{2}(y_{2k}, y_{2k+2})\right\} \\ &(\tau-\varphi)-(y_{2k+1}, y_{2k+2}).\end{aligned}$$

By (9.5), we get

$$(\tau - \varphi) - (y_{2k+1}, y_{2k+2}) \leq$$

$$(\tau - \varphi) - \{(y_{2k+1}, y_{2k+2}) - (y_{2k+1}, y_{2k+2})\}$$

And so $(\alpha(y_{2k+1}, y_{2k+2}) \leq 0)$ and $y_{2k+1} = y_{2k+2}$.
Taking $n_0 = 2k + 1$, $\Rightarrow y_{2k+2} = y_{2k+3}$, $\{y_n\}$ is constant.
So $\{P^{-1}, Q^{-1}\}$& $\{G,H\}$ have a point of coincidence in X.

If $(y_n, y_{n+1}) > 0$ for each n. we shall show that for each $n = 0,1,2,3,4,\ldots$,

$$\alpha(y_{n+1}, y_{n+2}) \leq (\tau - \varphi) - d_{G,H}(x_n, x_{n+1}) = (\tau - \varphi) - (y_n, y_{n+1}) \quad (9.6)$$

Using (9.6), we obtain that

$$\alpha(y_{2n+1}, y_{2n+2}) = (\tau - \varphi) - (P_{x2n}^{-1}, Q_{2n+1}^{-1})$$
$$\leq (\tau - \varphi) - \left(d_{G,H}(x_{2n}, x_{2n+1})\right)$$
$$(\tau - \varphi) - \left(d_{G,H}(x_{2n}, x_{2n+1})\right) \quad (9.7)$$
$$< (\tau - \varphi) - \left(d_{G,H}(x_{2n}, x_{2n+1})\right).$$

$$\alpha(y_{2n+1}, y_{2n+2}) \leq (\tau - \varphi) - \left(d_{G,H}(x_{2n}, x_{2n+1})\right) \quad (9.8)$$

Moreover, we have

$$\alpha(d_{G,H}(x_{2n}, x_{2n+1}))$$
$$= (\tau - \varphi) - \max\left\{\begin{matrix}(y_{2n}, y_{2n+1}), (y_{2n}, P^{-1}x_{2n},), (y_{2n+1}, Q^{-1}x_{2n+1}), \\ \frac{1}{2}\alpha - \psi - \left((y_{2n}, Q^{-1}x_{2n+1}) + (y_{2n+1}, P^{-1}x_{2n})\right)\end{matrix}\right\}$$
$$= (\tau - \varphi) - \max\left\{\begin{matrix}(y_{2n}, y_{2n+1}), (y_{2n}, y_{2n+1}), (y_{2n+1}, y_{2n+2}), \\ \frac{1}{2}(\alpha_1 - \psi_1) - \varphi(y_{2n+1}, y_{2n+2})\end{matrix}\right\}$$
$$(\tau - \varphi) - \leq \max\left\{\begin{matrix}(y_{2n}, y_{2n+1}), (y_{2n+1}, y_{2n+2}), \\ \frac{1}{2}(\alpha_1 - \psi_1) - \varphi\left((y_{2n}, y_{2n+1}) + (y_{2n+1}, y_{2n+2})\right)\end{matrix}\right\}$$
$$\leq (\tau - \varphi) - \max\{(y_{2n}, y_{2n+1}), (y_{2n+1}, y_{2n+2})\}$$

If $\alpha(y_{2n+1}, y_{2n+2}) \geq (\tau - \varphi) - (y_{2n}, y_{2n+1})$, then by using the last inequality and (9.7), we have $\alpha\left(d_{G,H}(x_{2n}, x_{2n+1})\right) = (\tau - \varphi) - (y_{2n+1}, y_{2n+2})$ and (9.8) implies that

$$\alpha(y_{2n+1}, y_{2n+2}) = (\tau - \varphi) - \left(d_{G,H}(Px_{2n}, Qx_{2n+1})\right)$$

$$\leq (\tau - \varphi) - (y_{2n+1}, y_{2n+2}) - (\tau - \varphi) - (y_{2n+1}, y_{2n+2})$$

This is only possible when $\alpha(y_{2n+1}, y_{2n+2}) = 0$. it is contradiction. Hence

$$\alpha(y_{2n+1}, y_{2n+2}) \leq (\tau - \varphi) - (y_{2n}, y_{2n+1}), \text{ and}$$

$$\alpha[d_{G,H}(x_{2n}, x_{2n+1})] \leq (\tau - \varphi) - (y_{2n}, y_{2n+1}).$$

$$\Rightarrow \alpha(y_{2n+3}, y_{2n+2}) \leq ((\tau - \varphi) - d_{G,H}(x_{2n+2}, x_{2n+1}))$$
$$= (\tau - \varphi) - (y_{2n+2}, y_{2n+1}).$$

So (9.8) will be true $\forall\, n \in N$.
$\Rightarrow \{\alpha(d(y_n, y_{n+1}))\}$ is nondecreasing and the limit

$$\alpha \lim_{n\to\infty} \alpha(y_n, y_{n+1}) = (\tau - \varphi) - \lim_{n\to\infty} d_{G,H}(x_n, x_{n+1})$$

exists. It can be denoted by l^*. & $l^* \geq 0$. suppose that $l^* > 0$.

$$\Rightarrow \alpha(y_{n+1}, y_{n+2}) \leq (\tau - \varphi) - (d_{G,H}(x_n, x_{n+1})) - (\tau - \varphi) - (d_{G,H}(x_n, x_{n+1})).$$

when $n \to \infty$, *it will* $\dashrightarrow$ *upper limit*

$$\alpha(l^*) \leq (\tau - \varphi) - \varphi(l^*) - (\tau - \varphi) - \lim_{n\to\infty} \inf (\tau - \varphi) - (d_{G,H}(x_n, x_{n+1}))$$
$$\leq \alpha(l^*) - (\tau - \varphi) - \varphi(l^*), \quad (9.9)$$

i.e., $(\tau - \varphi) - (l^*) \leq 0$. By control functions, we get $l^* = 0$, this will be contradiction. $\Rightarrow$ $(\tau - \varphi) - \lim_{n\to\infty} (y_n, y_{n+1}) = 0$. It is necessary to prove that $\{y_n\}$ is a CS in X. It is enough to prove that $\{y_{2n}\}$ is a CS.

If it is not so $\Rightarrow \in > 0$, $\exists$ subsequence $\{y_{2n(k)}\}$, $\{y_{2m(k)}\}$ of $\{y_{2n}\}$: $n(k)$ is the smallest index satisfying

$$\alpha(n(k)) > (\tau - \varphi) - m(k)$$

And

$$(\tau - \varphi) - (y_{n(k)}, y_{m(k)}) \geq \in.$$

In particular, $(\tau-\varphi)-\left(y_{n(k)-2}, y_{m(k)}\right)<\epsilon$.

$$|d(x,z - d(x,z)| \leq d(x,z),$$

$$\begin{aligned}\alpha \operatorname*{limit}_{k\to\infty}\left(y_{2n(k)}, y_{2m(k)}\right) &= (\tau-\varphi)-\lim_{k\to\infty}\left(y_{2\mathrm{n(k)}}, y_{2\mathrm{m(k)}-1}\right)\\ &= (\tau-\varphi)-\lim_{k\to\infty}\left(y_{2\mathrm{n(k)}+1}, y_{2\mathrm{m(k)}}\right)\\ &= (\tau-\varphi)-\lim_{k\to\infty}\left(y_{2\mathrm{n(k)}+1}, y_{2\mathrm{m(k)}-1}\right)=\epsilon. \quad (9.10)\end{aligned}$$

By using the previous limits, we get that

$$(\tau-\varphi)-\lim_{k\to\infty} d_{G,H}\left(x_{2\mathrm{n(k)}}, x_{2\mathrm{m(k)}-1}\right)=\epsilon.$$

Indeed,

$$(\tau-\varphi)-d_{G,H}\left(x_{2\mathrm{n(k)}}, x_{2\mathrm{m(k)}-1}\right)$$

$$(\tau-\varphi)-\max\left\{\begin{array}{l}\left(y_{2n(k)}, y_{2m(k)-1}\right),\left(y_{2n(k)}, y_{2m(k)+1}\right),\left(y_{2m(k)-1}, y_{2m(k)}\right),\\ \frac{1}{2}(\alpha_1-\psi_1)-\varphi\left(\left(y_{2n(k)}, y_{2m(k)}\right)+\left(y_{2n(k)}, y_{2m(k)-1}\right)\right)\end{array}\right\}$$

$$\to (\tau-\varphi)-max\{\epsilon,0,0,\frac{1}{2}(\epsilon+\epsilon)\}=\epsilon.$$

Applying (9.9), we obtain

$$\alpha\left(y_{2\mathrm{n(k)}+1}, y_{2\mathrm{m(k)}}\right)=(\tau-\varphi)-\left(P^{-1}_{x_{2\mathrm{n(k)}}}, Q^{-1}_{x_{2\mathrm{n(k)}-1}}\right)$$

$$\leq \alpha\left(d_{G,H}\left(x_{2\mathrm{n(k)}}, x_{2\mathrm{m(k)}-1}\right)\right)-(\tau-\varphi)-\left(d_{G,H}\left(x_{2\mathrm{n(k)}}, x_{2\mathrm{m(k)}-1}\right)\right).$$

limit $k\to\infty$, $\Rightarrow \alpha(\epsilon) \leq (\tau-\varphi)-(\epsilon)-(\tau-\varphi)-(\epsilon)$, it is contradiction.

So $\{y_n\}$ is a CS in the complete metric (X,d). $\exists\, u \in X: (\tau-\varphi)-\lim_{n\to\infty} y_n = u$.

To prove the uniqueness property of u, suppose that u' is another point of coincidence of G and P, that is

$$u' = Gv' = P^{-1}v' \tag{9.11}$$

For some $v' \in N$. By (9.6), we have

$$\alpha(u',\mathrm{u}) = (\tau-\varphi)-(\mathrm{P}v',\mathrm{Q}^{-1}\mathrm{u}) \le (\tau-\varphi)-\left(d_{G,H}(v',\mathrm{u})\right)$$
$$-(\tau-\varphi)-\left(d_{G,H}(v',\mathrm{u})\right)$$

Where

$$\alpha\left(d_{G,H}(v',\mathrm{u})\right) = (\tau-\varphi)$$
$$-\max\left\{(u',u),0,0,\frac{1}{2}\left(d_{G,H}(v',u)\right)-(\tau-\varphi)-\left(d_{G,H}(v',u)\right)\right\}$$

It follows from (9.11) that $u' = u$.

⇒ u is the unique point of coincidence of $\{P^{-1},Q^{-1}\}$ & $\{G,H\}$.

If $\{P^{-1}, Q^{-1}\}$ & $\{G, H\}$ are $(\tau - \varphi)$ – contractive compatible mapping, then by (9.10) and (9.11), we have

$P^{-1}u = P^{-1}(Gv) = G(P^{-1}v) = Gu = w_1$ and $Q^{-1}u = Q^{-1}(Hu) = H(Q^{-1}u) = Hu = w_2$. by (9.6), we have

$$\alpha(w_1,w_2) = (\tau-\varphi)-(P^{-1}\mathrm{u},Q^{-1}\mathrm{u}) \le (\tau-\varphi)-\left(d_{G,H}(\mathrm{u},\mathrm{u})\right)$$
$$-(\tau-\varphi)-\left(d_{G,H}(\mathrm{u},\mathrm{u})\right),$$

Where

$$\alpha\left(d_{G,H}(\mathrm{u},\mathrm{u})\right) = (\tau-\varphi)-\max\left\{(w_1,w_2),0,0,\frac{1}{2}(w_1,w_2)+(w_1,w_2)\right\}$$

It follows that $w_1 = w_2$, that is,

$$P^{-1}u = Gu = Q^{-1}u = Hu. \tag{9.12}$$

By (9.6) and (9.12), we have

$$\alpha(P^{-1}\,\mathrm{u},Q^{-1}\,\mathrm{u}) \leq (\tau-\varphi)-\left(d_{G,H}(\mathrm{u},\mathrm{u})\right)-(\tau-\varphi)-\left(d_{G,H}(\mathrm{u},\mathrm{u})\right),$$

Where

$$\alpha\left(d_{G,H}(\mathrm{u},\mathrm{u})\right)=(\tau-\varphi)-\max\begin{Bmatrix}(Gv,Hu),(Gv,P^{-1}v),(Hu,Q^{-1}u),\\ \frac{1}{2}(Gv,Q^{-1}u)+(P^{-1}v,Q^{-1}u)\end{Bmatrix}$$

$$(\tau-\varphi)-(P^{-1}v,Q^{-1}u).$$

Therefore, we deduce that $P^{-1}v = Q^{-1}u$, that is, $u = Q^{-1}u$. It follows from (9.12) that

$$u = P^{-1}u = Gu = Q^{-1}u = Hu.$$

Then u is the unique CFP of P^{-1}, G, H and $Q^{-}1$.

References

1. Barve S, K., Kabir A. Q., Daheriya, R. D, Unique common fixed point theorem for weakly compatible mappings in digital metric space, *International Journal of Scientific Research and Reviews,* Vol. 8, No. 1, 2019.
2. Boxer, L. (2005). Properties of Digital Homotopy. *Journal of Mathematical Imaging and* Vision, 22, 19–26.
3. Boxer, L. (2010). Continuous Maps on Digital Simple Closed Curves. *Applied Mathematics,* 1, 377–386.
4. Chauhan, M, S, Shrivastava, R, Verma, R, Kabir Aftab. Qazi, Fixed point theorem for quadruple self-mappings in digital metric space, *International Journal of Scientific Research and Reviews,* Vol. 8, No. 1, 2019.
5. Ege, O. and Karaca, I. (2013). Lefschetz Fixed Point Theorem for Digital Images. *Fixed Point Theory and Applications,* 2013, 13 pages.
6. Ege, O. and Karaca, I. (2015). Banach fixed point theorem for digital images, *J. Nonlinear Sci. Appl.,* 8, 237–245.
7. Han, S.E. (2015). Banach fixed point theorem from the viewpoint of digital topology, *J. Nonlinear Sci. Appl.,* 9, 895–905.
8. Jungck G, (1976). Commuting mappings and fixed point, *Amer. Math. Monthly,* 83, 261–263.

9. Jungck, G., Compatible mappings and common fixed points (2), ibid, 11, 285–288.
10. Kabir, Q. A., Mohammad, M., Jamal, R. and Bhardwaj, R. (2017) Unique fixed points and mappings in hyperconvex metric spaces, *International Journal of Mathematics and Its Applications*, Vol. 5, No. 3-B, pp. 171–177.
11. Mohammad, M., Jamal, R., Mishra, J., Kabir, Q.A., and Bhardwaj, R. Coupled fixed point theorem and dislocated quasi-metric space, *International Journal of Pure and Applied* Mathematics, 10(119), (2018), 1249–1260.
12. Rosenfeld, A. (1979). Digital Topology. *American Mathematical Monthly*, 86, 621–630.
13. Rosenfeld, A. (1986). Continuous functions on digital pictures, *Pattern Recognition Letters*, 4, 177–184.

10

Role of Mathematics in Novel Artificial Intelligence Realm

Kavita Rawat* **and Manas Kumar Mishra**

SCSE Department, VIT University Bhopal, India

Abstract

Artificial Intelligence has gained wide popularity in our day-to-day life. AI has progressed miraculously to deliver profits in fields such as healthcare, finance, education, manufacturing, etc. AI tools and Chatbots have been very recently a discovery in the speedily advancing tech domain. AI is not mystical; it is all about mathematics. The intellectual concept is to think like humans and copy their behavior with the assistance of mathematics. Both Artificial Intelligence and mathematics are designed for optimization purpose. But the main question that arises here is, What are the basic concepts of mathematics that lead to the imitation of human behavior? How is AI associated with mathematics? The mathematical concepts like Linear Algebra, Calculus, Probability, Statistics, etc., help to understand the logical reasoning behind Chatbots and the self-driving car. Though all the data are not numerical in nature it is often helpful to consider data in numerical format. In artificial intelligence the numerical data is represented as vectors and for representing them in tabular format matrices are used. The study of vectors and matrix comes under linear algebra. Likewise usually, when data are collected some sort of noise get introduced in data or observation. While dealing with such noisy data, uncertainty occurs so there is a need to quantify the uncertainty, which is the domain of probability theory. In this way artificial intelligence and mathematics are closely related with each other.

Mathematics helps to solve the challenging task of hypothetical problems in artificial intelligence using traditional methods and techniques. Mathematical concepts improve our skills of thinking and enhance our mental strength.

Keywords: Artificial Intelligence, mathematics, linear algebra, calculus, probability, statistics

**Corresponding author*: Kavita.rawat2018@vitbhopal.ac.in

Ramakant Bhardwaj, Jyoti Mishra, Satyendra Narayan and Gopalakrishnan Suseendran (eds.) Mathematics in Computational Science and Engineering, (211–232) © 2022 Scrivener Publishing LLC

10.1 Introduction

Artificial Intelligence is the branch of computer science that deals with developing a model that possesses human intelligence. AI has a broad range, from a ride-sharing app like Uber to self-steering car technology.

Whatever was in our imagination in childhood—what if a car could work without a driver, what if there was a machine that could do our work, what would happen if there was a machine that answered our questions, or a machine or computer we could talk to, or a car we could fly?—has now became reality with the help of Artificial Intelligence. Now there are self-driving cars that work without a driver, chatbots are always there to answer our questions, robots can assist us in our household and other activities, and a flying car can also be expected in the future. Mathematics is the key element of AI. Artificial intelligence mainly works on data, which is why data, probability, and statistics are some of the key parts of AI. The task of AI includes data analysis, Data Mining, Predictive Analysis and forecasting, and so on [10].

As an example, a restaurant may analyze the food ordering behavior of a customer, which means the trend that the customer follows each time while ordering food, and this helps to create the recommendation system of dishes that a customer might also ask for.

Second, we can take the example of a self-driving car in which machine learning techniques are used—supervised learning and reinforcement learning for learning the route as well as the texture of the road.

Google Maps is also a very good example of AI. In this app, Traffic analysis, shortest route prediction and the required time to cover the selected route helps the person to follow the path with ease.

10.2 Mathematical Concepts Applied in Artificial Intelligence

Mathematics plays an important role in the artificial intelligence domain. Without mathematics either the prediction of any problem or the analysis and visualization of dataset is not possible. There are many mathematical concepts used in artificial intelligence but this chapter covers some of the topics of mathematics which are needed for basic prediction problems as shown in Figure 10.1. The following areas of mathematics will be covered in this chapter:

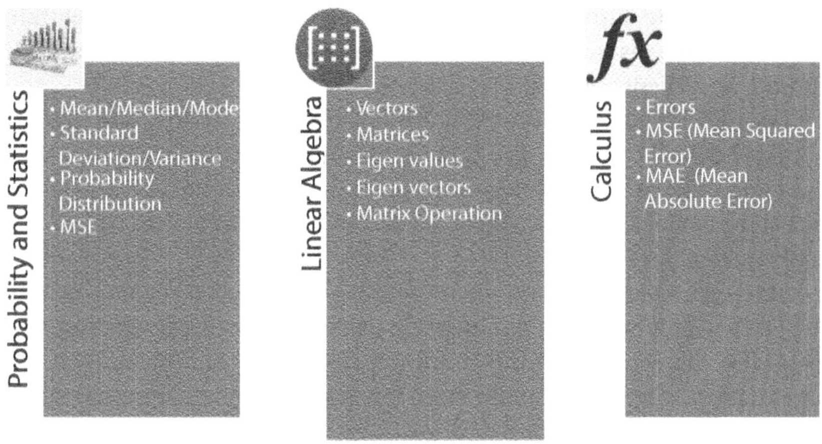

Figure 10.1 Mathematical topics covered.

1. Linear Algebra
2. Calculus
3. Probability & Statistics

10.2.1 Linear Algebra

Linear algebra is the core part of mathematics used in machine learn ing with the concept of matrices, vectors, linear transformations for the implementation of algorithms or performing operation in machine learning. Linear Algebra contributes in the majority of exploratory data analysis tasks, i.e., Data preprocessing, Data transformation, Dimensionality reduction and model evaluation. To understand the mathematical principle behind the linear algebra please refer to the book [1].

10.2.1.1 Matrix and Vectors

In Artificial Intelligence, there is a need to fit a machine learning model on a given dataset. Actually the dataset can be of any type, like numerical, categorical, time series, text, images.

Numerical data is also known as quantitative data and it is used to display the data points as numbers. These types of data are helpful in measurement types of problems like counting the number of restaurants in any particular area, counting how many houses were sold in the last two years, and so on. Numerical data is again characterized as continuous numerical data and discrete numerical data. Continuous data takes any value within a range while the discrete takes a distinct value or number, for example, the number of students present in a class.

Categorical data represents the data by its characteristics like green as 1 and yellow as 0. We can also take an example as representing women as 1 and men as 0. Also, property is of two types, commercial and residential; take commercial as 1 and residential as 0 but these numbers don't have any mathematical meaning.

Time series data is a sequence of numbers gathered at regular intervals over some timeframe. It is significant, particularly specifically for handling things like finance or neuroscience brain signals. We can understand time series data as recording of brain electrical activity of all the neurons simultaneously to predict any kind of disorder in the brain.

Text data is a data in word, sentence, and paragraph form. We can't process this type of data as it is so we need to convert it into numerical form for further processing.

Images and Photographs: When we are working with image type of data in computer vision applications, each image on which we will work is itself a tabular structure containing the value of each pixel according to its width and height value as black (1) and white (0). The image or photo is yet another example of matrix in linear algebra.

So, all the above types of dataset are needed to store in table-like structure where each row represents a sample or observation and each column represents an attribute or feature of the observation.

Below is displayed the snippet of Eye open shut (time series) dataset

Eye Open Shut (time series) Dataset

This type of data is actually a matrix, which is a very important data structure of linear algebra. Later on when we will split the above data into inputs and outputs to fit a model for machine learning, you will remain with the inputs as matrix (X) and outputs as vector (y). The vector is also another key data structure after matrix.

10.2.1.2 Eigen Value and Eigen Vector

When we are interested in finding the characteristics of a matrix, Eigen value and Eigen vector come into the limelight. Eigen values and Eigen vectors is helpful in dimensionality reduction techniques. For dimensionality reduction transformation and compression plays a vital role.

Transformation is used to convert matrix data of high dimensionality into low dimensionality by retaining all the resembling intrinsic features of

1352	4420	4015.9	4297.95	4127.18	4321.54	4596.41	4077.95	4624.1	4227.69	4256.92	4285.64	4337.95	4748.21	4512.31	eye close
1353	4402.56	4005.13	4287.18	4120	4320.51	4598.97	4082.05	4625.13	4226.15	4257.44	4278.46	4323.08	4736.92	4496.92	eye close
1354	4386.67	3998.46	4276.41	4112.82	4321.54	4598.46	4083.08	4628.72	4225.13	4254.87	4267.18	4313.85	4727.69	4486.15	eye close
1355	4378.97	3990.77	4271.79	4105.13	4323.08	4594.36	4082.56	4627.18	4222.56	4254.87	4270.77	4314.87	4734.36	4485.13	eye close
1356	4376.41	3984.1	4273.33	4102.05	4321.03	4593.33	408[illegible].03	4615.9	4204.1	4248.21	4273.85	4317.95	4731.79	4481.03	eye open
1357	4374.36	3984.1	4276.92	4106.15	4318.97	4591.28	4077.95	4604.1	4188.21	4236.41	4263.08	4317.44	4716.41	4471.28	eye open
1358	4372.82	3983.08	4278.97	4108.72	4322.56	4586.15	4073.85	4596.41	4191.79	4236.41	4255.9	4315.9	4709.74	4464.1	eye open
1359	4369.23	3980.51	4274.87	4105.13	4328.21	4586.67	4076.92	4597.44	4197.44	4242.05	4261.03	4313.85	4710.77	4460	eye open
1360	4362.56	3981.03	4266.15	4100	4326.67	4596.41	408[illegible].54	4606.67	4200	4242.56	4266.15	4310.77	4709.74	4457.44	eye open
1361	4357.44	3976.92	4262.05	4097.44	4323.59	4601.54	4082.05	4613.33	4205.64	4240.51	4262.56	4306.67	4709.74	4457.44	eye open
1362	4355.9	3973.85	4262.05	4095.38	4327.18	4598.97	4082.05	4612.82	4208.21	4244.1	4257.44	4303.59	4706.67	4452.82	eye open
1363	4354.87	3980	4266.67	4096.41	4329.23	4595.9	4083.08	4614.36	42[illegible]2.82	4248.21	4255.38	4304.62	4698.46	4446.15	eye open
1364	4355.9	3985.13	4275.38	4103.08	4326.67	4591.79	4084.1	4619.49	4222.56	4249.74	4257.95	4310.77	4700	4450.26	eye open
1365	4359.49	3983.08	4282.05	4106.67	4326.67	4589.74	4083.59	4617.44	4213.85	4251.79	4260.51	4313.85	4707.69	4456.41	eye open

high dimensional data. Transformation is of two types: features or attributes selection and features or attributes reduction.

The following equation shows a relationship between Eigen value and Eigen vector:

$$AV = \lambda V \tag{i}$$

Where A is a square matrix, V is a vector also known as Eigen vector and λ is a scalar termed as Eigen value. An Eigen value (λ) is a value or number which is a multiplied to an Eigen vector and produces the same result as any matrix get multiplied with any vector. Eigen values are helpful to identify the attributes of large data sets and reduce the dimensionality of datasets to achieve good accuracy. Eigen values are used to prioritize the features and identify those features that have highest priority to solve artificial intelligence problems and eliminate those features that have less priority and are not that much useful in solving a problem.

Eigen vector is the list of values (magnitude) and direction which can be plotted in a chart. Eigen vector is an important data structure of matrix which can be represented as array with n entries where n is the number of columns (rows) of a square matrix. When an Eigen vector is multiplied by a scalar (or applied liner transformation) its direction will not get affected; it will not change. Eigen vectors are used to give rank to each feature in a dataset.

Eigen values is a collection of scalar values which are applied on Eigen vector. It can change the shape (like stretch/compress) but can't change the direction of vector. The size of Eigen values are dependent on the size of actual matrix (A as shown in above equation i). After calculating the Eigen values we can calculate the Eigen vector using Gaussian Elimination.

As discussed above we will use Python for working in artificial intelligence models so we don't need to calculate Eigen values and Eigen vector by hand as we can use one of the module of NumPy library which is linalg.eig.

In Python we will import the above library as:

```
from NumPy import linalg as LA
```

This library will take square matrix as the input and it will return the Eigen values and Eigen vector.

```
Input values = np.array ([[1, 3], [2, 4]])

Eigen values, Eigen vector = LA. eig (input values)
```

10.2.1.3 Matrix Operations

The different matrix operations that are used to create a model in artificial intelligence are transpose, inversion, trace, determinant, rank. For more detailed study please refer to [2].

Transpose is used to create a new matrix by changing row to column and column to row. This is a flipping technique. The operation has no effect if a matrix is symmetric matrix; it will not change its dimensionality because the number of rows and number of columns are the same, but if a number of rows and number of columns are different then the dimensionality gets changed. Sometimes the datasets on which we are working need to change rows into columns and columns into rows; at that time transpose helps a lot. We can work on this using Python as discussed below.

In Python for transpose operation NumPy is used and array is used to display matrix

```
from NumPy import array
Data = array ([[2, 5, 3], [-1, 4, 6]])
Print (Data)
Transpose_Data = Data.T
Print (Transpose_Data)
```

Inversion is used to generate an identity matrix when a given matrix is multiplied by its inverted matrix. A matrix can be inverted in Python using the code

```
from NumPy import array
from NumPy.linalg import inversion
Data = array ([[2, 5, 3], [-1, 4, 6]])
Print (Data)
Inverted_Data = inversion (Data)
Print (Inverted_Data)
I = Data.dot (Inverted_Data)
Print (I)
```

10.2.1.4 Artificial Intelligence Algorithms That Use Linear Algebra

Many Artificial Intelligence beliefs are linked with linear Algebra. Supervised machine learning as well as unsupervised machine learning algorithms both use linear algebra [8]. Supervised learning deals with the tools to classify the problems using labelled data. Algorithms which use linear algebra are support vector machine (SVM), Decision tree, linear regression, and Logistic regression. Unsupervised learning deals with the problems having unlabeled data. Unsupervised learning algorithms that use linear algebra for dimensionality reduction is PCA. Principal component analysis (PCA) concept for exploratory data analysis uses Eigen values. Other areas are decomposition algorithms SVD, and clustering also uses linear algebra.

Artificial Intelligence mainly deals with high dimensionality data, which means data having many variables or attributes; this type of data is portrayed using matrices.

Linear algebra arithmetic is needed to process the data. Matrix Factorization is required for doing operations with higher order matrices and linear least squares is necessary in Artificial Intelligence models. Eigen values and Eigen vectors are used in noise reduction, noise generally introduced by human intervention while collecting the data. Eigen values also support to reduce the over fitting problem of machine learning.

10.2.2 Calculus

Artificial Intelligence uses calculus for almost every model. It helps in applying optimization technique in artificial intelligence by using differential and integral calculus. Optimization means to solve the problems of maximization and minimization with the help of derivative concept.

The subfield of artificial intelligence – machine learning and deep learning applications – generally deals with objective function, loss function and cost function. These functions express that the model performs well or not while applying data for fitting. With the help of calculus a scalar value is obtained which is used to optimize the parameters of the model and perform well with new samples of training set. But applying any random global minimum value or maximum value for optimizing each sample is a very cumbersome task and is very costly and time consuming, which is also why using iterative optimization technique makes the task easier. Calculus is used to predict the optimal values corresponding to the parameters of the cost function iteratively. One of the important mathematical concept of calculus is 'derivative' for minimizing the cost function which helps to get the minimum point along the curve. The term derivative refers to the slope of the function at any given point. We need to know the derivative for knowing the direction of coefficient values to achieve the lowest cost on the next iteration. To understand the mathematical principle behind the calculus please refer to the book [3].

10.2.2.1 *Objective Function*

The function is termed as objective function or criterion function where we want to either minimize or maximize it. When we focus on minimizing a function it is called as cost function or error function.

10.2.2.2 *Loss Function & Cost Function*

Loss function and cost function can be used interchangeably but there is a difference between them; loss function computes the error for single training sample and the cost function calculates the mean squared error (loss) between the predicted output and the target output for the entire training set. We can also say that it is the average of loss functions of the whole training dataset. With the help of loss function the model is predicted, which means if the loss is high then the model is not going to give a satisfactory result on the test dataset. Loss suggests how awful or satisfactory your model is. Cost function is displayed in Figure 10.2.

Cost function - "Half of MSE (Mean Squared Error)"

$$C(\theta_1,\theta_2)=\frac{1}{2M}\Sigma_{j=1}^{M}(H_\theta(Y^j)-T^j)^2$$

Objective function -

$$\text{Min } C\,(\theta_1, \theta_2)$$

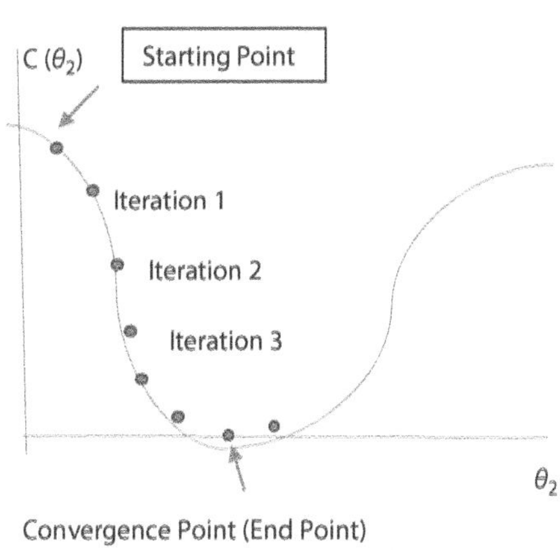

Figure 10.2 Cost function.

Derivatives -

$$\frac{d}{d\theta_1}C(\theta_1,\theta_2)=\frac{1}{M}\Sigma_{j=1}^{M}(H_\theta(Y^j)-T^j)^1$$

$$\frac{d}{d\theta_2}C(\theta_1,\theta_2)=\frac{1}{M}\Sigma_{j=1}^{M}(H_\theta(Y^j)-T^j)^1.X^j$$

θ_2 reaches to final point from the starting point after each iteration in search of a minimum value. The cost function is calculated for machine learning algorithms on the training dataset at each iteration. In the above equations for calculating the cost function, partial derivative and chain rule is applied.

10.2.2.2.1 Types of Loss Functions

Loss function is used to calculate the model error, which is helpful in optimization and decision theory. Different types of losses used in machine learning and deep learning are shown in Figure 10.3. Loss function or cost function is the difference between the actual output and expected output.

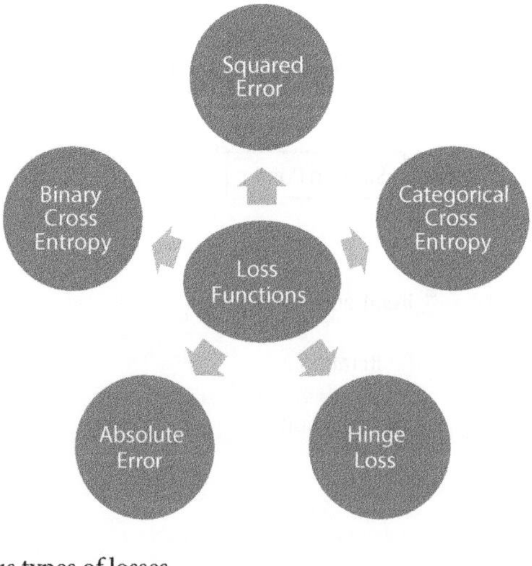

Figure 10.3 Various types of losses.

i. *Mean Squared Error*

Squared error loss is the square of the difference between the actual output and the predicted output. The mean of these squared loss is termed as MSE loss or L_2 error which has been discussed earlier.

ii. Mean Absolute Error

Absolute error is the difference between the actual output and the predicted output for each training set. For calculating the mean of this absolute error there is a need to sum up all the absoluter errors for each training set and divide the total sum by the total number of training sets.

Suppose number of training sets=n, MAE stands for mean absolute error

MAE = (Absolute Error 1+Absolute Error 2 +...................... Absolute Error (n-1)+Absolute Error n) / n

Absolute Error = Actual Output–Predicted Output

$$\textbf{MAE} = \sum_{i=1}^{n} |Y_i - Y_{pred(i)}| / \text{n}$$

$$\text{Actual Output} = Y_i \text{ Predicted Output} = Y_{Pred(i)}$$

iii. *Binary Cross Entropy*

This loss function is generally used to predict the performance of the classification models. This loss function is used for binary classification problems where there are only two classes which are labelled as either 0 or 1. In this error, probability values between 0 and 1 are given to the predicted output to know how the predicted probability value deviates from the actual output value. Binary cross entropy is also called sigmoid cross entropy.

Suppose number of training sets=n, Actual Output = Y_i Predicted Output = $Y_{Pred(i)}$

$$\text{Binary Cross Entropy Loss} = -1/\text{n} \sum_{i=1}^{n} Y_i \times \log Y_{pred(i)} + (1 - y_i) \times \log(1 - Y_{pred(i)})$$

iv. *Categorical Cross Entropy*

Categorical cross entropy is the generalization of binary cross entropy. This entropy is not confined only for two class labels but it is used for multiclass classification tasks. This is also termed as Softmax cross entropy. In this error the probability value can be 0 or 1, which means 1 for high probability of correct predicted output value and 0 for low probability of correct predicted output value.

Mathematical Formula:

Suppose number of training sets=n, Actual Output = Y_i Predicted Output = $Y_{Pred(i)}$

$$\textbf{Categorical Cross Entropy Loss} = -1/\text{n} \sum_{i=1}^{n} Y_i \times \log Y_{pred(i)}$$

v. *Hinge Loss*

Hinge loss is also called classification loss; it is basically used in support vector machine classifier for calculating the loss. Simply we can understand this loss as the score of the correct class must be greater than the score of all the incorrect class by some margin; that is why hinge loss is also known as maximum-margin classification. This loss is not at all differentiable; it uses convex function to work with convex optimizers. To understand the mathematical principle behind the calculus in depth please refer to the book [4].

10.2.2.3 Artificial Intelligence Algorithms That Use Calculus

Calculus is more useful in Machine Learning and Deep Learning. Optimization techniques, i.e., Gradient Descent, SGD (Stochastic gradient descent), Adam, Rmsprop, Ada Delta, use calculus and perform partial calculus differentiation and will equate the minimum 0 value, i.e., 0 slope value. Back propagation algorithm also uses the concept of calculus in predicting the error between calculated output and actual output and the error is propagated back to learn it and correct the calculated output.

10.2.3 Probability and Statistics

Probability and statistics are the main ingredients of the most evolving technology, Artificial Intelligence. This mathematical area helps in analyzing the vast amount of data that are generated continuously every day. There is a need to extract meaningful data with the help of statistical analysis. The collected and analyzed data can be visualized using statistical tools. For in-depth study of mathematical concept, probability please refer to book [5].

Statistics is the branch of mathematics that deals with all the important steps of model creation in artificial intelligence like data collection, data analysis, interpretation and visualization. The data are collected for the purpose of solving some real-life complex problems. The topic will be covered is shown in Figure 10.4.

Some examples of statistics used in exploratory data analysis are:

1. Latest example is the detection of people wearing a mask or not; here, by getting the image data of n number of people, the model will classify how many people are wearing a mask and how many are not wearing one.
2. Number plate detection model uses the data from traffic analysis.
3. The pharma companies working on making medicine for coronavirus, how would they conduct a test to confirm the medicine's effectiveness?
4. Sleep problem analysis by statistics – What is the sleep pattern followed by the average number of people? Do differences in age/gender affect the sleep pattern? Do the different

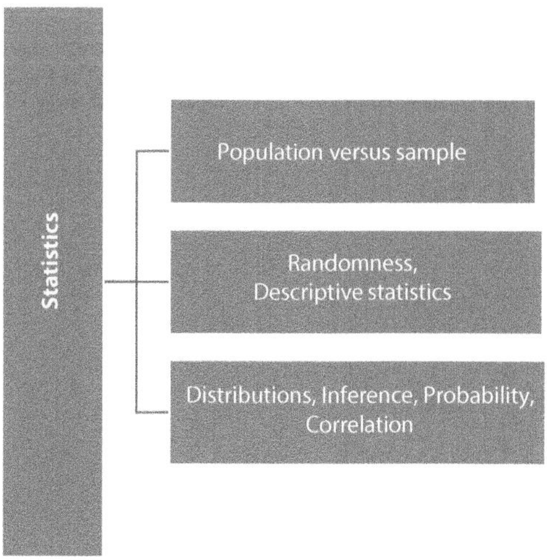

Figure 10.4 Statistical concepts.

lifestyles of people also influence the quality and quantity of sleep?
5. Business analysis by statistics – Profit gain in a year? How many products are there with defects? What is the trend of defect of each product? Analysis of employee satisfaction by changing the process of working?

10.2.3.1 Population Versus Sample

Population contains all the possible values of datasets; it means that for the cancer dataset it has all the possible values for each attribute.

A sample contains only a part or subset of values of a population. The size of a sample is always less than the population from where it has been taken. We can understand that if the size of a dataset (population) is 1000*24 it means 1,000 samples with 24 attributes, then the sample can be something like 400*24, it takes only 400 samples from the whole population. To understand statistics in detail please refer to the book [6].

10.2.3.2 Descriptive Statistics

This terminology focuses only on taking the summarized information from the whole dataset so that it can only showcase the important aspects of the

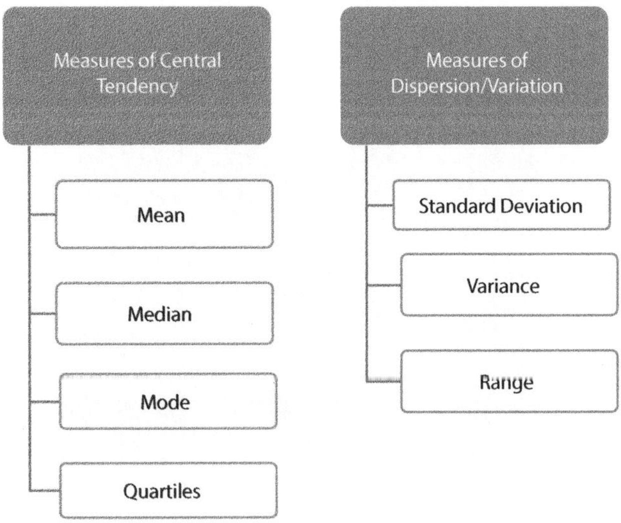

Figure 10.5 Descriptive statistics.

dataset. Two methodologies are used to do the summarization and highlight only the useful information from the dataset, and these are shown in Figure 10.5:

- Measures of central tendency
- Measures of dispersion/variation

10.2.3.3 *Distributions*

Distributions provide a way of observing all the values of a sample using visualizations, i.e., in tabular form or in histogram termed as frequency distribution. A distribution is described as a mathematical function for achieving the probability of happening of any particular event.

10.2.3.4 *Probability*

Probability is a foundation topic of artificial intelligence. Statistics is of no use without probability. Probability and statistics are an interconnected field; probability is a mathematical concept which helps in statistical analysis. In machine learning we are training our model on training dataset and fit the model on test datasets, which is not possible without probability. So the understanding of probability is very necessary in the artificial intelligence field for solving all these above issues and also for performance analysis. Actually the software developer is not much aware of probability while coding on deterministic programs, but for the data scientist or artificial intelligence engineer probability is a much-needed technique for solving predictive problems.

Probability is used in AI to express all kinds of uncertainty and the noise introduced in our model. Inverse probability (Bayes rule) deduce the unknown values in our dataset, modify the model with respect to the dataset, learn from data and make the prediction problems easier. Learning and prediction together act as an inference.

Probability is a statistical measurement tool for finding of the likelihood of any particular event or problem. Probability helps in proper prediction of happening of any event; if the value is 0 it means that the event is not going to happen, and if the value is 1 then the event certainly will happen. For better understanding of probability and statistics mathematically, please refer the book listed in reference [7].

10.2.3.5 Correlation

Correlation is a statistical measurement to measure the relationship in terms of association between events or dependence between them. It helps in analysis of the dataset; which attribute affects the other attributes; what is the relationship trend between the attributes. When two groups of data have a strong association then they have a "high correlation".

Bias is a statistical term used to measure the overfitting or underfitting the values of a population. With proper planning we can reduce the sources of bias, i.e., from where it can be introduced.

10.2.3.6 Data Visualization Using Statistics

Visualization is the representation of data graphically. Data visualization depends on the type of data we are working on. There are two categories of data on which data visualization is performed: Quantitative data or numerical data (e.g., height of a building, cutoff marks of any exam, age of a person) and Qualitative data or categorical data (grades allotted to students, i.e., A, B, C; gender of a person). Types of visualization can be understood by looking into Figure 10.6. Visualization plays an important role in displaying the relation, pattern and trends between the different features available in the dataset. Visualization works on finding out the complex relationship within the data and helps in easy understanding

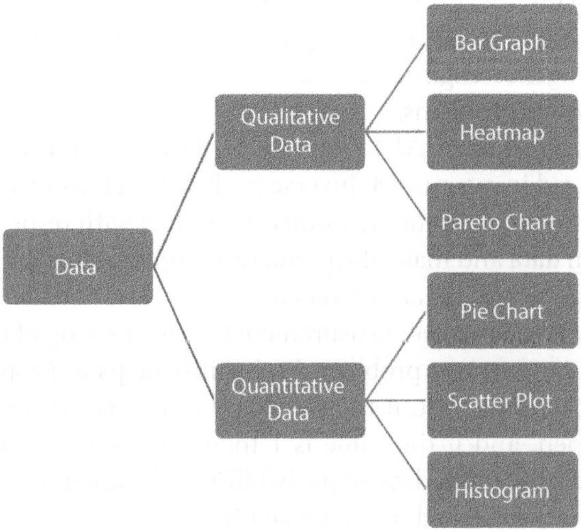

Figure 10.6 Data visualizations.

graphically. Data visualization can be performed using graphs, maps and charts. It contributes in analysis task of exploratory data analysis.

10.2.3.7 Artificial Intelligence Algorithms That Use Probability and Statistics

Naïve Bayes is a predictive classification model that uses conditional probability for the class label of a given dataset. Naïve Bayes is used for simplifying the calculations performed on datasets. Linear regression also uses probabilistic approach for minimizing the mean squared error of predictions [9]. Logistic regression is based on probability for minimizing the negative log likelihood for predicting the positive class. Hidden Markov model is another example of probability-based algorithms; it finds the probability of some hidden attributes based on some other observed values.

10.3 Work Flow of Artificial Intelligence & Application Areas

The subfield of artificial intelligence is machine learning and deep learning, which adopts the following steps in creating a model for prediction, analysis and forecasting types of problems. The basic work flow steps are shown in Figure 10.7.

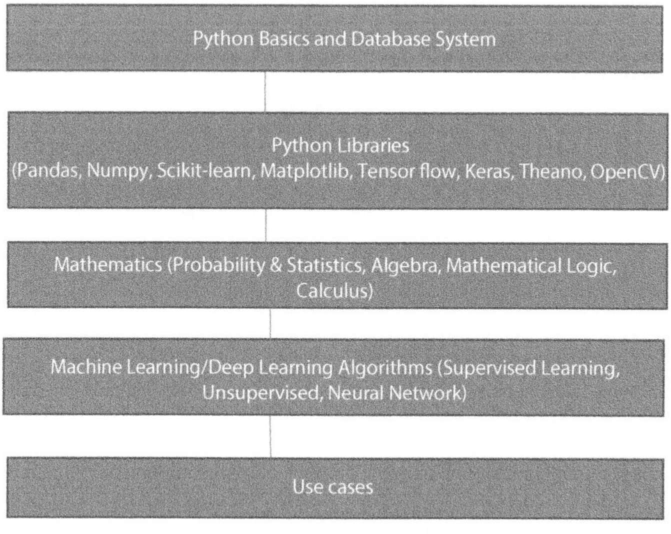

Figure 10.7 Work flow of building an artificial intelligence model.

Initially we need to select the language that we will use for building a machine learning or deep learning model. In this chapter we discuss the mathematical concepts using Python. In Python, for conducting all the operations like analysis, optimization, fitting, prediction, visualization there is a need to import inbuilt libraries, i.e., Pandas, NumPy, Scikit learn, Matplotlib, Tensor flow, Keras, Theano, Open CV which is shown in the above figure. Using these libraries mathematical concepts can be imported and helps in improvising the model. Later on different algorithms of artificial intelligence are applied in building an actual model.

For building any machine learning or deep learning model there are some steps to follow, like gathering the dataset of your problem. The dataset can be in any form, i.e.. numeric, text, images, as discussed earlier. Then the dataset is represented in such a way that it can be easily understood and prepared in matrix form to be able to feed to the model for performing classification task, regression task, etc. When the model is created on training set then it is fitted against the test dataset for prediction and forecasting purposes. Finally, the model is visualized using different visualization techniques.

The whole process can be seen diagrammatically in Figure 10.8 below.

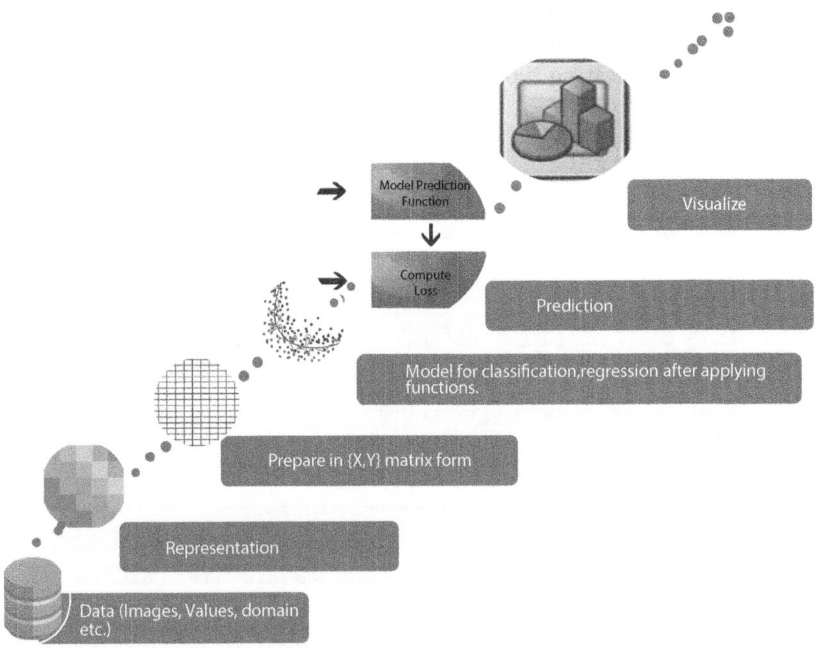

Figure 10.8 Steps of processing the dataset.

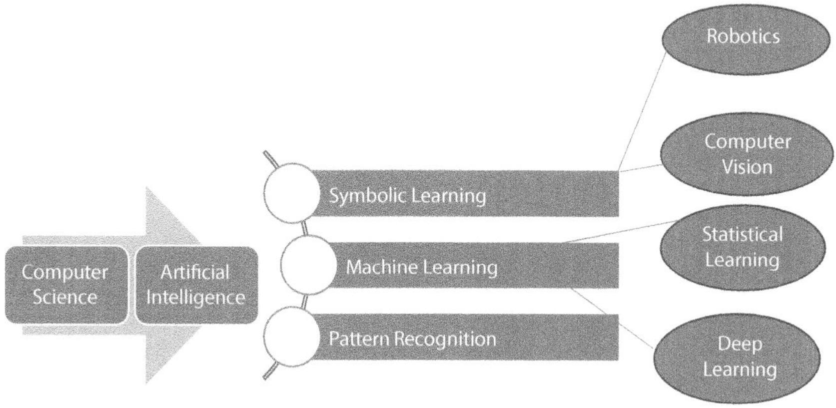

Figure 10.9 Application areas.

10.3.1 Application Areas

There are lots of application areas where artificial intelligence is used. The following diagram shows the different fields where artificial intelligence is used. Artificial intelligence is the subfield of computer science which encompasses symbolic learning, machine learning and pattern recognition which helps to perform intelligently in robotics, computer vision, statistical learning and deep learning areas. How AI can be characterized in respect of application areas is shown in Figure 10.9.

10.3.2 Trending Areas

Nowadays the trending areas where artificial intelligence is used is of great concern. Predictive analysis and forecasting artificial intelligence play an important role in prediction of diseases, prediction of stock market and forecasting of estimates, for forecasting of rainfall using past trends of data, etc. Content recommendation is another emerging field where AI is of great importance. With the help of content recommendation people can purchase those products that may be on their interest list. Voice search is also one of the trending areas nowadays; voice facility is also used in chatbots. All the trending areas are shown in Figure 10.10 below.

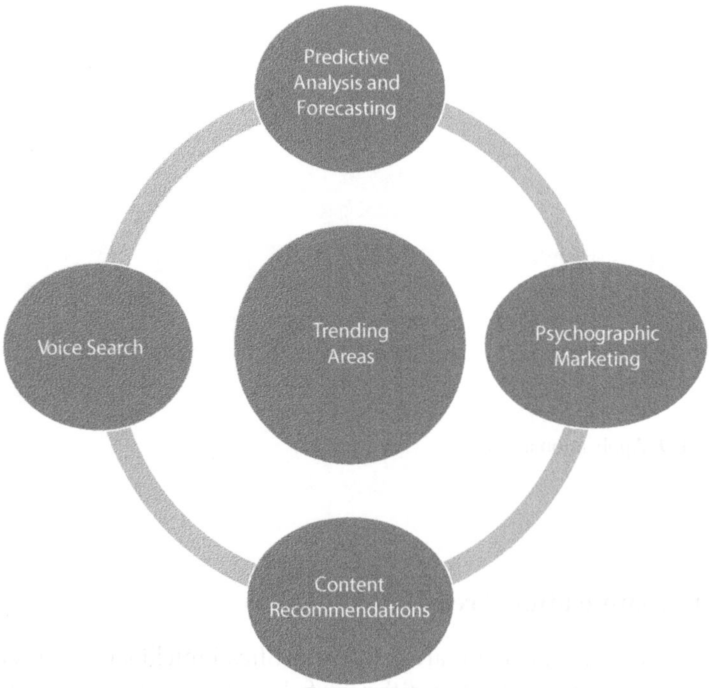

Figure 10.10 Trending areas.

10.4 Conclusion

Artificial Intelligence has acquired great significance over the past few decades in terms of implementation and integration of AI in our day-to-day lives. The progress that AI has made is surprising with the introduction of self-driving cars, disease diagnostics, and also in the playing of strategy games like chess.

The future for AI is remarkably encouraging. The day when we can have our own automated robots is not far off. This has pushed developers and scientists to start writing codes and begin the development of AI and ML programs. But writing codes is not that easy as it may require to figure out how to compose calculations for AI and ML and also requires broad programming and numerical information. The integration of mathematics and artificial intelligence builds the foundation for programming and in developing these real-life models. Each of the mathematical concepts has its own relevance in developing the artificial intelligence models like Bayesian algorithm and hidden Markov model that use probability, and exploratory data analysis task is made easier by linear algebra while the optimization and feature reduction is by calculus.

References

1. Carl D. Meyer, *Matrix Analysis and Applied Linear Algebra*, SIAM, 2010.
2. David C. Lay, Steven R. Lay, and Judi J. McDonald, *Linear Algebra and its Applications*. Pearson, 5th ed., 2014.
3. Silvanus P. Thompson, *Calculus Made Easy*. 3rd ed., St. Martin's Press, 1970.
4. Michael Spivak, *Calculus*. 4th ed., 2008.
5. Walter Rudin, *Principles of Mathematical Analysis*, 3rd ed., McGraw-Hill, 1976.
6. Neil A. Weiss *Introductory Statistics*. 9th ed., Pearson, 2011.
7. Vijay K. Rohatgi, A. K. Md. Ehsanes Saleh, *An Introduction to Probability and Statistics*. Wiley, 2015.
8. Angel Garrido, "Mathematics and Artificial Intelligence, two branches of the same tree" Elsevier *Procedia Social and Behavioral Sciences* 2010, 1133-1136.
9. Kotaro Ohori, Hirokazu Anai, *Mathematical Technologies and Artificial Intelligence Toward Human-Centric Innovation*. Springer, 2018.
10. Greenberg, H. J. A. prospective on mathematics and artificial intelligence: Problem solving=Modeling+Theorem proving. *Annals of Mathematics and Artificial Intelligence* 28, 17–20 (2000).

11

Study of Corona Epidemic: Predictive Mathematical Model

K. Sruthila Gopala Krishnan[1]*, Ramakant Bhardwaj[1], Amit Kumar Mishra[2] and Rakesh Mohan Shrraf[2]

[1]Department of Mathematics, Amity University, Kolkata, West Bengal, India
[2]Department of Computer Science and Engineering, Sagar Institute of Science and Technology, Bhopal, Madhya Pradesh, India
[3]Department of IT, Sallalah College of Technology, Thumrait Rd. Salalah, Dhofar, Sultanate of Oman, Oman

Abstract
The outbreak of SARS-CoV-2 (Covid-19) is one of the most unprecedented and devastating events that the world has witnessed so far. It was manifested in Wuhan, China in December 2019 and has spread worldwide. The rapidity at which Covid-19 is transmitted has become one of the major concerns regarding the safety of mankind. The similarity of symptoms between Covid-19 and normal flu, like cough, body ache and headache, makes it difficult to ascertain a case to be of normal flu or of Covid. Consequently, many Covid cases are unreported which further increases the risk of spread of infection. In the present chapter, by using three mathematical models, we aim to give an outline of the spread of Covid-19 in West Bengal and how lockdown has helped to reduce the number of Covid cases. The first model is an exponential model; the second model is based on Geometric Progression which shows spread of coronavirus using a tree chart. The third model, named as Model for Stay at Home, shows that due to lockdown, the number of cases is gradually attaining a constant level instead of growing exponentially; thus urging each citizen to stay at home during lockdown unless an unavoidable situation arises.

***Keywords*:** Mathematical modelling, predictive model, differential equations

**Corresponding author*: k.sruthila112@gmail.com

Ramakant Bhardwaj, Jyoti Mishra, Satyendra Narayan and Gopalakrishnan Suseendran (eds.) Mathematics in Computational Science and Engineering, (233–256) © 2022 Scrivener Publishing LLC

11.1 Mathematical Modelling

In the real world, it is believed that for each problem there exists a solution. Mathematical Modelling is one of the techniques used to solve real-world problems. Scientifically speaking, the term 'model' refers to representing something. Mathematical Modelling is a way of representing the real-world problems under some conditions in the language of mathematics with the help of mathematical concepts and scientific, social and economic laws [2]. Mathematical modelling involves:

- Understanding the problem at hand
- Identifying the essential and non-essential characteristics of the concerned problem
- Taking into consideration the essential characteristics of the problem and ignoring the non-essential ones
- Converting the problem into a mathematical one by the application of concepts of mathematics and relevant social, economic or scientific laws.

This leads to the construction of a structure like mathematical equations, called mathematical model, whose solution is then interpreted, validated and verified against the observations of the actual problem. If these observations and predictions of the model are not comparable, then the problem is further scrutinized in order to either improve the model or to develop a more suitable one.

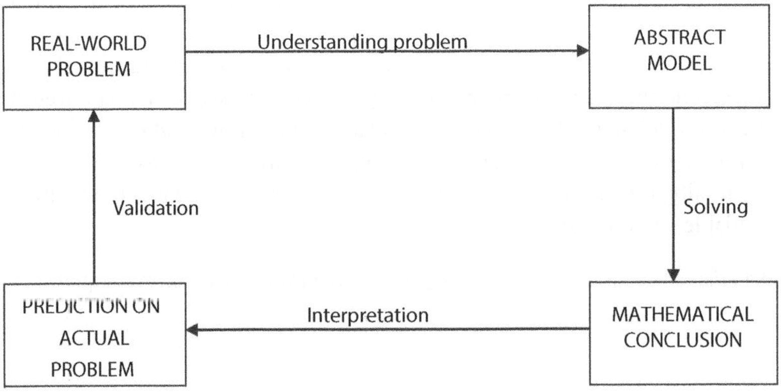

Process of Mathematical Modelling

11.2 Need of Mathematical Modelling

A collection of entities present in the real world is called a system. A system can be considered to be comprised of two aspects – one is the theoretical and mathematical aspect and the other is the experimental and observational aspect. When a system cannot be analyzed by physical, chemical, and linguistic models, then mathematical models come into play.

As mathematics is a concise and precise language, a simplified abstract model is obtained such as mathematical relations which can be of the form of algebraic, transcendental, differential, integral, difference, integro-differential, differential-difference or even inequations. The equations of the model can be solved analytically, numerically or by simulation.

Through mathematical modelling the physical behaviour of a system can be understood very clearly. It also proves to be very economical for measurement. By considering only the relevant features of the system and ignoring the irrelevant ones, parameters involved in the model can be controlled. Also, based on the past or present observations, one can predict about the future behaviour of the system which hasn't been observed so far.

For example, if we want to know the population of India after 100 years (say) assuming the death and birth rates to be dependent on the total size of the population, a mathematical model expressing the population as a function of time can be constructed. Then with the help of past or present population data, prediction about the future can be made without actually waiting for the future!

With the help of Mathematical modelling, direct and cumbersome methods to find the solutions to real-world problems can be avoided. The yield of rice in India from the standing crop can be estimated by first calculating the area under the cultivation of rice and then computing the yield per acre by cutting and weighing the crop from selected plots, instead of cutting and weighing the whole crop at once!

Let us consider another example. Suppose a person standing at one side of a river wants to find the breadth of the river without actually crossing it. Now the question arises, how to measure the breadth?

Let the person be standing at a position **P**.

Let **MN** be the required breadth of the river. Let **Q** be the point on the bank of the river on the side where **N** is situated.

Let **O** be the midpoint of the line **NQ** such that the points **M, O** and **P** lie on the same straight line.

So, **NO=OQ=s** (say)

Now we consider triangles MON and POQ

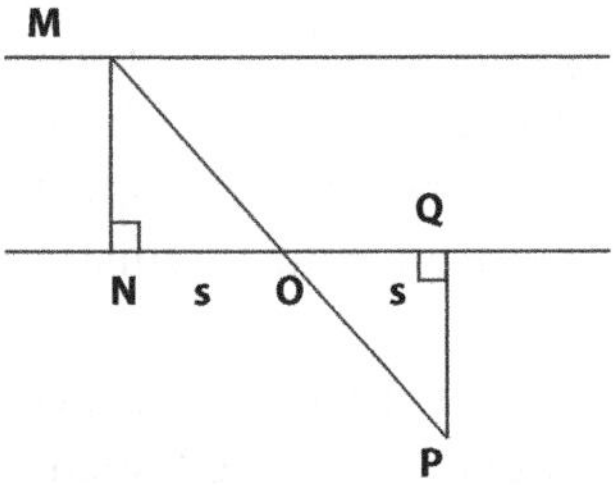

$$\angle MON = \angle POQ$$

(Since, vertically opposite angles are equal)

$$NO = OQ$$

(Since, O is the midpoint of line NQ)

$$\angle MNO = \angle PQO(\text{Right-angles})$$

$\therefore \Delta MON \cong \Delta POQ$ *{By Angle Side Angle(ASA)condition of congruency}*

So, $MN = PQ$(∵*Corresponding parts of congruent triangles are equal*)

Thus, the width (breadth) of the river can be evaluated just by measuring the length of **PQ**.

Hence, by looking upon these examples, it can be understood that how real-world problems can be solved in simple ways with the help of mathematical models.

11.3 Methods of Construction of Mathematical Models

11.3.1 Mathematical Modelling with the Help of Geometry

Mathematical modelling has been in utility even at the time of determining the path of the planets. Earlier it was believed that all the planets revolve round the earth. So, simple curves like circles, ellipses were not considered. Then it was found possible to describe the paths as a combination of some epicycloidal curves. An epicycloid is a curve which is traced by a point on the circumference of a circle rolling on the exterior of another fixed circle. This combination of epicycloidal curves or epicycloids is the **geocentric model** for planetary motion.

But it was observed by Nicolaus Copernicus that the sun, not the earth, is the centre of the universe. On the basis of this observation, Kepler proved that each planet moves in an elliptical path, with the sun at one of its foci. This is the **heliocentric model** of planetary motion.

Though both geocentric and heliocentric models are correct, the latter one is simpler than the former one. Later it was observed by Newton that the universal law of gravitation holds true in elliptical orbit. In 1957, elliptical orbits were considered as orbits of satellites, thus the model became a model of understanding and optimization.

11.3.2 Mathematical Modelling with the Help of Algebra

Through mathematical modelling the revolution of the planets round the sun can be studied. The planets revolve round the sun in an elliptical orbit with the sun at one focus. The ellipticities, the alternative measure of non-circularity, of the orbits are very small. So at first approximation, the orbits are considered as circles with the sun at the centre.

Let m_s and m_p denote the mass of the sun and the planet, respectively.

Let F_c denote the force acting on the planet due to the sun.

This force helps the planet to move in the circle with a velocity

say V, and is called **centripetal force.**

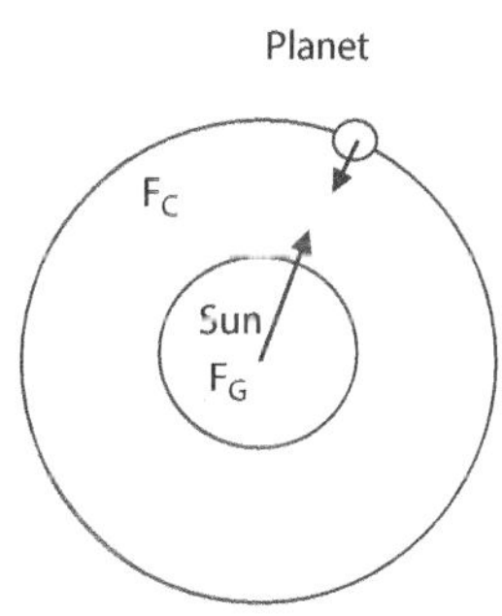

$$F_c = \frac{m_p V^2}{R}$$

where R is the distance of the planet from the centre of the sun.

Let F_G denote the force acting at the centre of the sun due to the planet. This force is called gravitational force.

$$F_G = \frac{G\, m_S m_p}{R^2}$$

where G is the gravitational force.

Both F_c and F_G are equal in magnitude but opposite in direction.

$$F_c = F_G$$

$$\frac{m_p V^2}{R} = \frac{G m_s m_p}{R^2}$$

$$\therefore V^2 = \frac{G m_s}{R} \qquad \text{-(i)}$$

If we denote the time period by T, then we get

$$V = \frac{distance}{time} = \frac{2 \sqcap R}{T}$$

Substituting the value of V in equation (*i*), we have

$$\left(\frac{2 \sqcap R}{T}\right)^2 = \frac{G m_s}{R}$$

$$\therefore T^2 = \frac{4 \sqcap^2 R^3}{G m_s}$$

If T' and T" denote the time period of two planets with orbital radii R_1 and R_2 respectively, then

$$\frac{T'^2}{T''^2} = \frac{4 \sqcap^2 R_1^3}{G m_s} \times \frac{G m_s}{4 \sqcap^2 R_2^3}$$

$$\therefore \frac{T'^2}{T''^2} = \frac{R_1^3}{R_2^3}$$

This implies that the squares of periodic time of two planets are proportional to the cubes of the radii of their orbits.

11.3.3 Mathematical Modelling Using Trignometry

By using mathematical models constructed with help of trigonometry, the distance of the moon from the earth, distance of a star from the earth, duration of length of a day and latitudes at which the sunset does not occur and many such real-world problems can be solved.

11.3.4 Mathematical Modelling with the Help of Ordinary Differential Equation (ODE)

There are many applications of mathematical modelling through ordinary differential equations such as demography, pharmacology and many other fields [1].

When a drug is administered to a person, it enters the bloodstream and is absorbed by the body over time. Through medical research, it has been proved that the greater the presence of the drug in the bloodstream, the greater is the rapidity at which it is absorbed by the body.

In order to translate this problem into a mathematical model, suppose that x=x (t) denote the amount of the drug present in the bloodstream at time t. So x is a function of t. At each time t, the rate of change in x with respect to time t is dx/dt which is proportional to the amount x in the bloodstream.

$$\therefore \frac{dx}{dt} = -px$$

where p is a positive constant of proportionality which depends on the drug and can be determined through experimentation.

The negative sign is considered as x decreases with time. If the initial dosage is known say $x=x_0$ at time t=0, then it is possible to obtain a general formula for x (t) by solving the initial value problem

$$\frac{dx}{dt} = -px \quad , \quad x(0) = x_0$$

11.3.5 Mathematical Modelling Using Partial Differential Equation (PDE)

Mathematical models can be constructed with the help of partial differential equation to solve a system dependent on more than one independent factor.

Let d(x, t) and v(x, t) denote the traffic density which is the number of vehicles, say, cars, per unit length of the highway, and the velocity of the vehicle, say, car on a highway at a distance x from the origin at time t.

If no cars are allowed to enter and leave the highway, using continuum model, we obtain continuity equation for traffic flow on a highway which is

$$\frac{\partial d}{\partial t}+\frac{\partial}{\partial x}(dv)=0$$

Since there are two dependent variables and one equation showing relation between them, if one more relation is taken into consideration both d(x, t) and v(x, t).

11.3.6 Mathematical Modelling Using Difference Equation

With the aid of difference equations, modelling change and modelling approximations can be done [3].

Suppose there are 500 students in a hostel out of which one or more students are in critical condition due to contagious disease flu (say).

Let s_n denote the number of infected students after some time period n. It is assumed that some interaction between infected and non-infected students is required to spread the disease.

If all the students are vulnerable to the disease, $(500\text{-}s_n)$ represents those who are vulnerable to the disease but not yet infected.

We can model the change of those infected as

$$\Delta s_n = s_{n+1} - s_n = p s_n (500 - s_n)$$

In the above model, s_{n+1} denotes the number of infected students after time period (n+1); the product s_n $(500\text{-}s_n)$ implies the number of possible interactions between the infected and non-infected students. A small factor p of these interactions would result in increment in the number of infected students represented by Δs_n.

So, the spread of congruence diseases can be studied through mathematical modelling using difference equations. It is highly recommended herein to study mathematical modelling using the reference books listed below.

1. Fox, William P., Frank R. Giordano, and Stephen B. Horton (2014), "A First Course in Mathematical Modeling", Fifth Edition, Cengage Learning, pp. 10-14.
2. Kapur J. N, *Mathematical Modelling*, Second Edition (New Age International Publication).
3. Michael D Alder, An Introduction to Mathematical Modelling, Heaven for book.com (2001).

11.4 Comparative Study of Mathematical Model in the Time of Covid-19 - A Review

11.4.1 Review

It is well known that the novel coronavirus (nCoV) was found in Wuhan, China, in December 2019 and has been transmitted to many countries across the world. In the present paper, we have reviewed different studies and research work through mathematical modelling for Covid-19.

Epidemics, a threat to the very existence of human lives, have been a concern of great importance for mankind, as we were reminded by Mimmo Lannelli [5] in his research work titled 'The Modelling of Epidemics' published in UNITEXT in January 2005. Lannelli referred to the 'Black Death,' which was the plague that caused the death of almost 25 million people across Europe in the 14th century, the plague in Bombay that occurred in 1905, measles in Trento, Italy, from 1977 to 1978 and in Trentino (Italy) from 1949 to 1999 and Influenza in Basilicata (Italy) from 2003 to 2005.

In order to describe any infectious disease, the following are the primitive or basic variables used:

- Number of susceptible at time t (say) denoted by S(t) (say), which implies number of healthy people but prone to infectious disease.
- Number of infective at time t denoted by I(t) (say), which implies number of people infected due to a contagious disease and are capable of spreading the disease to susceptible if came in contact with them.

- Number of immunes at time t denoted by R(t) (say), which represents the number of people immune to the disease through innate immune system after getting infected who are not capable of transmitting the infection to the susceptible through any interaction between them.

Lannelli has also explained that if the infected person after recovering is again vulnerable to that disease, that is, no lifelong immunity is imparted to the once infected person, then the growth and effect of disease cannot be studied using SIR model; instead SIS model should be used.

In SIR model, the susceptible once get infected and then recovered become immune to that particular disease for their lifetime. But in SIS model, once the infected recovers from the particular disease, again becomes susceptible to that disease getting no immunization against that disease for life-long.

Then explaining the impact of the outbreak of a single epidemic and also endemic states of the disease, it has been shown that reproductive number is an important parameter determining the disease progression. Reproductive number denotes the number of susceptible who can be infected due to contact with one single infected person.

For epidemicity of a disease and endemicity of a disease in SIR model

$$\textit{Reproductive number} = \frac{(\textit{Number of interactions per unit time}) \times (\textit{Effect of infection per interaction with one infected person})}{(\textit{Rate at which infectives become immune})}$$

For endemicity of a disease in SIS model,

$$\textit{Reproductive number} = \frac{(\textit{Number of interactions per unit time}) \times (\textit{Effect of infection per interaction with one infected person})}{(\textit{Rate at which infectives become immune}) + (\text{Rate of fertility or mortality})}$$

Here, it is assumed that rate of birth=rate of death

Furthermore, it has been observed that if

Reproductive number<1 → No epidemic outbreak or increase in intensity of epidemic

Reproductive number=1 → Stable equilibrium of the disease, won't result in outbreak
Reproductive number>1 → Outbreak of epidemic disease

Chayu Yang and Jin Wang [8] through their research article titled 'A Mathematical model for the novel coronavirus epidemic in Wuhan, China,' published in the *Journal of Mathematical Biosciences and Engineering* on 11th March 2020, have proposed a mathematical model to investigate the course of the coronavirus progression in such a way that the model takes into consideration two important aspects responsible for the spread of the disease. First is the contribution of the environment, as environment is considered to act as a reservoir of the coronavirus, thus resulting in transmission of the disease. Secondly, the variable rates of transmission which are subjected to changes due to the change in the status of the epidemic and conditions prevalent in surroundings due to different levels of precautions implemented to control the disease progression.

The total human population has been divided into four compartments:

- Susceptibles who are healthy people but are prone to the coronavirus.
- Exposed who do not show any symptoms of the disease but are capable of transmission of the disease to others.
- Infectives who show fully developed symptoms of the disease and are potentially capable of infecting other people.
- Recovered people who are those infectives who have recovered from the disease and are incapable of spreading the disease.

The basis reproductive number is considered to have three parts which represent three different routes of transmission: first, from exposed-to-susceptible; second, from exposed-to-infective; third, from environmental store of coronavirus-to-susceptible.

It has been observed that the reproductive number representing the first route is maximum due to inability to understand the exposed ones as they do not show any symptoms of the disease, resulting in higher chances of interaction between exposed and susceptible people. The reproductive number representing the second route is the least, indicating that the healthcare precautions help to curb the disease. The reproductive number corresponding to the third route of transmission is also much greater than 1, which shows significant contribution of the environment reserve in this epidemiological state.

Thus the strategies for curbing the coronavirus transmission should consider the above-mentioned three possible transmission routes.

Yichi Li, Bowen Wang, Ruiyang Peng, Chen Zhou, Yonglong Zhan, Zhuoxum Liu, Xia Jiang and Bin Zhao [6] through their research article titled 'Mathematical Modeling and Epidemic Prediction of COVID-19 and Its Significance to Epidemic Prevention and Control Measures' published in *Annals of Infectious Disease and Epidemiology* on 11th March, 2020, have used SEIQDR-Model which is good enough to completely describe the preventive measures taken by the government such as inspection detention, treatment in isolation and isolation of cities. Here, the total human population are divided into six categories: class of Susceptibles (S), class of exposed (E) but not showing any symptoms of the infection, class of infectives (I) who are highly infectious but they are not under quarantine, class of people who are diagnosed and are quarantined (Q), potential victims (D) and class of people recovered from the infection (R). The results obtained solving this model shows the effectiveness of preventive measures such as implementation of lockdown and social distancing between people, which have helped to restrain the propagation of coronavirus to a great extent.

Sayan Nag [7] in his research paper titled 'A Mathematical Model in the Time of COVID-19' pre-printed on 15th March, 2020, has explained how the spread of Covid-19 is determined by factors such as rate of immigration (including both healthy people and infected people), death rate and cure rate, by modifying and using the classic Lotka-Volterra model for simulating predator-prey dynamics as healthy-infected dynamics model.

By simulating the results, six cases were considered which explains the following (assuming that number of healthy people = number of infected people):

More the rate of protection, lesser will be the rate at which people get infected and hence more rapidly the infected population will become zero.

The spread of the infected can further be controlled by increasing the rate of cure of infected individuals.

If immigration is not allowed, then the transmission of coronavirus can be greatly curbed. But even if there is a slight increase in immigration rate (especially of infected people) then there occurs instability in the process of controlling the pandemic outbreak of Covid-19.

Zafer Cakir and Hasan Basri Savas [3] through their research titled 'A Mathematical Modelling Approach in the Spread of the Novel 2019 Coronavirus SARS-CoV-2 (COVID-19) Pandemic' accepted in *Electronic*

Journal of General Medicine on 27th March, 2020, have investigated how the pandemic coronavirus has taken its course by considering that the rate of spread of the disease is directly proportional to the interactions between the infected people and the people who are prone to get infected with the disease.

The product term consisting of number of infected and number of susceptibles indicates the possible interactions between these two classes of infectives and susceptibles.

So,

Rate of spread of disease = (a parameter) x (number of infectives) x (number of susceptibles)

Here, a parameter is taken to consider all other possible factors of spreading the disease. Taking the parameter as a function of time and considering two conditions at $t=0$ and $t=t_1$, an initial value problem is obtained.

The results obtained show the change in behaviour of the course of growth of the disease for different levels of precautions in a 30-day future time interval.

As the spread of coronavirus represents an exponential growth, so in order to curb this disease, every single precaution ranging from personal hygiene to social isolation plays a vital role.

B. Ivorra, M.R. Ferrandez, M. Vela Perez, A.M. Ramos [4] in their research article titled 'Mathematical modelling of the spread of the coronavirus disease 2019 (COVID-19) taking into account the undetected infections. The Case of China' published on 3rd April, 2020, have used θ-SEIHRD model instead using of SIR and SEIR or other models for general purpose. The total human population is divided into the following compartments:

- Susceptible - Healthy individuals but vulnerable to coronavirus
- Exposed - People who got infected with coronavirus but show no symptoms and can infect other people; in other words they are in the incubation period, after which they are considered in the Infectious compartment.
- Infectious - Individuals who have undergone incubation period, show clinical symptoms and are capable of infecting other persons.
- Infectious but are not detected by the authorities.

- Hospitalized patients or quarantined at home and are detected by the authorities. But they can still spread disease to other nearby people.
- The hospitalized people who will die.
- The dead Covid-19 patients.
- Recovered after being previously recorded as infectious people. They become naturally immune to the coronavirus.
- Recovered from the disease but were not earlier detected as infectious. They gain natural immunity against the virus and are not further capable of infecting other people.

In this θ-SEIHRD model, θ represents the fraction or ratio between detected and reported cases over real total infected cases. So, undetected cases are very much responsible for the propagation of Covid-19.

Jyoti Bhola, Vandana Revathi Venkateshwaran and Monika Kaul [1] in their article titled 'Corona Epidemic in Indian context: Predictive Mathematical Modelling' preprinted in *medRxiv* on 7th April, 2020, have explained the scenario of coronavirus transmission by the following basic differential equation system using SIR-Model. Here S denotes susceptibles, I denotes class of infectives and R denotes class of recovered (including the recovered people, people who died due to the disease or are naturally immune to the coronavirus).

By solving the mathematical model, the results obtained indicate that the more is the interaction between the susceptible and infectives, the more is the chance of spreading the disease.

11.4.2 Case Study

Rebecca E. Morrison and Americo Cunha [2] through their research article titled 'Embedded model discrepancy: A case study of Zika modelling' published on 4th May, 2020, has thrown light upon the fact that a mathematical model is a simplified form of the real problem and sometimes mathematical models fail to give the solutions consistent with the actual data.

To make these models more reliable, either of the two options can be chosen:

- to modify the model in direct way by including additional information, or
- to be accountable and represent the discrepancy of the model

Though first option is preferable but it may not be possible due to lack of knowledge to improve the model. So, second option has to be considered.

So to make the model consistent with the available data related to pandemic Covid-19 the model equations once used for study of Zika epidemic occurred in Brazil in 2016, are modified. For Zika Disease Modelling, SEIR-SEI model is used, which explains coupled rates of growth of susceptibles, exposed, infectives and recovered human beings as well as susceptibles, exposed and infected vectors.

For modelling of Covid-19, a discrepancy operator has been embedded in the system of differential equations of SEIR-SEI model, thus obtaining an enriched model.

The results show that solutions of the enriched model and the available data almost coincide with each other, thus enabling the model to be more reliable.

Also, the effects of under-reporting can be observed from the modified model which thus describes the whole dynamical behaviour of the outbreak of coronavirus even in highly under-reported situations. Thus embedded discrepancy operator is generally applicable to most epidemiological diseases and improves the agreement between the output and available data.

11.5 Corona Epidemic in the Context of West Bengal: Predictive Mathematical Model

11.5.1 Overview

The outbreak of SARS-CoV-2 (Covid-19) is one of the most unprecedented and devastating events that the world has witnessed so far. It was manifested in Wuhan, the capital city of Hubei Province, China, in December 2019 and has spread to different countries worldwide. The rapidity at which Covid-19 is transmitted has become one of the major concerns regarding the safety of humans. The diversity in habits, awareness, level of acceptance of the guidelines and many other socio-economic factors varies over districts of a state. Thus the number of infected people and number of recovered people vary from district to district. In addition, as the number of Covid cases is changing daily, it is very difficult to predict the exact course followed by the pandemic. Also, as the symptoms of Covid-19 is similar to that of normal flu, like cough, body ache, dry cough and headache, it is difficult to ascertain whether a person has normal flu or is covid-affected. As a result many Covid cases are not reported, which increases the risk of spread of infection at a larger scale. To tackle this unpredicted situation, research works are being carried out at various interdisciplinary levels. Also the higher authorities are implementing many measures for the safety of the citizens. One such example is implementation of lockdown. In the present chapter, with the help of mathematical modelling, we aim to give an outline of how Covid-19 is spreading in West Bengal and also show how lockdown has helped to reduce the number of Covid-infected people. Mathematical modelling plays a vital role to solve the real-world problem into a mathematical one whose solution is then interpreted, validated and verified against the observations of the actual problem. In the current paper, three mathematical models have been used. One is the exponential model which shows that the pandemic would follow an exponential course without the implementation of lockdown. One is a model based on Geometric Progression (G.P.). The third one is constructed using first-order ordinary differential equations. The model based on G.P. shows how coronavirus is spread using a tree chart, assuming that an infected person is capable of infecting two persons who come in contact with him or her. It can also be observed that there is a significant difference between the number of Covid patients with and without the implementation of lockdown. The third model, which is named as Model for Stay at Home, shows that had there been no lockdown, the number of Covid cases would have been increasing exponentially. But due to lockdown, the number of cases is gradually attaining a constant level; in other words the number of cases is approaching a constant value, instead of showing an exponential growth. Thus each citizen should stay at home during lockdown unless any unavoidable situation arises.

11.5.2 Case Study

Earlier, viral diseases were not considered seriously as these diseases could be overcome by the natural immunity system in humans, except for those who have an auto-immune disorder or a compromised immune system. As a result, not much attention was paid by researchers and funding agencies to the field. But the outbreak of the pandemic Covid-19 has made people realize the huge impact that viral disease outbreaks can have on all sectors of the society.

SARS-CoV-2 which is generally known as novel coronavirus is a single, positive stranded RNA. It belongs to the family Coronaviridae of order Nidovirales. Coronaviridae also includes viruses SARS-CoV responsible for SARS (Severe Acute Respiratory Syndrome) which emerged in 2002 and MERS-CoV, which is responsible for MERS (Middle East Respiratory Syndrome) which emerged in 2012.

The rapidity with which Covid-19 has been spread worldwide is so tremendous that the World Health Organization (WHO) declared it as public health emergency of international concern (PHEIC) on 30th of January, 2020, and a pandemic on 11th of March, 2020.

The novel coronavirus was introduced in West Bengal with the arrival of people from different Covid-affected countries. The first Covid-19 case reported in West Bengal was on 17th of March, 2020, in Kolkata. The state government has left no stone unturned to control the situation. The state government has issued many guidelines for the safety of the people which include social distancing, which has proved to be an effective measure to curb the disease. In this pandemic outbreak of Covid-19, researches are carried out at various interdisciplinary levels. Doctors, nurses and other medical and paramedical staffs are risking their own lives to cure the covid-infected people. Researchers in West Bengal have developed Covid-19 testing kits which got approval from Indian Council of Medical Research (ICMR). Scientists are working very hard to develop vaccines to eradicate the disease. Mathematicians have developed various mathematical models to study the spread of coronavirus and predict the future course of this pandemic.

But unfortunately, the super cyclone Amphan, which arrived on 20th of May, 2020, in Bengal, marred the efforts for effective containment of the virus to a great extent. Many people had to be shifted to shelter homes where social distancing was a matter of luxury. This has led to an increase in the number of Covid cases. Also, inter-state movement through special trains and domestic flights has caused an increase in the risk of spread of coronavirus.

As education, awareness and understanding of the people regarding the virus is different for different districts, the number of infectives (people infected with Covid) and the number recovered ones will also vary over districts. In addition, as the number of Covid cases is changing daily for different districts, it is very difficult to predict the exact course followed by the pandemic. By knowing the course of this pandemic, some effective measures can be implemented to control the spread of coronavirus.

Also through the models, it has been observed that implementation of lockdown proves to be an effective measure to curb the spread of coronavirus. There is a significant difference in the number of Covid cases with and without the implementation of lockdown.

11.5.3 Methodology

Mathematical modelling plays a vital role in the study of the transmission of pathogens causing an epidemic. The following shows how the spread of the coronavirus can be studied through modelling.

11.5.3.1 Exponential Model

Let P denotes the number of Covid patients at time t.

Then, change in the number of patients per day is $\left|\frac{dP}{dt}\right|$

Since the number of patients is increasing day-by-day,

$$\left|\frac{dP}{dt}\right| = \frac{dP}{dt}$$

As the coronavirus spreads through social contact, so new cases that will be reported in the upcoming days will directly depend on the number of patients already existing.

So,

$$\frac{dP}{dt} \propto P$$

$$\Rightarrow \frac{dP}{dt} = cP$$

Here **c** is proportionality constant.

$$\Rightarrow \frac{dP}{P} = dt$$

Integrating with limits, we get

$$\int_{P_0}^{P} \frac{dP}{P} = c \int_{0}^{t} dt$$

$$\log\frac{P}{P_0} = ct$$

$$P(t) = P_0 e^{ct}$$

Initially suppose, there was 1 patient $\therefore P_0 = 1$
Suppose, on 11th day, total number of covid cases was 15, so

$$1e^{10c} = 15$$

$$\therefore e^{10c} = 15$$

Therefore on 21st day,

$$P(20) = 1e^{20c} = 1 \times (e^{10c}) \times (e^{10c}) = 1 \times 15 \times 15 = 225$$

On 61st day,

$$P(60) = 1e^{60c} = 1 \times (15)^6 = 1,13,90,625$$

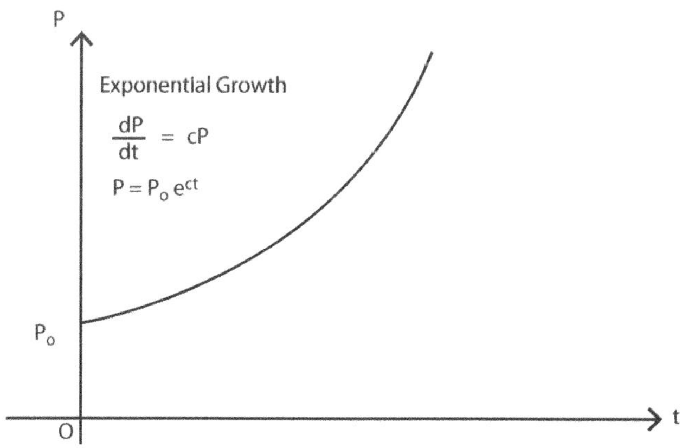

So it can be observed that the spread of coronavirus is growing exponentially (without implementation of lockdown). Here the concept of reproductive number (number of healthy people infected due to a single infective) is not used.

11.5.3.2 Model Based on Geometric Progression (G.P.)

Let it be considered that a single infected person can infect two healthy persons (reproductive number=2). Then, the transmission of coronavirus can be explained with the following tree chart.

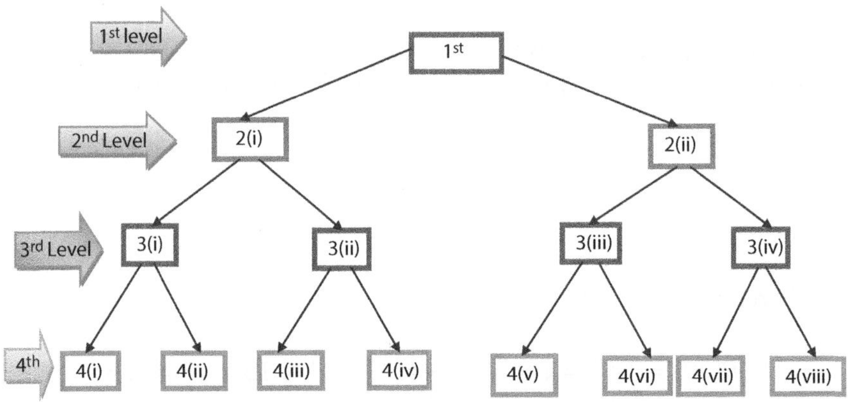

TREE CHART SHOWING SPREAD OF NOVEL CORONAVIRUS

So, number of persons infected daily is following the G.P series 1,2,4,8,...

So, till nth level, the GP series is 1,2,4,8,........, 2^{n-1} (n is any natural number)

So, total number of Covid cases after n levels

$$= \text{Summation of G.P. series}$$

$$= a\left(\frac{r^n - 1}{r - 1}\right)$$

$$= 1\left(\frac{2^n - 1}{2 - 1}\right)$$

$$= 2^n - 1$$

Let us consider two scenarios: One without the implementation of lockdown and the other with lockdown implementation:

11.5.3.2.1 Without Implementation of Lockdown

Let it be assumed that it takes 3 days for the transmission of virus from one level to another level.

So, it can be written as

On 1st day → 1st level

On 4th day → 2nd level

On 7th day → 3rd level

On 10th day → 4th level

and so on.

So, on 61st day → 21st level, the number of covid cases will be $= 2^{21}-1$

$= 20,97,151$

11.5.3.2.2 With the Implementation of Lockdown

Let it be assumed that it takes 6 days for the transmission of virus to next level, due to less interaction between infectives and susceptibles.

So, it can be written as

On 1st day → 1st level

On 7th day → 2nd level

On 13th day → 3rd level

On 19th day → 4th level

and so on.

So on 61st day → 11th level, the number of covid cases will be $= 2^{11} - 1$

$= 2047$

So, it can be seen that how lockdown proves to be an effective way to control the spread of Covid-19.

11.5.3.3 Model for Stay At Home

Let it be considered that after the lockdown for a long period, the number of cases will become constant. Let the value of that constant be **C**. So now the rate of will not depend on the total number of cases at time t= 0, instead it will depend on the difference between the final number of cases after the lockdown and the initial number of cases.

Let N denotes the number of cases at time t.

Let initially (at t=0), the number of cases be N_0

At time dt, the number of cases be N_0+dN

Rate of change in number of cases = $\frac{dN}{dt} = C - N$

$$\Rightarrow \frac{dN}{C-N} = dt$$

Integrating with limits, we get

$$\int_{N_0}^{N} \frac{dN}{C-N} = \int_{0}^{t} dt$$

$$\Rightarrow log \frac{C-N}{C-N_0} = -t$$

$$\therefore N = C + (N_0 - C)e^{-t}$$

So, it can be obtained that,

when t=0, $N=N_0$

and

when $t \rightarrow \infty$, $N \rightarrow C$

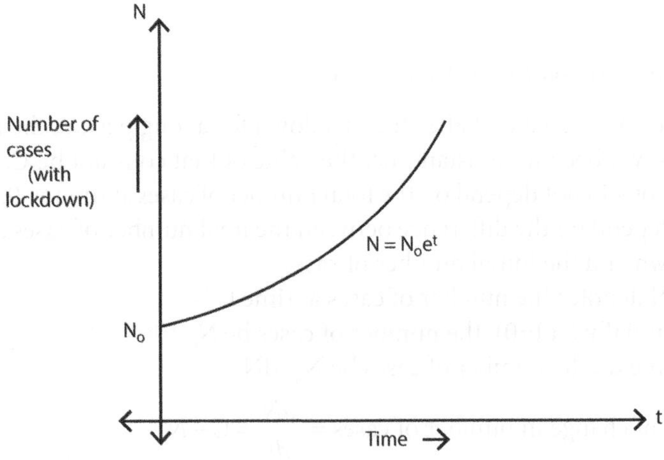

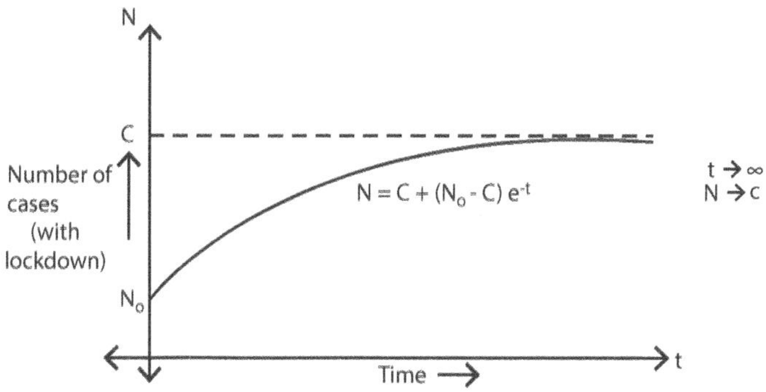

NOTE: All the calculations are made based on the assumption Lockdown is followed with 100% efficiency.

11.5.4 Discussion

The three models discussed above show that coronavirus is spreading day by day but due to implementation of lockdown the cases are not increasing in such a manner as it would have been without any social distancing. Thus each person should stay at his or her home to be safe from coronavirus instead of being a carrier of the virus, so that others do not suffer due to him or her. Let us all be the soldiers by following the guidelines issued by the higher authorities and protect each other by staying at home unless unavoidable circumstances arise.

References

1. Bhola Jyoti, Koul Monika, Venkateswaran Revathi Vandana (2020) "Corona Epidemic in Indian context: Predictive Mathematical Modelling", *MedRxiv*, pp. 1-14, Doi No 10.1101
2. Cunha Americo, Morrison .E Rebecca (2020) "Embedded model discrepancy: A case study of Zika modelling", *Chaos* 30, pp. 1-8, Doi No 10.1063
3. Cakir Zafer, Savas Basri Hasan (2020), "A Mathematical Modelling Approach in the Spread of the Novel 2019 Coronavirus SARS-CoV-2(COVID-19) Pandemic", *Electronic Journal of General Medicine*, pp. 1-3. Doi No 10.29333
4. Ferrandez, R.M, Ivorra .B, Ramos .M.A, Vela-Perez .M (2020), "Mathematical modelling of the spread of the coronavirus disease 2019 (COVID-19) taking into account the undetected infections. The case of China", MOMAT Research Group, Interdisciplinary Mathematics Institute, Complutense University of Madrid, pp. 1-5, Doi No 10.1016

5. Iannelli Mimmo (2005), "The Mathematical Modelling Of Epidemics", UNITEXT, pp. 1-10, Doi No 10.31219
6. Jiang Xia, Li Yichi, Liu Yonglong, Peng Ruiyang, Wang Bowen, Zhan Yonglong, Zhao Bin, Zhou Chen (2020), "Mathematical Modeling and Epidemic Protection of Covid-19 and Its Significance to Epidemic Prevention and Control Measures", *Annals of Infectious Disease and Epidemiology*, Volume Number 5, pages 1-9
7. Nag Sayan (2020), " A Mathematical Model In the Time of Covid-19", pp. 1-5, Doi No 10.31219
8. Yang Chayu, Wang Jin, "A Mathematical model for the novel coronavirus epidemic in Wuhan, China", *Mathematical Biosciences and Engineering*, pp. 2708-2710, DOI: 10.3934/mbe.2020148

12

Application of Mathematical Modeling in Various Fields in Light of Fuzzy Logic

Dr. Dhirendra Kumar Shukla

Department of Education in Science and Mathematics, Regional Institute of Education, NCERT, Bhopal (M.P.), India

Abstract

In this chapter, the author explains the concept of "fuzzy logic" and various approaches. In the present scenario, fuzzy logic is a widely accepted and used term in the light of development for applications, tools, and techniques as Fuzzy Cognitive Maps, Fuzzy Cluster Means, etc. Here Fuzzy Logic Concept has been studied and tried to explain applications of the concept in various fields as Mathematics, Science, Business, Finance, Controller of Temperature, Home appliances, Aeronautics, Defence, Medical Science and Bioinformatics, and Engineering Fields such as Mechanical, Industrial, Production, Electronics, Chemical, Automotives, Signal Processing and Communication, and Robotics.

Keywords: Fuzzy logic, fuzzy cognitive maps, induced fuzzy cognitive maps, fuzzy cluster means, fuzzy logic control, mathematical modeling

12.1 Introduction

12.1.1 Mathematical Modeling

"The model represents something in a thumbnail; an illustration or image; a conception or analogy for the observation of something that can not be detected; a system of axioms, details & assumptions that are provided by a mathematical explanation of the object or state of the business." Different methods can characterize devices and behaviors. Physical object words,

Email: dhirendrashukla1982@gmail.com

Ramakant Bhardwaj, Jyoti Mishra, Satyendra Narayan and Gopalakrishnan Suseendran (eds.) Mathematics in Computational Science and Engineering, (257–286) © 2022 Scrivener Publishing LLC

drawings, or sketches, computer programs, or maths. We also should conclude that the model operations can be conducted in many languages, often simultaneously [4].

A mathematical model demonstrates the conduct of real devices and objects in mathematical terms. It is required to understand that,

- How can mathematical images or models be constructed?
- How can they be approved?
- How will they be used?
- How and why is the use of mathematical modeling limited?

However, there is a good debate about why mathematical modeling should be used before discussing these crucial problems.

A mathematical design is a representation of a structure. Mathematical modeling is called the period of the progress of a mathematical model. Mathematical models are used in the investigation of the discovery of operations and financial matters. A model helps to reveal a structure and ponder the impacts of different segments and make predictions for normal sciences, such as physical chemistry, chemistry, geology, and order design. Numerous constructs may be taken up by mathematical models, including but not limited to dynamic systems, empirical models, differential conditions, or game-hypothetical models. A model that includes a variety of theoretical structures may cover these and various types of models [4].

Mathematical models, by and large, can grasp logical models. As a rule, a logical field's existence depends on how the mathematical models built on the hypothetical side are in harmony with the studies' reproducible aftereffects. If the speculation grew better, the absence of understanding between the hypothetical mathematical models and the test quantifies periodically prompts enormous advances. Mathematical modeling (Figure 12.1) is a conceptual report, where a "real world" and a "theoretical world" are recognized. The rest of the universe is known as the real world; here, we experience numerous wonders and activities, regardless of whether the ancient rareties are of normal beginning or delivered. The measured universe is the soul's world. In three steps, the theoretical universe can be considered: perception, modeling, and forecast. Mathematical modeling is an operation that relies on the values that help it and the methods that can be modified with development. The principles are general or objective principles detailed as inquiries about the mathematical modeling objectives and destinations. Such ideals are practically metaphysical. We will diagram the concepts in the following area and communicate about a portion of the techniques quickly [4, 10].

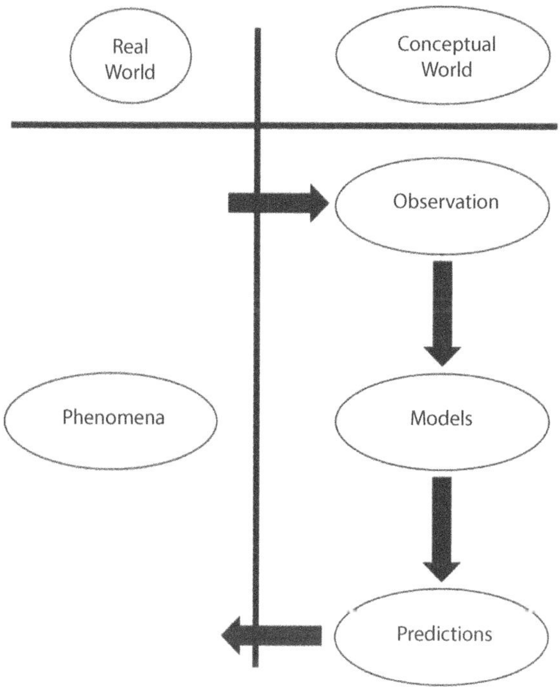

Figure 12.1 An elementary representation of the world's scientific method is connected to discoveries made within the real world [4].

12.1.2 Principles of Mathematical Models

Mathematical modeling is an activity focused on its supporting principles and successful implementation strategies. The concepts are common or objective, formulated as questions about mathematical modeling purposes and objectives. These definitions are almost philosophical. In the following section, we will outline the concepts and discuss some of the methods briefly.

Figure 12.2 shows a visual illustration of the fundamental philosophical approach. These methodological model principles are also included in the following discussion:

- What's our question? Recognize the necessity for that model.
- What are we going to know? A list of the data.
- Does knowledge matter? The value of available data can be established.

- What are we to take for granted? Determine the circumstances under which this is concerned.
- How can the model be taken into account? Identify the guiding principles of mathematics.
- What is our prediction model? Identify the equations and the answers that will be used in the calculations to be carried out.
- Are the predictions true? Determine experiments to test the model, i.e., do they follow its principles and hypotheses?
- Do the projections work? Determine the tests to evaluate the model, i.e., what is useful for what was achieved in the initial reason?
- Can the model be enhanced? Identify the values of undefined parameters of variables to be included and/or future

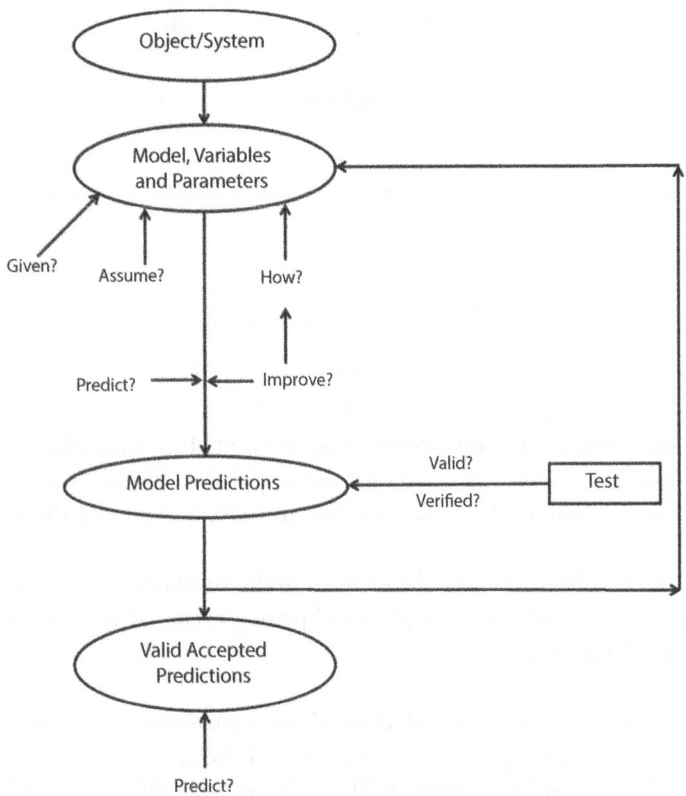

Figure 12.2 The conceptual approach to creating a model relates to the development of that model [4].

assumptions/restrictions. Apply the iterative loop we call Model-Validate-Verify-Predict.
- How do we use the pattern? How can we use the design?

This discussion isn't an algorithm for constructing a suitable mathematical model. The underlying principles are the keys to mathematical modeling and the formulation of the problem as a whole. Therefore, we can hope that individual issues are always repeated during the modeling of business processes. This list may be taken as a very general guideline for thinking in mathematical modeling. To understand the flow chart of preparing a mathematical model see the Figure 12.2.

12.2 Fuzzy Logic

The concept of fuzzy logic is influenced by human perception and cognition, on the principle of relatively measured association. In 1965, the first exploratory paper of Lotfi A. Zadeh on fuzzy sets was delivered. Fuzzy reasoning, i.e., doubtful, ambiguous, or half way apparent, can handle com putational observation and cognition data without stringent limits. The fuzzy logic takes account of dubious human judgments in defining problems. It also provides a suitable procedure for breaking various laws and for better testing alternatives. New figure techniques based on fuzzy logic can be used for improving knowledgeable frameworks for complicated, recognizable proof, design identification, creation, and control.

For others, those concerned with pioneering work, such as engineers, mathematicians, PC-programmers and professional designers, regular researchers, clinical analysts, social researchers, fuzzy logic are serious, including those who have been engaged in groundbreaking work, such as electrical engineers (Electrical, Mechanical, Civil Chemicals, Aviation, Agricultural, and Biomedical).

There is a fuzzy logic in many design and logical works, where it is treated as a numerical interest. Fuzzy logic is being utilized in several applications; facial recognition, air conditioning control systems, clothing washing machines, vacuum cleaner, skid stoppage, transmission frameworks, autonomous helicopters and metro frameworks, information framework for multi-objective frameworks to enhance strength, climate guessing, and new product modeling. In many ways, fuzzy logic has been efficiently used, including managing concept structures, image preparation, power design, state-of-the-art robotics, mechanical technology, customer hardware, and progress. This part of science has incorporated new life into logical fields which have been lethargic for quite a while.

A significant number of analysts work with fuzzy logic and product licenses and exploration documents. The hypotheses or the use of the fuzzy logic are recorded in two discovery diaries; there are 89,365 hypothesis distributions or service in the INSPEC database of fuzzy logic; there are 22,657 hypothesis distributions in the MathSciNet database of furrowed logic or use, there were 16,898 patent applications as recorded by Zadeh on 4 March 2013 for "fuzzy logic." The number of commitments to exploration increases daily and is growing more and more. In support of fleeting reasoning and responsive figurative ideas and usages, Zadeh started a soft-computing project for Berkeley (BISC), a well-known research center at the University of California, Berkeley.

Fuzzy logic, in other words, is a kind of multi-worth logic that can have any actual numbers between 0 and 1. It is used to confront the idea of partial truth, in which the value of truth ranges from absolutely right to entirely false. By comparison, in boolean logic, only 0 or 1 integer values contain values that are true variables. Fuzzy logic is founded on the observation that people make decisions based on imperfect and non-numerical data. Mathematical means for communicating vague and imprecise information are fuzzy models or sets. These models may classify, represent, manipulate, interpret, and use ambiguous and uncertain data and expertise. Fuzzy logic has been extended to several areas, from artificial intelligence to control theory.

12.2.1 Fuzzy Cognitive Maps & Induced Fuzzy Cognitive Maps

We should define the "Cognitive Maps" (CM) before the "Fuzzy Cognitive Maps" (FCM) debate. Edward Tolman initiated the CMs in 1948 to establish and research social science expertise in decision-making in foreign policy–related activities. CM is a signed digraph designed to reflect a framework of beliefs or declaration of causality about a specific area by an individual or a group of people and to use that statement to evaluate the effect of a selection on a specific target. In 1994, Julie A. Dickerson and Bart Kosko developed a "Fuzzy Cognitive Map" (FCM), which expands the idea of Cognitive Maps to enable the concepts to be linguistically interpreted in a game of fuzzy logic. The FCM links causal events, actors, values, and trends to a dynamic feedback blurring system. The FCM lists the fuzzy law or the causality movement routes of the reporting events.

For example, a Socioeconomic model can be built through various social issues as Population, Crime, Economic condition, Poverty, and Unemployment. According to the study in the real world, a relation can

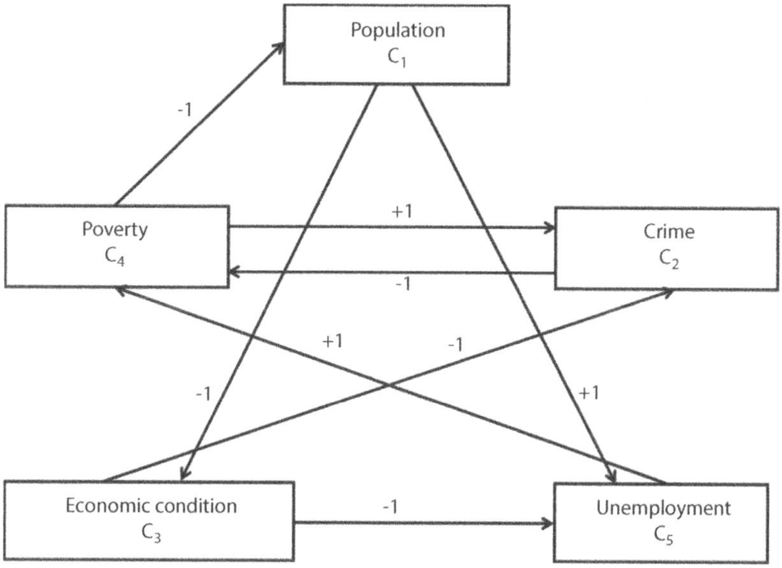

Figure 12.3 A socioeconomic model can be constructed based on issues related to society [7]

be made to understand the problem's scenario in society, as shown in Figure 12.3.

12.2.2 Fuzzy Cluster Means

"Fuzzy Cluster Means" or "Fuzzy C-Means" is a data agglutination technique in which a particular degree of each point belongs to a cluster. Jim Bezdek originally proposed this technique in 1981 to improve the methods of the cluster. Data clustering is the method of separating data objects into classes or groups from the original images. The same class items are as similar as possible, and the things in the different categories are as different as possible. Different similarity measures can be used to position elements of classes, depending on the nature of the data and the purposes for which the clustering is used; in the similarity measure, the way groups are combined is managed. In clusters, distance, connectivity, and intensity are examples of measures that can be used [1].

This unusual problem explores various fields in the developments in fuzzy logic–based applications and commercial goods. Although fuzzy logic has submissions in different regions, people without new, creative frameworks do not yet know how well it can be incorporated in various products currently accessible to the consumer. For certain persons, the philosophical and logical meaning of the word fuzzy remains unpleasant. These individuals need to figure out when and how they can use fuzzy logic.

12.3 Literature Review

All over the world in various countries, scientists and researchers in multiple fields tried to use the concept of fuzzy logic to invent something new to help society. A few details are given below to understand the strength of fuzzy logic.

S. no.	Researchers	Title	Details
1.	M. S. Dattathreya *et al.*	*"Detection and elimination of a potential fire in engine and battery compartments of hybrid electric vehicles."*	A new fuzzy deterministic non-controller type (FDNCT) and an FDNCT inference algorithm (FIA) were defined. The FDNCT is used in an intelligent system to detect and prevent potential fire in a hybrid electric vehicle's engine and battery parts. The result of the simulation Comparison and identification of activities for the removal of potential fires between the FIA and singleton inference algorithms.

(*Continued*)

S. no.	Researchers	Title	Details
2.	R. Dixit and H. Singh	*"Comparison of detection and classification algorithms using boolean and fuzzy techniques."*	For a Hypothetical Target Classification Scenario (HTCS), the researchers compare various logic analysis approaches and display the findings. Working of preprocessing can consistently retain confidence in results and compare Boolean, multiquantization Boolean, and fuzzy techniques has been shown.
3.	R. Dixit and H. Singh	*"BDD, BNN, and FPGA on fuzzy techniques for rapid system analysis."*	To streamline the data processing of large multivariate military sensor systems, the researchers look at fuzzy techniques.
4.	F. Kaleem *et al.*	*"A fuzzy preprocessing module for optimising the access network selection in wireless networks."*	The researchers present a fuzzy multicriteria scheme of estimating the necessity of vertical handoff design and implementation. When considering the service's quality and consistency and customer satisfaction, their strategy determines the best vertical delivery time.

(*Continued*)

S. no.	Researchers	Title	Details
5.	A. M. Dixit and H. Singh	"*A soft computing approach to crack detection and impact source identification with field-programmable gate array implementation.*"	The researchers present a fuzzy inference method for automating the Crack Detection and Impact Source Identification (CDISI) and delivering the work on an automated CDISI microchip.
6.	A. K. Dash	"*Analysis of adaptive fuzzy technique for multiple crack diagnosis of faulty beam using vibration signatures.*"	Using a fuzzy Gaussian method, the resracher proposes a method for multi crack structure detection.
7.	D. Pal and D. Bhattacharya	"*Effect of road traffic noise pollution on human work efficiency in government offices, private organisations, and commercial business centres in agartala city using fuzzy expert system: a case study.*"	The researchers are discussing the decrease in the productivity of human labor due to increasing road noise emissions. They track and model road traffic disruptions & the performance of personal work using fuzzy logic.
8.	A. D. Torshizi and J. Parvizian	"*A Hybrid approach to failure analysis using stochastic petri nets and ranking generalised fuzzy numbers.*"	The researchers are advancing a failure analysis approach that the flexibility of fuzzy logic with structural characteristics of the Stochastic Petri Nets. This algorithm includes a wide range of industrial uses.

(*Continued*)

S. no.	Researchers	Title	Details
9.	K.- Y. Song *et al.*	"*Excluded-mean-variance neural decision analyser for qualitative group decision making.*"	In uncertain cases, the mean-variance neural approach for group decision-making is a revolutionary step. The excluded mean-variance approach shows that this approach increases qualitative decision-making efficiency by giving the decision maker a new cognitive tool to aid in the reasoning process.
10.	A. M. G. Solo	"*Warren, McCain, and Obama needed fuzzy sets at presidential forum.*"	The investigators have shown how the moderator and the presidential candidates in a presidential forum needed a fuzzy logic to pose a discussion question and address it correctly. The author explains how to grasp the fuzzy logic to correctly pose and answer questions about the meaning of imprecise linguistic terms. Then the researchers distinguish among qualitative definitions and quantitative definitions of approximate language terms and between tangible quantitative definitions of imprecise linguistic terms.

(*Continued*)

S. no.	Researchers	Title	Details
11.	M. H. F. Zarandi *et al.*	*"A fuzzy rule-based expert system for evaluating intellectual capital."*	The researchers described fuzzy expert system for the evaluation of intellectual capital. It helps managers to identify and assess the degree to which intellectual activities produce each asset.

12.4 Applications of Fuzzy Logic

Fuzzy logic is a generally utilized idea in different fields as Mathematics (Pure and Applied), Medical Science, Nano Technology, Micro-Biology, Bio-Technology, Bio-Informatics, and especially in Engineering. Here barely any utilization of fuzzy logic is examined for comprehending the significance of the idea [14].

Lotfi Askar Zadeh, an engineer in the University of California at Berkeley in 1965, recommended a factual diagram of those classes which had no characterized membership models for the most generally utilized words in ordinary dialects, for example, "high temperature," "round face," or "oceanic creature." They've been called Zadeh's fuzzy assortments. Any number from 0 to 1 is an association of a fuselage, which speaks to a continuum from "not in the set surely" through "somewhat in the set" to "totally in the set." For "Fuzzy sets," nonetheless, a few properties of the normal Set-Operations are not, at this point, legitimate. For instance, the intersection of a "fuzzy subset" and its complement may not be void in the "Fuzzy sets" [14].

Consequently, the principle of the excluded middle in a logic based on Fuzzy sets isn't pertinent. The functionalities are described by the hypothetical just as expository strategies, which rely upon a specific application. They may require the utilization of showing strategies and enhancement strategies. Fuzziness is non-measurable, recognized by Zadeh, showing an unclearness reliant on the human creative mind, not a vulnerability in a probabilistic setting.

"Fuzzy control" alludes to innovation projects or calculations that utilize the fuzzy logic to permit apparatus to settle on choices dependent on practical information on the human administrator. The essential issue with programmed controls is deciding the system's satisfactory reaction or plant under some random set of terms. The first fuzzy logic controller was dispatched in the mid-1970s by E. H. Mamdani, a teacher at Queen Mary College, in London, intrigued by the learning systems' plan. Mamdani and his understudy, Seto Assilian, have recorded 24 heuristic guidelines for directing a little steam engine and boiler combination activity.

The business utilization of fuzzy logic started in the mid-1980s and was before long turned into a center point for a scholarly and mechanical examination of Japan's fuzzy systems. Fuzzy logic has been utilized, for example, to manage concrete assembling/cement manufacture, and the maker of "fuzzy controllers" has worked water sanitization measures/water purification processes and robotized tram-trains from the Japanese city of Sendai.

The Japanese furor has inevitably been sponsored for fuzzy products. Be that as it may, fuzzy logic is still extremely present, yet less recognizably, with different purchaser items, for example, the programmed transmissions of certain vehicles containing a fuzzy component, which detects driving style and cars to choose the best apparatus. Master programs with the guide of fuzzy logic can help specialists to analyze diabetes. Management science, securities exchange investigation, extraction of information, phonetic, and conduct testing are just a few different fields in which productive utilization of fuzzy logic ideas has been made. It involves courses for the fields of engineering and related fields.

Crossbreed systems were created in the last part of the 1990s that join the benefits of at least two computational methods. The fuzzy logic (Figure 12.4) and neuro-fuzzy systems have incorporated certain types of learning. Systems with neuro-fuzzy components can be utilized in fields, for example, financial exchange expectations/stock market, prediction/theoretical studies, information system, and data mining.

12.4.1 Controller of Temperature

1. Identify the problem.
2. Adjust the heater fan's speed based on the temperature and humidity of the room.
3. Four Cold, Cool, Wet, and Hot settings are required for a temperature control system.
4. Low, medium, and high can describe humidity.

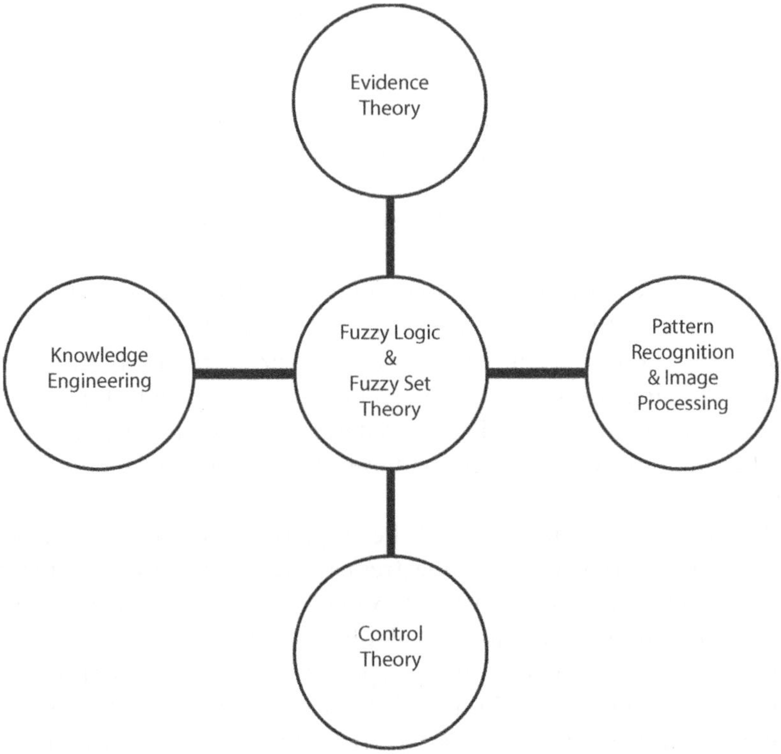

Figure 12.4 Some fields related to fuzzy logic and fuzzy set theory [2, 3, 5].

12.4.2 Usage of Fuzzy Logic in a Washing Machine

1. Fuzzy logic washing machines are becoming more common. These machines have the advantages of performance, productivity, simplicity, productivity, and lower costs. Sensors perceive variations continuously [14].
2. The machine conditions and adjusts processes for the finest wash efficiency accordingly. Since there is no norm for fuzzy logic, different machines behave in different ways.
3. Fuzzy logic controls the washing process, intake of water, the temperature of the water, washing time, rinsing output, and spin-time. It increases the life cycle of the washing machine.
4. Machines also learn from previous experience, memorize, and modify programs to reduce operating costs.
5. "One-touch power" is featured on most fuzzy logic machines and built with energy savings features.

6. The fuzzy logic checks the amount of dirt and fat, soap and water, the direction in which the spin is applied, etc.
7. The machine rebalances the washing load to ensure proper spinning. If the excess is detected, the spinning speed would otherwise decrease. Also, load washing prevents noise from spinning. Neuro fuzzy logic involves optical sensors that detect dirt and a tissue sensor in water to detect the tissue shape and to adjust the cycle accordingly [11].

12.4.3 Air Conditioner

Due to the use of fuzzy logic, an automatic decision-making system fas been made for air conditioners to work properly, which can be expressed as follows [21] also see Figure 12.5:

1. The air conditioner provides cool air through the pipe in a central heating and cooling system through the house's interior. It provides a mechanism to filter hot air interior and prevent heat.
2. The compressor condenses and converts a refrigerant over and done with the external unit from one gas to a liquid. The liquid is then inserted into the inner evaporator or refrigeration bowl. The indoor fan circulates in the air at the fins of the evaporator. Thermal energy exchange with the atmosphere through the evaporator's metal fins. The coolant

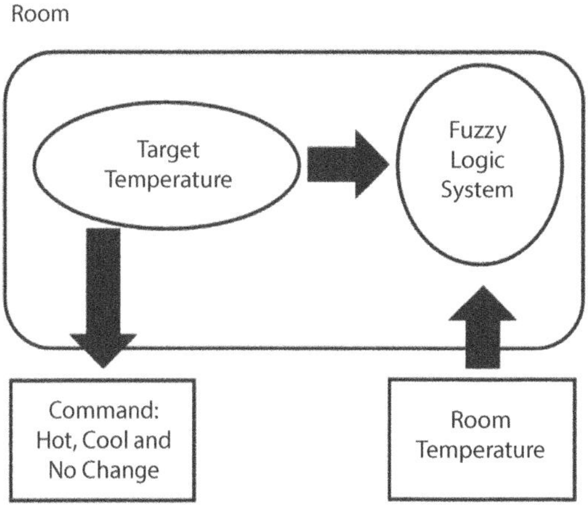

Figure 12.5 The working process of the air conditioner [21].

transforms from the fluid into vapor, which prevents heat from surrounding air. The air is cooled and blown back into the home as the heat is taken from the air.

Follow the figure mentioned to understand the working process of the air conditioner.

12.4.4 Aeronautics

1. Spacecraft and Satellite altitude control, flow control and mixture in deicing aircraft vehicles.
2. Applications focused on fuzzy logic assist in the design and validation of a morphing wing. In actuation theory, the shaped morphing mechanism used intelligent materials like shape memory alloy (SMA).
3. Two major applications of the technique of fuzzy logic are:
   a. Identification of a device model, starting from some input-output experimental results.
   b. An automatic machine control.
4. The neuro-fuzzy network and numerical values help find results (forces, waves, temperatures, and elongations) during SMA experimental research.
5. Four Fuzzy Inference Systems (FIS) are used to acquire four neuro-fuzzy controllers:
   a. First for controller for the current increase.
   b. Second, For a continuous current.
   c. Third, the current decrease.
   d. Fourth, Null Current Controller.
6. In designing the SMA actuators controller, fuzzy logic techniques have been used, starting with the model of the SMA actuators developed.

In other words, logical methodologies based on fuzzy logic for wing morphing or aircraft morphing research in multidisciplinary research is beneficial [6].

12.4.5 Automotive Field

For the simulation of several different vehicle dynamics research scenarios, a full vehicle dynamics and driver model is important, since the mathematical models used to estimate the proposed control systems' performance are highly cost-effective. Fuzzy reasoning has proved to be an effective method for addressing imprecision and ambiguity, both critical driving environment characteristics, and can take vague human judgments into account. A few apps are as follows [18]:

1. A fuzzy logic controller can describe driver behavior for a non-linear vehicle dynamics problem by providing the opportunity to establish rules that make intuitive sense. Since the 1980s, the fuzzy methods in driver modeling originated from the fuzzy set theory have been investigated.
2. Antilock braking system (ABS): ABS was developed in 1971 to be used in vehicles. The antilock braking system intends to avoid wheel locking and additional yaw time while preserving the vehicle's directional stability during braking because of the control operation. The ABS's key control goal is to maintain the usable friction coefficient as high as possible during braking to have the greatest possible brake force while preventing excessive wheel slip. Choosing an acceptable control methodology is necessary for any dynamic vehicle system to get an effective response from the system.
3. Vehicle Stability Control Systems (VSC): It is possible to identify VSC systems as direct yaw moment (DYC) or steer dependent. The most successfully commercialized device, Electronic Stability Control (ESP), relies on the DYC foundation. The ESP has a closed-loop algorithm built by programmed intervention in the braking system and/or drivetrain to enhance vehicle braking and handling response.
4. Fuzzy logic helped human beings to develop their decision-making system, and it works in two ways (Figure 12.6):
   a. Two fuzzy systems are required
      - Infer Driving Style
      - Select gear
   b. Gear selection based on
      - Sensor data
      - The fuzzy judgment of current driving style

It is known that fuzzy logic is very promising for many areas of vehicle dynamics. In particular, logic is very efficient based on practice for systems with uncertainty surrounding complex vehicle dynamics. Some of the "fuzzy" automotive applications include trainable fuzzy systems for

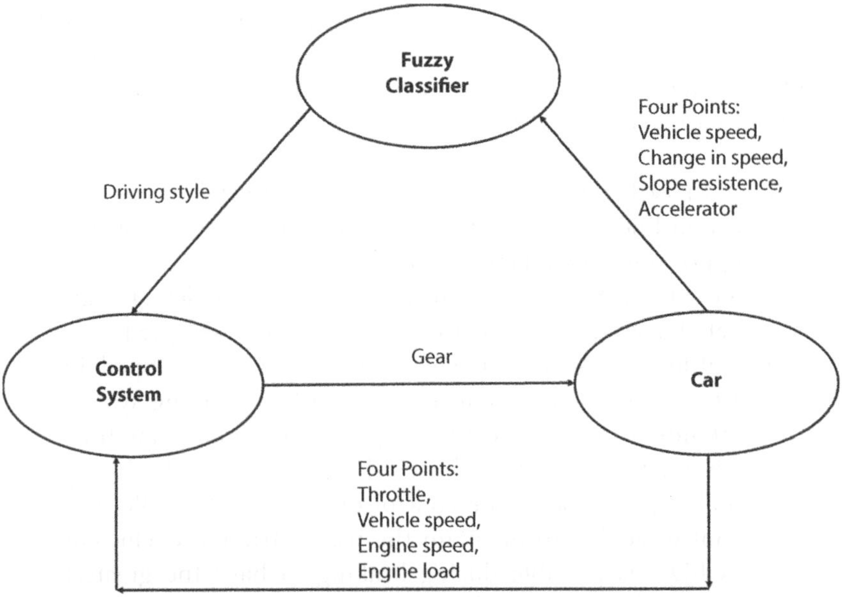

Figure 12.6 Use of fuzzy logic for decision making during car driving [18].

idle speed control, automatic transmission method shift scheduling, smart road systems, traffic control, improved automatic transmission efficiency as shown in the Figure 12.6.

12.4.6 Business

Businessmen and traders are now obligated to cross a database of products to be sold. Some characteristics of a product are more important than others for the consumer. The salesperson must take this information into account. Salespeople use computers as well, and all these things can be easily done using furtive business and business logic to represent their customers effectively. It shows a simple and very fast display system according to the needs and expectations of the customers [13].

The consumer and fulfillment of his desires and expectations should be the most critical factor in the industry. If the dealer has just a few things in stock, the situation is straightforward. He possibly explains all the products perfectly and can tell the consumer quickly. However, we live in an age where supermarkets and hypermarkets sell hundreds, if not thousands of items. Consequently, the vendor can not recall all the products for sale or their parameters. He does not immediately and thoroughly advise and serve the consumer. Therefore, the computer is responsible for remembering all data, but the computer works by the standard two-way reasoning, only determining "true" or "false." For example, the application "reasonable pricing mowers of an average range" should be translated into specified selection criteria. This problem can be solved concerning the selection criteria by using fuzzy logic. The computer (like a human being) makes decisions and control systems based on incorrect information by implementing it. The application of flouted logic reduces problems with sharp limit variables and leads to a more natural "perception" of the universe [13].

Therefore, shoppers can be assured with fuzzy reasoning that a product is selected according to all their desires and requirements. Moreover, an appropriately designed system may define unrealistic customer requirements. The customer and the dealer will identify and verify them. The system can be expanded to include every product variety offered in any store; the wider the product range, the greater the benefits of using a system. The application of fuzzy logic in retail, as one of the opportunities for multicriteria decision making, has its undeniable advantages [13].

In other words, by using the fuzzy logic principle, decision-making support systems and staff assessment in a large organization can be achieved very effectively and quickly.

12.4.7 Finance

Buckley introduced the fuzzy logic which explores financial mathematical and applies it to study money value-time in the financial field. Since it can resolve imprecise, incomplete, and ambiguous knowledge, fuzzy logic is used effectively in finance. This technique was applied in risk management and credit rating in the financial sector, though only somewhat. But the footprint of the fuzzy logic was nearly non-existent about the particular case of banking crises [13].

Due to the global financial crisis, Banking Crisis Prevention and Management (BCPM) for bank escape using public money for rescue banks has been the main financial priority. This priority highlights the importance of coping with this problem in various ways. The use of fuzzy logic to analyze bank crises can be an essential step towards more efficient planning and control of bank crises [13].

In general, significant adverse effects are correlated with the banking crisis, which reduces tax revenues, raises budget expenditures and raises government debt on productivity, production, and asset prices [13].

The financial sector's need to establish new instruments for preventing and managing bank crises must be acknowledged. Four fields have been illustrated in the banking crisis in research into the proper use of fuzzy logic [13]:

1. Banking crisis management.
2. Banking crisis management and assessment of its economic effects.
3. The idea of a strong regulatory and institutional environment.
4. Banking crisis resolution.

12.4.8 Chemical Engineering

In the concept of both chemical components and gas recognition, chemical engineering has used fuzzy logic. The process control, the batch distillation column, separation process and the cinematics are also expanded. Chemical engineering focuses mainly on the growth, development and maintenance of procedures concerning chemical and biological transformations. A few submissions of fuzzy logic are mentioned as follows [20, 21]:

1. **Piping risk assessment:** Risk management of road and rail tubing was the safest and most cost-efficient transport procedure, but numerous accident databases with transported medium releases into the atmosphere reported many tube failures and significant damage. Adam and al. discuss how long-pipeline transport of flammable substances can be carried using the fuzzy logic of risk management.
2. **Safety analysis:** Safety analysis processes can expose different hazards and process large quantities of hazardous chemical compounds in processes and chemical plants. Adequate awareness and good evaluation of dangers from such a facility are practices known as process safety analysis (PSA) for the efficient management of these installations. PSA is a highly competitive practice committed to the chemical industry's decision-making methods, with complexities associated with the lack of expertise that can give rise to numerous plant safety overviews. The use of fuzzy sets that boost the process of data acquisition solves this problem.
3. **Furnace control:** Various furnaces have long constant durations, and no single variable might be used for process control. Regulated expert systems and oven are commonly used for such furnaces. These devices use inputs from several furnace parameters observed.
4. **Fluidized Catalytic Cracking:** FCC is a significant oil refining process that makes high molecular oils lighter. Optimizing gasoline production in the petrochemical refinery presents

a significant challenge that provided the non-linear and interdependent input and output variables. These processes are considered difficult to manage and control cracking and coke burning due to their large-scale, complex hydrostatic dynamics, and complex kinetics. The fuzzy logic control was used efficiently to increase the FCC process control as a promising control technology in the refining field.

5. **Separation Process:** A separation process is used for chemical engineering to convert a substance mixture into two or more separate products. Chemical or some physical characteristics of separate products, size, crystal changes/other separations of components may differ. In the process of separation, fuzzy logic played an important part.
6. **Food Production:** The non-linearity, diversity, and heterogeneity of the sensitive parameters do not make a traditional approach adequate for the food industry. For example, it is complex and incomplete to model the cooking process. Fuzzy logic makes the use of established "professional" skills and the consideration of consistency variables. In combination with other methods, fuzzy rule bases have been used.
7. **$_p$H Controller:** Benz *et al.* Developed a computerized self-adjustable pH measurement and control system and tested in a laboratory reactor for regulating fermentation and neutralization of wastewater streams. The controller was defined and based on advanced logic as experience and non-theoretical models, which required applying subjective information. From the conventional PID controller, the pH value can be changed more easily by the fuzzy control [9].
8. **Reactor control:** In chemical engineering, fuzzy logic played an important role in reactor control-vessels designed to hold chemical reactions in chemical reactors. To maximize the net reaction present value, chemical engineers construct reactors.
9. **Batch crystallizer:** Batch and semi-batch procedures are appropriate for treating goods such as special polymers, fine substances, and pharmaceuticals. Unseeded crystallization was used to examine the efficiency of the Adaptive Neuro-Fuzzy Logic Controller. The controller effectively accomplished the "straight" aims of both test sets, which maintained supersaturation and the chord length gap for default ranges.

10. **Combustion process:** Combustion is a thermal sequence of processes. Due to the characteristics of each combustion stage, the process is subject to substantial dynamic changes. Performance and lower emission measures should be focused on knowledge of combustion quality in real time. Ruusunen and Leivisk proposed a foggy model-oriented strategy for estimating carbon dioxide content in a wood process. In real time and low caloric values of wood fuel, model outputs for combustion efficiency measurements, including the estimate of oxygen content, are used.

12.4.9 Defence

The future of war can be decided by technological progress instead of brute destructive forces. Artificial intelligence systems based on fuzzy logic can operate in complex fighting environments and produce improved outcomes compared to existing control systems. Within the battlefield's complexities and uncertainty, fuzzy controllers have achieved promising military applications like mobility, cybersecurity, and target detection and interoperability. Fuzzy controllers are at ease to implement and adjust, in contrast to other complex approaches. Logic-based artificial intelligence can transform war in an intelligenized form from its present "informatised" form. For instance, a landmine is pre-programmed to detonate if a certain heaviness is added to its pressure plate. Autonomous systems probably justify a particular set of inputs/data. There are many advantages of using fuzzy logic as a control algorithm over standard Boolean logic for autonomous military platforms. The following are the fuzzy logic security applications [17]:

1. The NATO reference mobility model (NRMM) is one of these traditional models. It predicts military vehicles' mobility in off-road and on-road conditions, but it's not enough for a complex environment. According to user feedback, the above-mentioned fuzzy model is more comfortable to improve, highly detailed, and upgradable to suit complex and to change realistic environments. For instance, flat tires, or reduced traction by rain, may consider a flush-based model.
2. The adaptive Neuro-Fuzzy Inference System (ANFIS) is to create a competent method for effective combat ground operations to sustain firing. Fire support is used to support a

battle or combat program by using indirect weapons or area guns. These military systems have a fugitive rule-based mortar, artillery, air defense, etc. In cybersecurity, fuzzy logic helps develop indicators intended to alert system managers of potential or expected risks to cybersecurity.

3. Unmanned Combat Aerial Vehicles (UCAVs) are aviation that can operate independently to carry out combat missions. It has been shown that fading logical control systems have been applied most extensively and effectively. Learning Enhanced Tactical Handling Algorithm (LETHA) was designed to implement fuzzy control logic systems in challenging environments with excellent results.
4. The learning-enhanced tactical algorithm utilizes a Genetic Fuzzy Tree (GFT) and a Hoplological Autonomous Defend and Engage Simulation (HADES). HADES is an environment for the simulation that LETHA uses to train its fuzzy tree and is an adaptable, efficient, and realistic simulator.

12.4.10 Electronics

Fuzzy electronics is an electronic technology that uses fuzzy logic rather than the two-state Boolean logic, in general, used in digital electronics. Fuzzy electronics have a broad range of uses, such as control systems and artificial intelligence. A few examples can explain the use of fuzzy logic in electrical engineering [5, 15]:

1. As ordinary logic of action is used to drive a metro train, it speeds up to high speed and then jams on the brakes, speeds up again, slows, accelerates, slows, speeds, and finally jams the brakes to enter the station with a shouting after knocking down all the passengers. In contrast, using fuzzy logic, you can smoothly accelerate, sustain a steady high speed and slowly decelerate to the station.
2. The fuzzy logic, which was better than the conventional dedicated IC controller, was able to control a switch-mode voltage regulator.
3. A pressure controller in a steam boiler had a significantly better response, including an increase of at least twice as fast, when a Fuzzy Controller has been added.
4. Fuzzy controllers mounted in Japan in coolers, washing machines, and cameras provide inherent benefits.

12.4.11 Medical Science and Bioinformatics

Conventional quantitative approaches to diagnosis are unsuccessful due to the uncertainty of medical practice. There are common facts about the absence of knowledge and often its imprecise and conflicting existence in medicine. The causes of uncertainty may be defined as follows [11, 16]:

1. A patient's records.
2. The patient's medical history, which is normally given by the patient and/or their relatives. Generally, this is rather subjective and unprecise.
3. The Inspection Physique. The doctor typically gets objective evidence, but the distinction among normal and pathological status is not sharp in some instances.
4. Laboratory and other medical test findings are often prone to such anomalies and even to the patient's inappropriate behavior before the examination.
5. Simulated, exaggerated, understated symptoms can occur in the patient or even fail to mention any of them.
6. About the lack of natural classification, we emphasize the paradox of the growing number of mental disorders. Classification is difficult in critical (i.e., borderline) situations, particularly when considering a categorical diagnosis type.
7. Help devise for hospital diagnosis, arterial pressure control during anesthesia, multivariable anesthesia control.
8. Modeling of neuropathological findings, radiological diagnosis, in patients with Alzheimer's.
9. Fuzzy prostate cancer diagnosis and diabetes by implication.

In medicine, fuzzy logic plays an important part. The following are some examples that illustrate that several disease classes are crossed by fuzzy logic.

1. Predict the reaction to alcohol dependency treatment with Citalopram.
2. For diabetic neuropathy study and early diabetic retinopathy diagnosis.
3. Determining a sufficient dose of lithium.
4. To assess brain tissue volumes with magnetic resonance imaging (MRI) and to interpret functional MRI outcomes.

5. To characterize subtypes of stroke and co-existing ischemic stroke causes.
6. For radiation therapy to improve decision-making.
7. During anesthesia, to control hypertension.
8. Recognizing methods for flexor-tendon repair.
9. For the detection of cancer of the breast, prostate, or lung.
10. Helping to classify tumors of the central nervous system (astrocytic tumors).
11. To make a distinction between malignant melanomas and benign skin lesions.
12. To imagine the human brain's nerve fibers.
13. To explain drug use, quantitative figures.
14. To research the element of auditory P50 in schizophrenia.
15. Several other, to name a few, application fields are
    a. To discuss Fuzzy Epidemics.
    b. To make nursing choices.
    c. Electroacupuncture for conquering housing.

In bioinformatics, fuzzy logic and fuzzy technologies are still widely used. These examples are the following.

1. Improving the stability of motifs for proteins.
2. Polynucleotide Experiments on variants.
3. Using fuzzy adaptive resonance theory to test experimental speech results.
4. Centered on a fuzzy recast of a complex programming algorithm for arranging sequences.
5. DNA screening through fuzzy constructs of genetics.
6. To cluster genes from the results of microarrays.
7. Using the fuzzy k-nearest neighbor algorithm to infer protein subcellular positions from their dipeptide composition.
8. To model complex gene-influenced traits with fuzzy-valued outcomes in populations with a pedigree.
9. Using a method of fuzzy partitioning, fuzzy C-means are used to assign cluster membership values to genes.
10. Evolutionary comparison leads to hypothetical proteins' successful functional characterization to map complex sequence patterns to putative functional groups. The authors used a fuzzy alignment model.
11. To analyze the results of gene expression.

12. Using fuzzy alignment techniques to unravel functional and ancestral protein relationships or using the neural network's generalized radial basis architecture produces fuzzy classification rules.
13. To discover gene interactions and to decode a genetic network.
14. For the processing of microarray images with Complementary Deoxyribonucleic Acid (cDNA). Because of the large number of spots, the process should be automated, and it is achieved using a fuzzy vector filtering framework.
15. To categorize amino acid sequences into distinct superfamilies.

12.4.12 Robotics

A robot is described as a mechanical handler that can be programmed by the end-user, which moves materials or tools over-engineered trajectories to accomplish the desired task by application needs. A few applications are mentioned to understand the utility of fuzzy logic in robotics as follows [19, 20]:

1. A strongly non-linear coupling method is the robotic system. The robotic manipulator is a non-linear device. Traditional control methods can not be invented or approximated easily when the controller is built, fuzzy logic offers a viable way of managing non-linear structures. A fluorescent controller can be produced to simulate the output of a two-link manipulator.
2. For implementing the fuzzy rules if-then, the Fuzzy Inference Development Environment (FIDA) software is used designed by Aptronix. The two-link manipulator is simulated by c-language code.
3. Mobile robots with differential drives can be operated by fuzzy logic controllers. A fuzzy logic controller was thus applied in real time to control an independent mobile robot's movement. This type of control (with some amendments) can be utilized in many real-world applications, including interoffice mail, disaster area building investigations, autonomous vehicles, etc.
4. Fuzzy logic provides a modern robotic control solution that eliminates the two main problem areas of dynamic simulation and the whole operating system's time constraints.
5. Robots based on fuzzy logic all controllers can be approximated to biological behavior, such as eliminating barriers or the following of walls in a room.

12.4.13 Signal Processing and Wireless Communication

Wireless communication introduces seamless logical applications and concepts to explain productive use. It can be defined in two parts; channel estimation, channel equalization, and decoding areas. The first part uses fuzzy logic or hybrid fuzzy approaches and secondly, explains what is best suited to fuzzy logical techniques. Some descriptions of the fuzzy logic application in signal processing and communication are as follows [12, 19, 20]:

1. The channel estimate introduced fuzzy processes for fuzzy tracking based on Kosko's fuzzy associative memory models and TSK's. Fuzzy logic is also used in conjunction with adaptive algorithms such as LMS and RMS. The planning of membership function parameters in the fuzzy inference system is widely used in the neural network and adaptive algorithms.
2. In channel equalization, the two fuzzy Type-1 and, more recently, Type-2 systems were highly influential. To better represent the Bayesian decision solution, the Bayesian architecture, which incorporates fuzzy basis function and Gaussian membership functions, has been efficiently developed using human specialists' knowledge or training. Another category in channel equalization is the blind approach using Fuzzy-C means or Neural Network versions. The key advantages of using fuzzy logic methods are [8].
3. When adaptive techniques must be used, the fuzzy logic approaches work exceptionally well under non-linear conditions and time variations.
4. Complex and not fully understood models with minor degradation in effectiveness compared to standard methods may use fuzzy logic–based methods to achieve faster convergence and reduced complexity.

12.4.14 Transportation Problems

The transport problem can be seen as a very interesting area because it deals with many different decision-making alternatives. These criteria include assessments by experts in specific areas such as paving processes, transport management techniques, transportation planning, signal monitoring, etc. Most problems in the real world are dynamic, vague, or impressive, hardly numerical values understood by so-called crisp values. To deal with such problems, therefore, a tool is required. Fuzzy logic seems to be the most effective approach, and a few implementations are explained [3]:

1. Implementation of the fuzzy logic as the intersection control between various signals timing algorithms for passenger and motor vehicles. Fuzzy logic has helped to monitor the traffic signals situation with various solutions and vehicle movements.
2. The fuzzy logic was used to manage a signalized isolated pedestrian crossing in minimal waiting times, as well as to manage the multi-phase car controls to ensure the intersection between the signals-a combination of time value, effect, and traffic safety. The logic was used to control the signalized isolated traffic.
3. Fuzzy logic can be used as a module for the rehabilitation of the related fuzzy logic that can be used as a module for the rehabilitation of the related Life-Cycle Cost Analysis (LCCA) maintenance and rehabilitation problem in transport.
4. There are several factors in traffic accidents, such as the surroundings, driver conditions, road transport speeds, etc. It is an unpredictable and complex subject. Incremental cost-benefit analysis with elastic limitations and dynamic programming was used to address this problem.
5. Traffic regulation is also a dynamic social problem, like road, traffic, etc. For example, it can simulate the artificial intelligence policy framework for decisions that analyze innovation in the integration of infrastructures and land through transport corridor planning to explain highway alignment choices in environmental impact analysis.

According to the above explanation, several applications are available to solve problems in the real world, although, it is essential to know a few new ideas/subjects that are developed based on the concept of fuzzy logic and fuzzy set theory, such as:

- Alternative Set Theory
- Defuzzification
- Fuzzy Concept
- Fuzzy Mathematics
- Fuzzy Set Operations
- Fuzzy Subalgebra

- Interval Finite Element
- Linear Partial Information
- Multiset
- Neuro-Fuzzy
- Rough Fuzzy Hybridization
- Rough Set
- Sørensen Similarity Index
- Soft Fuzzy Set
- Type-2 Fuzzy Sets and Systems
- Uncertainty

12.5 Conclusion

In this chapter, an in-depth discussion on fuzzy logic is described with its applications in various fields such as Mathematics, Science, Medical, Bioinformatics, and Engineering, etc. It is required to understand that the concept of fuzzy logic has made a significant change in day-to-day life. It has helped researchers/scholars to develop new ideas in any specific field or multidisciplinary research.

References

1. Bezděk, Václav. "Using fuzzy logic in business." *Procedia Social and Behavioral Sciences* (2014).
2. Chen, Ching-Han, *et al.* "Fuzzy logic controller design for intelligent robots." *Mathematical Problems in Engineering* 2017 (2017).
3. Duong, Dieu Thi Xuan, and Veeris Ammarapala. "Fuzzy Logic Application in Transpotation Problems." *Proceedings of the 4th International Conference on Engineering, Project, and Production Management (EPPM 2013)*. 2013.
4. Dym C. Principles of mathematical modeling. Elsevier; 2004 Aug 10.
5. Emami, MR Sarmasti. "Fuzzy logic applications in chemical processes." *J. Math. Comput. Sci* 1.4 (2010): 339-348.
6. Erman, Maria, Abbas Mohammed, and Elisabeth Rakus-Andersson. "Fuzzy Logic Applications in Wireless Communications." IFSA/EUSFLAT Conf.. 2009.
7. Grigorie, Teodor Lucian, and Ruxandra Mihaela Botez. "New Applications of Fuzzy Logic Methodologies in Aerospace Field." *Fuzzy Controllers, Theory and Applications*. IntechOpen, 2011.

8. Karnik, Nilesh Naval, Jerry M. Mendel, and Qilian Liang. "Type-2 fuzzy logic systems." *IEEE Transactions on Fuzzy Systems* 7.6 (1999): 643-658.
9. Menzl, Stefan, Michael Stühler, and Roland Benz. "A self adaptive computer-based pH measurement and fuzzy-control system." *Water Research* 30.4 (1996): 981-991.
10. Novák, Vilém, Irina Perfilieva, and Jiri Mockor. *Mathematical Principles of Fuzzy Logic*. Vol. 517. Springer Science & Business Media, 2012.
11. Pérez-Neira, Ana, *et al.* "Neuro-fuzzy logic in signal processing for communications: from bits to protocols." International Conference on Nonlinear Analyses and Algorithms for Speech Processing. Springer, Berlin, Heidelberg, 2005.
12. Peri, Vamsi Mohan, and Dan Simon. "Fuzzy logic control for an autonomous robot." NAFIPS 2005-2005 Annual Meeting of the North American Fuzzy Information Processing Society. IEEE, 2005.
13. Sanchez-Roger, Marc, María Dolores Oliver-Alfonso, and Carlos Sanchís-Pedregosa. "Fuzzy logic and its uses in finance: A systematic review exploring its potential to deal with banking crises." *Mathematics* 7.11 (2019): 1091.
14. Singh Harpreet, *et al.* "Real-life applications of fuzzy logic." Hindawi Publishing Corporation. *Advances in Fuzzy Systems* (2013).
15. Stoian, Viorel, and Mircea Ivanescu. *Robot Control by Fuzzy Logic*. INTECH Open Access Publisher, 2008.
16. Torres, Angela, and Juan J. Nieto. "Fuzzy logic in medicine and bioinformatics." *Journal of Biomedicine and Biotechnology* 2006 (2006).
17. Tyagi, Lakshya, and Swati Singal. "Application of Fuzzy Logic Control Systems in Military Platforms." *2019 9th International Conference on Cloud Computing, Data Science & Engineering (Confluence)*. IEEE, 2019.
18. Uzunsoy, Erdem. "A brief review on fuzzy logic used in vehicle dynamics control." *Journal of Innovative Science and Engineering* (JISE) 2.1 (2018): 1-7.
19. Vashisth, H., and Peng-Yung Woo. "Application of fuzzy logic to robotic control." *Proceedings of the 1996 IEEE IECON. 22nd International Conference on Industrial Electronics, Control, and Instrumentation*. Vol. 3. IEEE, 1996.
20. Wakileh, B. A. M., and K. F. Gill. "Use of fuzzy logic in robotics." *Computers in Industry* 10.1 (1988): 35-46.
21. Zhang Jianwei. "Applications of fuzzy logic control in autonomous robot systems." *International Journal of Systems Science* 24.10 (1993): 1885-1904.

13

A Mathematical Approach Using Set & Sequence Similarity Measure for Item Recommendation Using Sequential Web Data

Vishal Paranjape, Dr. Neelu Nihalani, Dr. Nishchol Mishra and Dr. Jyoti Mishra*

Rajiv Gandhi Prodyogiki Vishwavidyalaya (RGPV), Jabalpur, Madhya Pradesh, India

Abstract

There has been an explosive growth of data and information in recent years with the coming of the World Wide Web. A major challenge in this arena is to serve the correct information to the correct person which adds up to a complex measure in efficient decision making. To solve these problems, the recommender system plays a vital role. Most of the e-commerce websites used today make use of recommender systems for effective decision making. Today's recommender system takes into account only the content information, ignoring the sequential details, which also play a vital role for recognizing the behavior of users. The present paper explores the different types of recommender techniques with their mathematical foundation and also discusses some of the problems in the prevailing system. Our proposed approach makes use of sequential patterns of web navigation along with the content information and is based on set and sequence similarity measure (S3M) for generating recommendations on web data. The paper makes use of mathematics involved in finding the set and sequence similarity for recommendation to user on CTI news dataset. To create suggestions for users, our proposed method uses the principle of upper approximation & singular value decomposition.

***Keywords*:** Recommendations, e-commerce, set similarity, sequence similarity, singular value decomposition

**Corresponding author*: jyoti.mishra198109@gmail.com

Ramakant Bhardwaj, Jyoti Mishra, Satyendra Narayan and Gopalakrishnan Suseendran (eds.) Mathematics in Computational Science and Engineering, (287–310) © 2022 Scrivener Publishing LLC

13.1 Introduction

Due to the vast growth of the World Wide Web it is difficult for users to choose relevant items due to information overload. A recommender system plays a major role in overcoming this problem to a great extent [1]. The Recommender Framework (RS) helps in suggesting best items to the users on the basis of their likings and allows in large information spaces to take decisions [2]. Based on their past purchasing habits, likes, feedback, demographics, etc., these systems are used to recommend items to users. Most of the companies today are deploying recommendation systems to boost their sales by suggesting items to their customers on the basis of their interests. Several companies like Amazon, NetFlix, YouTube and several other e-commerce-based companies are using this for efficient recommendations.

Recommender systems proposed in this era comprise both personalized and non-personalized recommendations. While a personalized recommender system uses information of user for making predictions, a non-personalized system does not take this into account [3].

Recommender systems have been in research since the mid-1990s as with the coming of the World Wide Web most of the things which went online and had any type of rating survey or reviews needed to provide best suggestions to their valuable customers [5]. Past ratings on a particular item serve as a tool for knowing the interests of the user and predicting items as part of future recommendations [4]. The more ratings being provided by the user the more will his likes or dislikes be known by the system and more accurate will be suggestions or recommendations. Most of the areas where this plays an important role include book recommendations, movie recommendations by Netflix, item recommendations by Amazon, etc. For movie recommendations, we present an example of user movie rating where the rating is given on a scale of 1-5 and is shown in Table 13.1 below.

In the above example five users have rated the movies with ratings on a scale from 1-5. Non-rated movies are shown by a '-' symbol. Now, on the basis of ratings given by the user, the recommendation system creates a profile by separating users on the basis of movie genre liking and ratings. Finally, a recommendation engine predicts a rating for an unrated movie and provides the user with suitable recommendation on that basis.

We will address the different types of recommendation systems in the next section. The following Figure 13.1 demonstrates a description of recommendation techniques:

Table 13.1 User-movie rating matrix.

User	Toy Story	Money Train	Othello	GoldenEye
A1	4	3	3	1
A2	5	-	2	1
A3	2	3	-	5
A4	1	4	1	-
A5	5	-	3	2

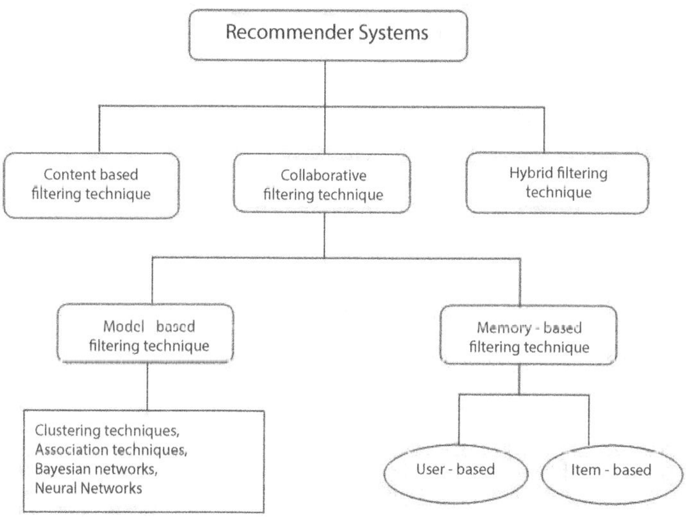

Figure 13.1 An overview of recommendation techniques [13].

(i) **Content-based filtering technique:** These systems suggest items based on the similarity of items liked by the same user in the past, irrespective of preference of other users [6]. For getting this information from the user, two types of feedback are taken into consideration, namely:

a) Explicit Feedback: This is provided by users in the form of ticking the "like"/"dislike" buttons or rating an item by number of stars, etc. Most of the time users do not provide explicit data which affects the accuracy of recommender systems.

b) Implicit Feedback: This is provided when a user visits a site, views a particular item out of several items, reads a manual related to the item and orders that item. These operations or likings of user are kept by the recommender engine for providing the next recommendations (Figure 13.2).

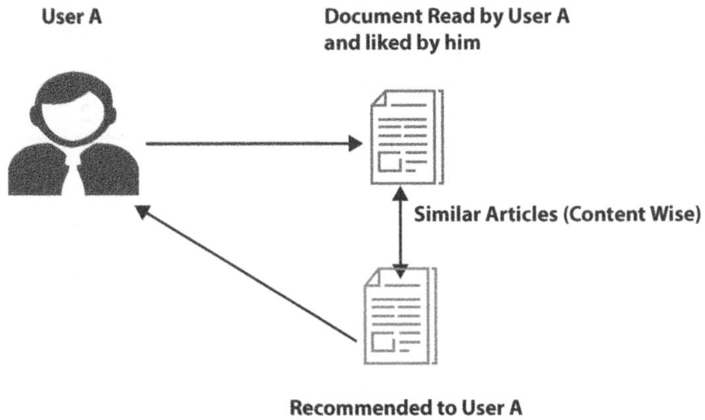

Figure 13.2 Content-based filtering technique [5].

They focus on past dealings of a given user and do not take other users into consideration [7]. Computation of attribute similarity of recent items and recommended similar items is the basic approach.

Based on the similarity between items for a particular user we find the utility of item in content-based recommendation methods based [8]. The root of the content-based filtering approach is information retrieval [9, 10] and information filtering [11], and mostly textual information like documents, websites (URLs) etc. are used by content-based filtering technique. One of the best IR techniques is Term Frequency & Inverse Document Frequency (TF-IDF) [10, 38, 39] and can be defined as:

Suppose N is the total number of documents to be suggested to users and the keyword ki appears in them, then the term frequency is given as

$$TF_{ij} = \frac{\text{fi, j}}{\max \text{fz, j}} \tag{13.1}$$

Where, $f_{i,j}$ = how many times k_i keyword occur in d_j document
$TF_{i,j}$ = term frequency of k_i keyword in d_j document
$\max_z f_{z,j}$ = max. over the frequencies $f_{z,j}$ of all keywords k_z

When we want to differentiate between relevant and non-relevant document Keywords play a major role in it. In association with the TF, the IDFi is used to define:

$$IDF_i = \log\left[\frac{N}{n_i}\right] \tag{13.2}$$

Then, in document dj, for the keyword kithe TF-IDF weight is specified as

$$w_{ij} = TF_{ij}\ X\ IDFi \tag{13.3}$$

The user's previously liked items are normally recommended by Content-based filtering technique [12–14]. Keyword analysis techniques are taken into consideration to create content-based profile (c) of the user on the basis of his taste and likings. These serve as a tool for recommending him item in the future. Content-based suggestions can be obtained by the Rocchio algorithm [15]. Utility function in content-based recommender system is given as:

$$\text{w(m,n)= score (CBP(m), content(n))} \tag{13.4}$$

where CBP= Content based profile

Recommendations for a specific item are now provided on the basis of the above-mentioned ContentBasedProfile(m) of users and document content(n). In addition, some scoring heuristics described in terms of Tc and Ts, such as cosine similarity measure [16, 17], provide the utility function.

$$\text{w}(m,n) = \cos\left(\vec{T}_C, \vec{T}_s\right) = \frac{\vec{T}_c . \vec{T}_s}{\|\vec{T}_c\|_2\ X\ \|\vec{T}_s\|_2} \tag{13.5}$$

$$= \frac{\sum_{i=1}^{k} w_{i,m} w_{i,n}}{\sqrt{\sum_{i=1}^{k} w_{i,m}^2} \Big/ \sqrt{\sum_{i=1}^{k} w_{i,n}^2}} \tag{13.6}$$

Where total number of keywords is 'k'.

There are some limitations with content-based filtering technique:

(i) **Problem with new user:** In order to really understand the interest and choices of the user, the user needs to give a sufficient number of ratings; only then will the technique be able to know choices.

(ii) **Analysis of limited content:** The filtering technique based on content suffers from the problem that if the same set of features is used it is difficult to differentiate between two items. Moreover, this technique works well when features are extracted from text documents, but since there is no automatic technique for feature extraction it is difficult to apply it to multimedia data [18].

(iii) **Overspecialization:** This problem recommends items to the user that are all of the same type. The consumer will not get the products outside the predefined limit, leading to poor recommendations [19].

(ii) **Collaborative filtering technique:** This approach (Figure 13.3) is based on a user's suggestion of an object based on reactions from similar users. This works by selecting a smaller collection of users from a wide community of individuals with tastes close to a single user. In this, the main recommendation principle is that other users offer ratings to a specific object.

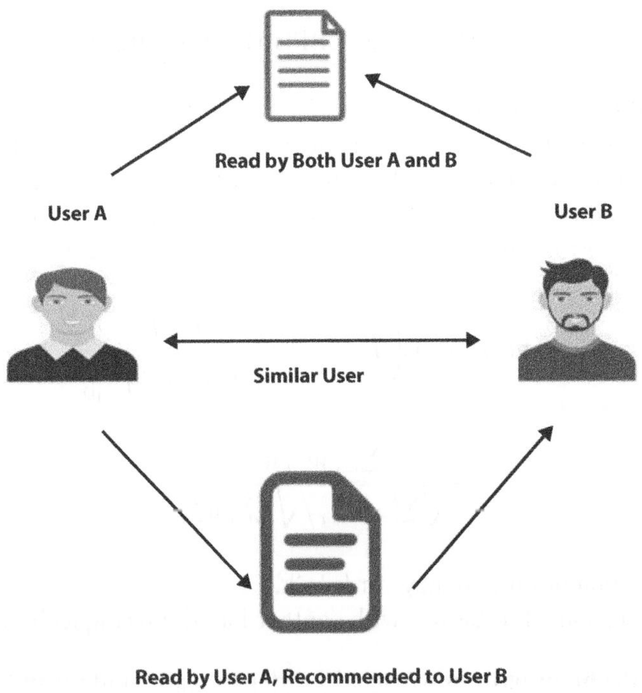

Figure 13.3 Collaborative filtering technique [5].

Measuring user similarity in Collaborative-based filtering technique:

(i) Pearson Correlation:

$$\text{sim (a,b)} = \frac{\sum_{P\in P}(ra.p-ra)(rb.p-rb)}{\sqrt{\sum_{p\in P}(ra.p-ra)2}\sqrt{\sum_{p\in P}(rb.p-rb)2}} \quad (13.7)$$

Where, a & b are users while $r_{a.p}$ is rating and P is set of items read by both users.

(ii) Cosine Similarity measure: It is measured by the angle between the vectors

$$\sin(\vec{m},\vec{n}) = \frac{\vec{m},\vec{n}}{|\vec{m}|*|\vec{n}|} \quad (13.8)$$

U: set of users who have rated both items m and n.

There are some challenges with collaborative-based filtering technique:

(i) **New user problem:** This is the same issue as addressed in content-based filtering, where the system is unable to suggest products if new users arrive, as it is not aware of the likes of new users. Often certain techniques based on item popularity, user personalization, item entropy, etc., are taken into account to give the best advice to new users in order to deal with these problems [20, 21].

(ii) **New item problem:** A main issue with recommending programmes is that they do not have any reviews when a new product comes to the system, so it is very difficult to recommend them to consumers.

(iii) **Sparsity:** Rating plays a very important role in efficient prediction but when users do not rate a particular item or if many ratings of an item are not done it leads to false prediction and hence recommendations of items are not up to the expectation of the customer. User profile information helps to overcome this problem to an extent; two users are treated as similar if they gave same rating to the items and they belong to the same demographic [22].

(iv) **Hybrid Methods:** A mix of both was taken into account and labelled as a hybrid approach in order to overcome the drawbacks of both the content-based filtering approach and collaborative filtering. Hybridization can be accomplished through the creation of a generic model integrating those characteristics of collaborative-based content and vice versa [23, 24].

Several hybrid recommendation systems propose collaborative and other approaches based on content. Gilbert Badaro *et al.* [25] suggested a hybrid approach to solve the problem of identifying in a user-item rating matrix the ratings of unrated objects through a weighted combination of collective filtering based on consumer and item.

13.2 Measures of Assessment for Recommendation Engines

In order to assess the efficiency of recommendation systems we need to measure some metrics like Recall, Precision, Root Mean Square Error, Mean Absolute Error, etc. [6, 8].

1. Recall: a measure of completeness, determines the fraction of relevant items retrieved out of all relevant items

$$Recall = \frac{T_P}{T_P + F_N} \tag{13.9}$$

   Here, T_P = No. of products suggested to a consumer and $T_P + F_N$ = Total products that consumer want.
2. Precision: a measure of exactness, determines the fraction of relevant items retrieved out of all items retrieved.

$$Precision = \frac{T_P}{T_P + F_P} \tag{13.10}$$

   Here, T_P = Number of products proposed to a consumer by RS and T_P + F_N = Total suggested products. The greater the precision, the better is the recommendation.
3. Root Mean Squared Error (RMSE): To measure the bias between actual and predicted value we use RMSE.

$$RMSE = \sqrt{\frac{\sum_{i=1}^{N}(Predicted_i - Actual_i)^2}{N}} \tag{13.11}$$

Here, the value given by the model is 'Predicted' and 'Actual' is the original value. The recommendation is to lower the RMSE value.

4. Mean Absolute Error (MAE): It is used to measure the difference between actual and predicted value.

$$MAE = \frac{1}{N}\sum|Predicted - Actual| \tag{13.12}$$

The lower the MAE rating, the better the recommendation would be.

13.3 Related Work

The system of recommendations was described as a means of supporting and enhancing the social process of using others' recommendations to make choices when there is insufficient personal knowledge or experience of alternatives [26]. The recommendation scheme is known as a user decision-making strategy in complex knowledge environments [27]. Recommender systems solve the issue of data overload that users usually face by presenting them with tailored, specific content and service recommendations. By the use of various available recommendation techniques different methods have recently been proposed [28–30]. The recommendation system has also been described as a tool from the perspective of e-commerce that allows users to search through records of information relevant to the interest and preference of users [31]. An online filtering framework using collective filtering for creating user-based profiles based on ratings on music album is Ringo [30]. Content-based filtering techniques typically base their predictions on user knowledge and, as in the case of collaborative techniques, they disregard inputs from other users [32, 33]. R. Mishra [34] proposed a technique for item recommendation using upper approximation & SVD. Programmes using content-based filtering to aid a user in finding information on the internet was proposed by Letizia [35]. It is the most mature and most widely applied collaborative filtering strategy. By finding other users with similar taste, mutual filtering recommends products; it uses their opinion to suggest items to the active consumer. In various technology fields, collaborative recommendation frameworks have been introduced. GroupLens is a news-based architecture that uses collaborative methods to help users find articles from broad news databases [36]. The system uses a user interface that lets users browse the internet; it is capable of monitoring a user's browsing pattern to predict the pages they may be interested in. An intelligent agent-based technique using naïve Bayesian classifiers to predict web pages of user interest was given by Pazzani *et al.* [37].

13.4 Methodology/Research Design

Our proposed recommender system is based on two prominent phases, clustering and classification. In the first phase we make clusters to acquire more and more knowledge to the system and then the classification technique is used later for better and enhanced recommendations. In our work we made use of similarity upper approximation technique for a better recommendation. S^3M technique is applied in our work to give better recommendations. After forming of clusters SVD technique was used to classify web user sessions.

The different steps involved in our work are as follows:

1. Web data collection through web logs.
2. Preprocessing involving user identification, data cleaning, conversion of categorical data, etc.
3. Clustering is done using clustering algorithm.
4. Identification of Top M clusters based on similarity.
5. Generation of response matrix using Top M clusters.
6. Creation of weight vector being filled using SVD.
7. Predictions and recommendation are obtained.

13.4.1 Web Data Collection Through Web Logs

Our proposed work is based on CTI news dataset obtained from China News channel which comprises topics based on 16 domains, the number coding for which is given below (Table 13.2):

Table 13.2 Number coding of web pages for CTI dataset.

Web page name	Number coding	Web page name	Number coding
Search	1	Resources	9
Programs	2	Authenticate	10
News	3	Cti	11
Admissions	4	Pdf	12
Advising	5	Calendar	13
Courses	6	Shared	14
People	7	Forums	15
Research	8	Hyperlink	16

The web navigational dataset of 10 users from CTI dataset is given below (Table 13.3) on the basis of their interests.

Table 13.3 Web navigational dataset.

T1	news people
T2	programs programs admissions programs courses
T3	resources forums
T4	courses courses courses courses
T5	courses people
T6	hyperlink news
T7	research research courses news news
T8	authenticate cti programs cti cti
T9	authenticate cti news
T10	people admissions cti cti admissions admissions people people

For recommendations it is very important to study similarity features either for content-based similarity where we are confined to similarity between items, or, as in this work, where we have to perform clustering based on sequential data to find similarity between users based on web sequential data. To find similarity in content-based model we use Jaccard & Dice similarity, whereas for sequence-based similarity we use Longest Common Sequence (LCS) and hamming distance, etc.

Our work is based on a hybrid similarity model where we used both a content-based as well as sequence-based approach to form a new one called S^3M for clustering. In S^3M approach both content as well as order or sequence is kept in mind to build similarity matrix.

The following steps of algorithm outline the formation of clusters from sequential data:

1. First we find similarity between 2 web user session S1 & S2 using function Sim(S1,S2)
2. Next we find the first similarity upper approximation using threshold $\delta = 0.2$
3. In this step we remove redundant subsets.
4. Finally we identify cluster centres and remove them from others.

To find Sim(S1, S2) we use the formula
For this we have to calculate, SeqSim(S1,S2) and SetSim(S1,S2)

$$SeqSim(s_1,s_2) = \frac{\text{LLCS}}{max(|\text{S1}|,|\text{S2}|)} \tag{13.13}$$

$$SetSim(s_1,s_2) = \frac{|\text{S1} \cap \text{S2}|}{\text{S1 U S2}} \tag{13.14}$$

Finally Set & Sequence similarity is given as:

$$S^3M = p \text{ X } SeqSim(S1,S2) + (1-p) \text{ X } SetSim(S1,S2) \tag{13.15}$$

For i^{th} element of S1 & j^{th} element of S2

Let us consider 10 navigation patterns T1,T2,……,T9 and T10 obtained from CTI dataset. S3M technique is used to find similarity with p=0.7. We can compute S^3M between two sequences as follows:

Suppose web navigation of 2 users are as given below:

T1= 1,4,5,8,12,15,18

T2= 3,4,5,12,18,20,22,26

|L1|= 7

|L2|= 8

SeqSim between T1&T2 can be calculated using eqn 13.12.

LLCS= 4 , max(L1,L2) = 8 , SeqSim = (4/8)= 0.5

SetSim (Jaccard Similarity) between T1&T2 can be calculated using eqn 13.13.

SetSim= (4/11) = 0.3636

Let us take, p=0.9 , q=0.1
We get, Set & Sequence Similarity Measure = 0.486

Step 1 : Similarity Table 13.4 below depicts similarity between users.

Table 13.4 Similarity matrix for the sequences.

	T1	T2	T3	T4	T5	T6	T7	T8	T9	T10
T1	1	0	0	0	0.45	0.45	0.21	0	0.30	0.16
T2	0	1	0	0.24	0.21	0	0.20	0.20	0	0.14
T3	0	0	1	0	0	0	0	0	0	0
T4	0	0.24	0	1	0.32	0	0.24	0	0	0
T5	0.45	0.21	0	0.32	1	0	0.21	0	0	0.15
T6	0.45	0	0	0	0	1	0.21	0	0.30	0
T7	0.21	0.20	0	0.24	0.21	0.21	1	0	0.20	0
T8	0	0.20	0	0	0	0	0	1	0.43	0.13
T9	0.30	0	0	0	0	0.30	0.20	0.43	1	0.13
T10	0.16	0.14	0	0	0.15	0	0	0.13	0.13	1

The diagonal entries in the above table is 1 as it is the similarity comparison of a sequence it itself. The values in the above table is 0 where there is no matching similarity in the sequence; other values are calculated using S^3M technique as explained above.

Step 2: Let us take similarity threshold δ=0.2, now we discard every row value which is less than the threshold to get upper approximation, so the resultant rows will be as follows: R(1) = T1,T5,T6,T7,T9 ; R(2) = T2,T4,T5,T7,T8 ; R(3) = T3 ; R(4) = T2, T4,T5,T7 ; R(5) = T1,T2,T4,T7 ; R(6) = T1,T6,T7,T9 ; R(7) = T1,T2,T4,T5,T6,T7,T9 ; R(8) = T2,T9 ; R(9) = T1,T6,T7,T8,T9.

Step 3: Various repeated subsets occur in upper approximation which is removed which gives the following: R(1) = T1,T5,T6,T7,T9; R(2) = T2,T4,T5,T7,T8; R(3) = T3; R(5) = T1,T2,T4,T7; R(7) = T1,T2,T4,T5,T6, T7,T9; R(9) = T1,T6,T7,T8,T9.

Step 4: Now discard cluster centers.

13.4.2 Web User Sessions Classification

Now response matrix for new user is generated. Here we have generated this matrix using CTI dataset. Matrix "A" contains sixteen columns as we have 16 categories in CTI dataset.

Response for the first user A1 is given by (4,0,0,2,0,7,3,6,0,0,1,4,5,6,8,5)

The response matrix is depicted by Table 13.5 which is of dimension (20 X 16). For determining recommendations we need to break our matrix using Singular Value Decomposition (SVD) technique.

Now we apply SVD on the given matrix using Matlab to break our matrix into 3 matrices,U, S & V, where U is user specific & SV are feature specific.

Recommendations for the web user

Now weight vector for new user is created from response matrix. Suppose for the first 5 visit sequential pattern is given by {2,7,6,1,5}.

Now $W_{i,j}$ is weight of page 'i' visited by user in 'jth' position & can be calculated as:

$$w_{ij} = \frac{|Vi,j|}{|Vi|} \tag{13.16}$$

Table 13.5 Sample response matrix for User.

	T1	T2	T3	T4	T5	T6	T7	T8	T9	T10	T11	T12	T13	T14	T15	T16
A1	4	0	0	2	0	7	3	6	0	0	1	4	5	6	8	5
A2	3	11	4	5	8	9	1	2	4	1	1	0	0	4	2	6
A3	3	3	3	1	1	4	6	7	0	0	2	0	3	17	3	7
A4	6	5	4	12	6	7	1	1	4	9	0	3	5	0	0	2
A5	7	9	5	1	4	3	7	8	18	4	8	0	11	8	5	6
A6	8	0	0	22	3	17	8	15	4	8	5	0	1	1	4	2
A7	0	5	7	1	1	3	16	8	5	7	2	5	7	7	9	4
A8	2	8	0	12	9	7	4	6	2	11	0	0	3	6	2	8
A9	0	7	6	3	1	1	7	4	9	24	8	2	1	0	0	4
A10	1	3	2	4	5	3	7	2	2	8	11	0	2	3	3	1

(*Continued*)

Table 13.5 Sample response matrix for User. (*Continued*)

	T1	T2	T3	T4	T5	T6	T7	T8	T9	T10	T11	T12	T13	T14	T15	T16
A11	7	2	3	5	1	0	0	3	1	2	3	1	1	13	11	7
A12	8	1	1	6	0	9	6	22	4	11	7	2	0	0	8	4
A13	9	1	0	0	3	5	6	7	22	0	7	5	9	12	5	9
A14	5	4	7	8	7	1	1	0	0	6	11	17	8	5	9	10
A15	3	8	8	1	6	7	6	9	2	4	1	5	5	21	1	0
A16	2	9	4	1	2	9	6	1	0	0	3	7	11	24	8	2
A17	5	0	5	5	7	1	2	6	5	4	8	11	0	7	3	1
A18	4	12	7	6	9	12	1	0	7	3	12	6	5	7	1	0
A19	2	3	2	8	4	6	7	8	19	8	6	4	9	1	0	11
A20	8	6	5	8	3	1	1	0	5	2	6	9	7	4	2	1

Where

$|V_{ij}|$ = No. of times page 'i' is in 'jth' position.

$|V_i|$ = No. of times 'ith' page been in all positions.

Weight calculation is explained as:

Let the S1,S2,S3,S4 be four sequences of length "6"

S1= 5,3,9,4,1,6

S2= 4,2,8,1,5,8

S3= 3,24,5,6,9

S4= 5,7,21,8,2

Suppose new user visits page in the order {5,2,8,1,9}

Now the weights can be calculated as follows :

$$W_{51}= 0.2,\ W_{22}= 0.5,\ W_{83}= 0.33,\ W_{14}= 0.5 \text{ and } W_{95}= 0.5$$

Now we need to find weight for next page visit Wk6 where, k= {1,2,3,........,16}. For new user weight vetor P1 of length 16 will be formed. Weight of corresponding page would be entered.

Initially P1= {0.5,0.5, X,X,0.25,X,X,0.33,0.5,X,X,X,X,X,X,X}. To calculate the unknown entries of 'P1' we use:

$$R_d - \Sigma_k U_{ik} S_{kk} V_{jk} \tag{13.17}$$

Where,

R_d = Length of dth page

U,S,V = decomposed matrix obtained by applying singular value decomposition of matrix 'A'.

i = ith user

k = kth feature

In the above case R_1= 0.5 , R_2= 0.5, R_5 = 0.25, R_8 = 0.33, R_9 = 0.5 are known for the new user.

The mathematical formulation can be shown as: {52819}

$$R_5 = U_5S_{55}V_{55} + U_2S_{22}V_{52} + U_8S_{88}V_{58} + U_1S_{11}V_{51} + U_9S_{99}V_{59} \tag{13.18}$$

$$R_2 = U_5S_{55}V_{25} + U_2S_{22}V_{22} + U_8S_{88}V_{28} + U_1S_{11}V_{21} + U_9S_{99}V_{29} \tag{13.19}$$

$$R_8 = U_5S_{55}V_{85} + U_2S_{22}V_{82} + U_8S_{88}V_{88} + U_1S_{11}V_{81} + U_9S_{99}V_{89} \quad (13.20)$$

$$R_1 = U_5S_{55}V_{15} + U_2S_{22}V_{12} + U_8S_{88}V_{18} + U_1S_{11}V_{11} + U_9S_{99}V_{19} \quad (13.21)$$

$$R_9 = U_5S_{55}V_{95} + U_2S_{22}V_{92} + U_8S_{88}V_{98} + U_1S_{11}V_{91} + U_9S_{99}V_{99} \quad (13.22)$$

Now we use matlab to solve the following set of equations in order to get the values of U5, U2, U8,U1,U9.

$$0.25 = U_5 \times 24.86 \times (-0.1023) + U_2 \times 40.10 \times (-0.0006) + U_8 \times (14.6993) \times (0.1878) + U_1 \times 92.40 \times (-0.1905) + U_9 \times 12.99 \times (-0.0055)$$

$$0.25 = U_5\,(-4.733) + U_2(-0.024) + U_8\,(2.760) + U_1\,(-17.602) + U_9\,(0.071) \quad (13.23)$$

$$0.5 = U_5 \times 24.86 \times (0.1904) + U_2 \times 40.10 \times (-0.1194) + U8 \times (14.6993) \times (0.0401) + U_1 \times 92.40 \times (-0.2342) + U_9 \times 12.99 \times (0.3926)$$

$$0.5 = U_5\,(4.733) + U_2\,(-4.787) + U_8\,(0.589) + U_1\,(-21.64) + U_9\,(5.099) \quad (13.24)$$

$$0.33 = U_5 \times 24.86 \times (0.0401) + U_2 \times 40.10 \times (0.2430) + U8 \times (14.6993) \times (0.1425) + U_1 \times 92.40 \times (-0.2944) + U_9 \times 12.99 \times (-0.0865)$$

$$0.33 = U_5\,(0.996) + U_2\,(9.744) + U_8\,(2.094) + U_1\,(-27.202) + U_9\,(-1.123) \quad (13.25)$$

$$0.5 = U_5 \times 24.86 \times (-0.3437) + U_2 \times 40.10 \times (0.0242) + U8 \times (14.6993) \times (0.2810) + U_1 \times 92.40 \times (-0.2113) + U_9 \times 12.99 \times (-0.0153)$$

$$0.5 = U_5\,(-8.544) + U_2\,(0.970) + U_8\,(4.130) + U_1\,(-19.52) + U_9\,(-0.198) \quad (13.26)$$

$$0.5 = U_5 \times 24.86 \times (-0.2438) + U_2 \times 40.10 \times (0.1449) + U8 \times (14.6993) \times (0.0740) + U_1 \times 92.40 \times (-0.3051) + U_9 \times 12.99 \times (-0.1569)$$

$$0.5 = U_5\,(-6.06) + U_2\,(5.810) + U_8\,(1.087) + U_1\,(-28.191) + U_9\,(-2.038) \tag{13.27}$$

Solving equation 13.23, 13.24, 13.25, 13.26 & 13.27 in matlab we get

```
>> A = [-4.733 -0.024 2.760 -17.602 0.071 ; 4.733-4.787 0.589
-21.64 5.099; 0.996 9.744 2.094 -27.202 -1.123; -8.544 0.970 4.130
-19.52 -0.198; -6.06 5.810 1.087 -28.191 -2.038];

>> B = [0.25;0.5;0.33;0.5;0.5]

B =
    0.2500
    0.5000
    0.3300
    0.5000
    0.5000

>> U3 = inv(A)*B                                    (13.28)

U3 =
   -0.0936 = U5
    0.0869 = U2
   -0.0921 = U8
   -0.0025 = U1
           = U9
```

13.5 Finding or Result

Now using the formula given below and placing the respective values we can calculate the weights (Table 13.6) based on equation 13.17 and then we number it in terms of descending order as the one having highest value is the top most prediction or suggestion.

$$R_d = U_5S_{55}V_{d5} + U_2S_{22}V_{d2} + U_8S_{88}V_{d8} + U_1S_{11}V_{d1} + U_9S_{99}V_{d9}$$

Table 13.6 Recommendations based on weight.

Page	Weight	Suggestion
R1	0.5	6
R2	0.5	7
R3	0	13
R4	1.35	4
R5	0.25	10
R6	2.01	1
R7	0.12	12
R8	0.33	9
R9	0.5	8
R10	0	14
R11	1.39	3
R12	0	15
R13	0.20	11
R14	0	16
R15	1.68	2
R16	0.68	5

13.6 Conclusion and Future Work

As information on digital platforms increases due to vast growth of the internet it is important to keep updated with the users' browsing history. Several e-commerce applications deal with the preferences of users for a particular item, so that they can offer the customer the right product. A major challenge in this arena is to serve the correct information to the correct person, which adds to a complex measure in efficient decision making. The present paper explores the different types of recommender techniques with their mathematical foundation and also discusses some of the problems in the prevailing system. Our proposed approach makes use of sequential patterns of web navigation along with the content information and is based on set and sequence similarity measure (S3M) for generating recommendations on web data. The work was conducted on CTI dataset. Matlab tool was used for SVD matrix decomposition and equation solving. In future, web-based recommendations can also be obtained by machine learning techniques for better and more accurate recommendations.

References

1. John K. Tarus, Zhendong Niu, Dorothy Kalui (2017), "A hybrid recommender system for e-learning based on context awareness and sequential pattern mining", Springer-Verlag GmbH Germany, DOI 10.1007/s00500-017-2720-6.
2. Konstan, J.A., Riedl, J., Recommender systems from algorithms to user experience. User Model User-Adapt Interact 22, 101–123. Springer Science + Business Media (2012).
3. Ricci F, Rokach L, Shapira B, Kantor P (2011), *Recommender Systems Handbook*. Springer, Berlin.
4. K. N. Asha and R. Rajkumar, "A Comprehensive Survey on Web Recommendations Systems with Special Focus on Filtering Techniques and Usage of Machine Learning", School of Computing Science and Engineering, Vellore Institute of Technology, https://doi.org/10.1007/978-3-030-37218-7_106 Springer 2020.
5. M. J. Pazzani, "A framework for collaborative, content-based and demographic filtering," *Artif. Intell. Rev.*, vol. 13, nos. 5_6, pp. 393_408, Dec. 1999.
6. G. Adomavicius, and A. Tuzhilin, "Toward the Next Generation of Recommender Systems: A Survey of the State-of-the-Art and Possible Extensions", *IEEE Transactions on Knowledge and Data Engineering*, Vol. 17, No. 6, June 2005.
7. M. Kavitha Devi, P. Venkatesh, "An Improved Collaborative Recommender System", *2009 First International Conference on Networks & Communications-2009 IEEE* DOI 10.1109/NetCoM.2009.69.
8. Gediminas Adomavicius and Alexander Tuzhilin, "Toward the Next Generation of Recommender Systems: A Survey of the State-of-the-Art and Possible Extensions", *IEEE Transactions an Knowledge And Data Engineering*, Vol. 17, No. 6, June 2005.
9. R. Baeza-Yates and B. Ribeiro-Neto, *Modern Information Retrieval*. Addison-Wesley, 1999.
10. G. Salton, *Automatic Text Processing*. Addison-Wesley, 1989.
11. N. Belkin and B. Croft, "Information Filtering and Information Retrieval," *Comm. ACM*, vol. 35, no. 12, pp. 29-37, 1992.
12. K. Lang, "Newsweeder: Learning to Filter Netnews," *Proc. 12th Int'l Conf. Machine Learning*, 1995.

13. R.J. Mooney and L. Roy, "Content-Based Book Recommending Using Learning for Text Categorization," *Proc. ACM SIGIR '99 Workshop Recommender Systems: Algorithms and Evaluation,* 1999.
14. M. Pazzani and D. Billsus, "Learning and Revising User Profiles: The Identification of Interesting Web Sites," *Machine Learning,* vol. 27, pp. 313-331, 1997.
15. J.J. Rocchio, "Relevance Feedback in Information Retrieval," *SMART Retrieval System—Experiments in Automatic Document Processing,* G. Salton, ed., chapter 14, Prentice Hall, 1971.
16. R. Baeza-Yates and B. Ribeiro-Neto, *Modern Information Retrieval.* Addison-Wesley, 1999.
17. G. Salton, *Automatic Text Processing.* Addison-Wesley, 1989.
18. U. Shardanand and P. Maes, "Social Information Filtering: Algorithms for Automating 'Word of Mouth,'" *Proc. Conf. Human Factors in Computing Systems,* 1995.
19. D. Billsus and M. Pazzani, "User Modeling for Adaptive News Access," *User Modeling and User-Adapted Interaction,* vol. 10, nos. 2- 3, pp. 147-180, 2000.
20. A.M. Rashid, I. Albert, D. Cosley, S.K. Lam, S.M. McNee, J.A. Konstan, and J. Riedl, "Getting to Know You: Learning New User Preferences in Recommender Systems," *Proc. Int'l Conf. Intelligent User Interfaces,* 2002.
21. K. Yu, A. Schwaighofer, V. Tresp, X. Xu, and H.-P. Kriegel, "Probabilistic Memory-Based Collaborative Filtering," *IEEE Trans. Knowledge and Data Eng.,* vol. 16, no. 1, pp. 56-69, Jan. 2004.
22. M. Pazzani, "A Framework for Collaborative, Content-Based, and Demographic Filtering, *Artificial Intelligence Rev.,* pp. 393-408, Dec. 1999.
23. M. Balabanovic and Y. Shoham, "Fab: Content-Based, Collaborative Recommendation," *Comm. ACM,* vol. 40, no. 3, pp. 66-72,1997.
24. C. Basu, H. Hirsh, and W. Cohen, "Recommendation as Classification: Using Social and Content-Based Information in Recommendation," *Recommender Systems. Papers from 1998 Workshop Technical Report WS-98-08,* AAAI Press 1998.
25. Gilbert Badaro, Hazem Hajj1, Wassim El-Hajj, Lama Nachman, "A Hybrid Approach with Collaborative Filtering for Recommender Systems", 978-1-4673-2480-9/13/$31.00 ©2013 IEEE.
26. Resnick P, Varian HR. Recommender Systems. *Commun ACM* 1997;40(3):56–8. http://dx.doi.org/10.1145/245108.24512.
27. Rashid AM, Albert I, Cosley D, Lam SK, McNee SM, Konstan JA *et al.* Getting to know you: learning new user preferences in recommender systems. In: *Proceedings of the international conference on intelligent user interfaces; 2002.* pp. 127–34.
28. Acilar AM, Arslan A. A collaborative filtering method based on Artificial Immune Network. *Exp Syst Appl* 2009;36(4):8324–32.

29. Chen LS, Hsu FH, Chen MC, Hsu YC. Developing recommender systems with the consideration of product profitability for sellers. *Int J Inform Sci* 2008;178(4):1032–48.
30. Jalali M, Mustapha N, Sulaiman M, Mamay A. WEBPUM: a web-based recommendation system to predict user future movement. *Exp Syst Applicat* 2010;37(9):6201–12.
31. Schafer JB, Konstan J, Riedl J. Recommender system in ecommerce. In: *Proceedings of the 1st ACM conference on electronic commerce; 1999*. pp. 158–66.
32. Min SH, Han I. Detection of the customer time-variant pattern for improving recommender system. *Exp Syst Applicat* 2010;37(4):2911–22.
33. Celma O, Serra X. FOAFing the Music: bridging the semantic gap in music recommendation. *Web Semant: Sci Serv Agents World Wide Web* 2008;16(4):250–6.
34. R.Mishra, P. Kumar,B.Bhasker, "A web recommendation system using sequential information", Decision support system, 10.1016/j.dss.2015 Elsevier.
35. Lieberman H. Letizia: an agent that assists web browsing. In: *Proceedings of the 1995 international joint conference on artificial intelligence. Montreal, Canada; 1995*. pp. 924–9.
36. Adomavicius G, Tuzhilin A. Toward the next generation of recommender system. A survey of the state-of-the-art and possible extensions. *IEEE Trans Knowl Data Eng* 2005;17(6):734–49.
37. Pazzani MJ. A framework for collaborative, content-based and demographic filtering. *Artific Intell Rev* 1999;13:393–408, No. 5(6).
38. J. Mishra, Fractional hyper-chaotic model with no equilibrium, *Chaos, Solitons & Fractals* 116, 43-53.
39. J. Mishra, Modified Chua chaotic attractor with differential operators with non-singular kernels, *Chaos, Solitons & Fractals* 125, 64-72.

14

Neural Network and Genetic Programming Based Explicit Formulations for Shear Capacity Estimation of Adhesive Anchors

Tawfik Kettanah[1] and Satyendra Narayan[2]*

[1]Department of Architectural Technology, Sheridan Institute of Technology, Hazel McCallion Campus, Mississauga, ON, Canada
[2]Department of Applied Computing, Sheridan Institute of Technology, Brampton, ON, Canada

Abstract
This paper uses the Artificial Intelligence methodology of Neural Network (NN) and Genetic Expression Programming (GEP) to provide an explicit formula for estimating and predicting the shear capacity of a single adhesive anchor post. The adhesive anchor is usually installed into uncracked hardened concrete.

The American Concrete Institute (ACI) Committee has already developed a database for the adhesive anchors and it is available in public domain. An Article 355 has been received from Dr. Ronald A. Cook and filtered into a file of relevant records only. The filtered records are separated into training file and testing file for further use in the NN and GEP computerized models. The NN and GEP models are developed with eleven input parameters, and one hidden layer for output of shear capacity. The generated NN has a high correlation coefficient for testing (0.9725), whereas the error percentage is 2.03% for testing. The generated GEP model has a correlation coefficient of 0.9719 for testing, whereas the Mean Absolute Error (MAE) is 1.58% for testing. This model has been selected for genetic programming, which has recorded a very good "Best Fitness" value (827 out of 1000) despite the noisy nature of the input parameters.

Keywords: Adhesive anchors, genetic programming, neural network, shear capacity

**Corresponding author*: Satyendra.narayan@sheridancollege.ca

Ramakant Bhardwaj, Jyoti Mishra, Satyendra Narayan and Gopalakrishnan Suseendran (eds.) Mathematics in Computational Science and Engineering, (311–328) © 2022 Scrivener Publishing LLC

14.1 General Introduction

Adhesive anchors are commonly used as attachments to hardened concrete. These attachments transfer load in different formats such as shear, torque, and tensile to concrete member in which they are embedded. These attachments in some cases connect to two parts of the structure or carry a combination of variety of loads (e.g., tensile, shear, torque) which need to be estimated for design or assessment purposes. There are a variety of anchors currently in use that serve as attachments to structural concrete members. Anchors can be classified into two main categories: cast-in-place and post-installed. The cast-in-place anchors are installed into fresh concrete and post-installed anchors are installed into hardened concrete. The post-installed anchors are more popular due to the flexibility in location, type of anchor, strength capacity of the anchor, as well as the ability to be installed at any phase of a structure's life. The post-installed anchors are further classified into two categories: post-installed mechanical anchors and post-installed bonded anchors. The post-installed mechanical anchors are classified into expansion anchors and undercut anchors. The post-installed bonded anchors are classified into adhesive anchors and grouted anchors [1]. The difference between "adhesive" anchors and the "grouted" anchors is the size of the drilled hole and the materials used as a bonding agent. The diameter of the drilled holes for adhesive type anchors is 10-25% more than the diameter of the bolt. The diameter of the drilled holes for the grouted type anchors is at least 150% larger than the diameter of the anchor [2]. The bonding agent for these anchor bolts is mainly epoxies, vinylesters, and polyesters whereas the grouted anchors use cementitious or polymer grouts [3].

The adhesive anchors are made up of a reinforced steel bar or threaded steel. A hardened concrete usually contains drilled holes. The adhesive anchors are inserted into these drilled holes and then they are bonded with a bonding agent, which is filled into the gap between the anchor bolt and the drilled hole.

All of the aforementioned members of the adhesive anchor's assemblage make their own individual contribution to the overall performance of the anchor. The contribution is based on each member's individual quality, and the overall quality of the final assemblage. A bonding agent with a specific strength to bond steel to concrete will present different performance matrices under less favorable surface conditions pertaining to either the anchor bolt or the concrete. Considering all possible combinations, this makes estimating the performance with the use of a standard provision less reliable for cases of installation done under poorly controlled construction sites.

The adhesive anchor demonstrates variety of modes of failure based on both the strength of the components for anchor assemblage, as well as the geometric parameters. The main components include the anchor bolt, the bonding agent, and concrete, whereas the geometric parameters are embedment depth and the diameter of anchor. These modes of failure are steel failure and concrete breakout for anchors near an edge [ACI 355.2-11].

The document ACI 318-02 provides mechanism for computing shear strengths for both cast-in-place and the post-installed anchors in the cracked and uncracked concretes. However, this mechanism is not pertinent for adhesive type of anchors.

Precise shear capacity or shear strength estimating tools are currently not available. As a result, industry encourages to conduct product- and condition-specific testing as an alternate solution to this situation. The test results provide more reliable values for shear capacity or shear strength of adhesive anchors. These approaches are expensive and require expertise to perform tests which are commonly practiced in the construction industry.

The objective of this paper is to provide a mathematical tool to predict the shear capacity or shear strength of an anchor without any expansive laboratory testing and expertise required. The Artificial Intelligence (AI) techniques are well-suited for assessment and prediction purposes. This is well recognized in civil and structural engineering. AI develops a mathematical model and consequently presents a mathematical formula for solution [3].

Artificial Intelligence-based software (NN & GEP) has been used as a tool for predicting performance of structural concrete members over the last several years. AI has the ability to accept and process input of varying data-types (numeric and symbolic) to efficiently predict performance. This makes AI ideal methodology for use in situations involving the prediction of shear capacity in adhesive anchors due to the nature of the input parameters [4, 5]. The input parameters include the diameter and embedment depth which are numeric, and the bonding agent type (epoxies, vinylesters, polyesters, etc.) which is symbolic.

14.2 Research Significance

The document ACI 318-06 contains an appendix titled "Appendix D-Anchoring to Concrete". This appendix provides mechanism for computing load-carrying capacity for post-installed anchors. The post-installed anchor is a category of anchors that includes mechanical (Expansion and Undercut) and the bonded anchors (Adhesive and Grouted). The formulas used in this document are generic in nature and do not represent adhesive anchors appropriately.

The lack of provisions directly related to adhesive anchors (ACI 355.Y Draft is in public discussion still) and to issues involving the variety of the assemblage elements demand formulation of a new method that is more efficient in estimating shear capacity accurately.

The objective of this paper is to provide a mathematical formula for estimating shear carrying capacity of an adhesive anchor by using Artificial Intelligence (NN & GEP). The proposed tool is to generate a mathematical formula and a model which is built based on a range of material types and conditions.

14.3 Biological Nervous System

The Artificial Neural Network (ANN) simulates the nervous system of the human body. ANN program receives input and changes the state based input. After processing the input data, it produces output in the response.

To understand ANN fairly, it is required to explore the way in which the nervous system works with a main focus on neurons. The neurons are the basic units of the nervous system. By analyzing the structure and how the neurons are communicating with other neurons, one may acquire the understanding of the ANN.

Neurons are the main processing unit in the brain; the brain is composed of billions of these units, which are interconnected, and operate parallel to each other. The neurons of each layer function are independent and are not connected to the same layer's neurons. They receive input, process the data, and forward the information to the next layer or output [6]. Each neuron is composed of a body (called Soma), an input receiving element (called Dendrites), a forwarding data element (called Axon) and a connecting element to another neuron (called Synopsis).

The neurons receive, process data, and forward information to other neurons via Axons in the form of electrochemical pulses. The magnitude of these pulses (weights) varies from very weak that will not transmit to strong enough that will be transmitted to the next neuron.

The magnitude of these pulses (weights) changes as the nervous system works on processing certain tasks repetitively. The processing starts with forward modelling to produce an outcome and compare such outcome to the desired outcome. The result of the comparison dictates whether there is a need to repeat the process (forward modelling) or not. If the comparison is not good, the processing model parameters are adjusted to produce a new outcome until the comparison between the desired outcome and outcome produced by the forward model is fair. This process of repetition is usually called the learning process, or back propagation in the ANN [7].

There are two important tasks in the biological neural system modelling: the ability to learn to achieve certain tasks and the use of an existing methodology (called learned method) to produce an outcome for a given input or group of inputs.

The structure of the ANN system is normally modeled on the basis of biological neural system which is composed of three main layers: the input, hidden layer or processing layer, and the output layer for the result.

The input layer is defined by a group of parameters. This includes the properties and the features of the item and its surroundings which is presented by the adhesive anchor bolt and concrete. The ANN is capable of accepting all types of data as an input which helps to include all the related properties and features of a model. The proposed research takes into the consideration of all related features and properties that could possibly have an effect on the shear capacity (outcome) of an anchor such as anchor embedment depth, bonding agent type, concrete strength, etc., as parameters (see Table 14.1).

The input parameters are connected to the hidden layer nodes or processing elements. Each input parameter is connected to every processing unit in the hidden layer.

The hidden layer or processing layer must have a certain number of Processing Elements (PE). This number is determined through the trial and error method to select the appropriate number of PE for the model. The process of learning is repeated a number of times in this study until experimental and computed results are matched. The observation of the results and use of comparison tools such as Correlation Coefficient ("r"), Error Percent (E%), Mean Absolute Error (MAE), and Mean Square Error (MSE) when applicable are used to find the best combination of the model structure element's numbers. The model's structure elements are networks, PE, number of layers, learning method, Threshold size, and number of epochs required to constitute the most critical factor for the establishment of the optimum model [8]. Once the best model is developed, the mathematical formula is finalized for the implementation in predicting the shear capacity of the anchors.

Each processing element performs two main operations: weighted summation of input, which is calculated by summing the received input multiplied by the weight added to the bias as in the following equation:

$$Net_j = \sum_{i=1}^{n} w_{ij} * x_i + b$$

Here x_i is the input value of ith neuron, w_{ij} is the weight coefficient between i^{th} and j^{th} neurons, n is the number of input neurons that goes into a cell, and b is a bias value [3].

The second operation is the activation function which processes the net input derived from the summation operation and calculates output for the processing element (node). The following equation is a mathematical representation of this statement:

$$y_i = f(Net_j) = \frac{2}{1+e^{-2U_i}} - 1$$

Each PE has to go through the two above operations, which decides weight of the transmission of the information to the next PE (in the next layer) and sometimes it is weak enough to the extent that it may not be transmitted successfully to the next layer [3, 7].

In this study, the worldwide database compiled by the "ACI Committee 355" has been used and the adhesive anchor related records are filtered for relevance. Ninety-six useful records found from this process are applied here.

Based on the diameter size of the anchors, the database of records is classified into training and testing sets for the better representation of the given recodes. Based on this classification, the 20% of each category is moved to another file which is used for testing the remaining records. The given 96 records are divided into 73 training sets and 23 testing sets [8, 9]. These same two sets are also used for both NN and GEP models.

Eleven columns of these training records are used as input data for the ANN model developed here. The shear capacity column is used as output for the ANN model (Table 14.1).

The generated results from each run from the ANN mode are compared with the experimental laboratory results. The compared results are examined to decide whether next repetition of the process is required or not.

If the outcome is not satisfactory in term of accuracy, then the repetition is required. Such repetitions are called "Epoch". The number of epochs is usually decided before running the ANN model.

14.4 Constructing Artificial Neural Network Model

The worldwide database compiled by the ACI Committee 355 is used in this study. The database consists of 2,929 adhesive anchors laboratory conducted tests. The records are collected in this database from various sources including reports and technical papers published in countries such as Europe, USA and Japan. The worldwide database lists tensile and shear strength load testing in uncracked and cracked concretes with "single" anchors, and groups of "two" and "four" anchors. The records clearly indicate the distances for the tested anchors, the type of anchors, being threaded rod or not, insert sleeves, and reinforced bars [10].

A dataset of 96 records are selected from the original dataset of 2,929 records. The original dataset is currently maintained for "ACI 355 Committee by Dr. R. A. Cook" who thankfully passed on the dataset to us [8]. This study considers a set of criteria and the required specifications for the selected records are summarized below:

1. "Single anchor" test or "group of anchors" with the spacing of greater than 1.6 times the embedment depth.
2. The "anchors with the minimum edge distance" greater than 0.8 times of the embedment depth.
3. Tests are carried out in "uncracked" and "unconfined" concretes.
4. "Anchors" are installed into clean drilled holes. They are tested under the static shear.
5. "Anchors" are exposed to the "short-term loading" at room temperature.
6. "Anchors: show typical adhesive failure modes except for the "bolt breakage".

The input parameters are eleven items as indicated below:

- X_1 in mm is outer diameter of the anchor bolt
- X_2 is the injection system (e.g., cartridge injection)
- X_3 is the glass capsule
- X_4 is the type of bonding agent such as epoxy

- X_5 is the unsaturated polyester
- X_6 is the "anchor bolt type" (e.g., threaded rod)
- X_7 is the rebar
- X_8 in mm is the embedment depth
- X_9 in mm is the annular gap
- X_{10} in MPa is the concrete strength and
- X_{11} in mm is the clear clearance between the "hole" and the "anchor bolt" [3].
- Y (in kN) is called Shear capacity. It is the output of the models developed in this study (Table 14.1).

The non-numeric design variables (e.g., bonding agent type, anchor bolt type and the type of injection system) are assigned a dummy variable value of either 1 or 0. For example, if epoxy is used as the bonding agent, the epoxy (X_4) is assigned a value of 1 and unsaturated polyester (X_5) is

Table 14.1 Data range of model parameters.

Parameters		Input parameter min	Data range	
			Max	
Diameter [mm]		X1	8	25.4
Type of injection system	Cartridge	X2	0	1
	Glass Capsule	X3	0	1
Type of bonding agent	Epoxy	X4	0	1
	Unsaturated Polyester	X5	0	1
Type of anchor bolt	Threaded Rod	X6	0	1
	Rebar	X7	0	1
Embedment depth [mm]		X8	80	229.8
Annular gap [mm]		X9	0.79	4.76
Compressive strength [MPa]		X10	13.13	43
Edge distance		X11	38.1	262.5
Shear		Shear	6.49	188.75

assigned a value of 0. A similar approach has been followed for the anchor bolt type and the type of injection system.

Several ANN models are constructed and evaluated for adoption. The most precise and accurate ANN model has been chosen for this study. The adopted model has the following characteristics:

- MLP for type of network with 14 processing Elements PE
- Conjugate Gradient CG for learning rule
- one hidden layer
- thresholds of 0.00001, and
- 7000 epochs

The details of these parameters are shown in Table 14.1.

Graph 14.1 shows the comparison of test results conducted in the laboratory to the results of current study's model which was built by Neural Network. There is a good fit between the laboratory results and theoretical model results.

Table 14.2 shows statistical data extracted from the comparison between the laboratory results and the current study's model.

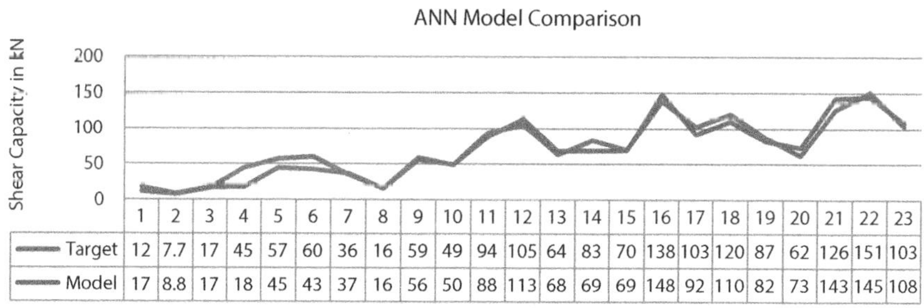

	1	2	3	4	5	6	7	8	9	10	11	12	13	14	15	16	17	18	19	20	21	22	23
Target	12	7.7	17	45	57	60	36	16	59	49	94	105	64	83	70	138	103	120	87	62	126	151	103
Model	17	8.8	17	18	45	43	37	16	56	50	88	113	68	69	69	148	92	110	82	73	143	145	108

Graph 14.1 The comparison of testing result for Neural Networks.

Table 14.2 Statistical data for the selected model.

Statistical tool	ANN
MSE	94.80
MAPE	0.12211
MAE	2.03
Correlation Coefficient " r " for testing	0.97258
Coefficient of Determination R-Sqr	0.94590

14.5 Genetic Programming (GP)

Genetic programming is a tool that operates group of programs/functions to achieve a specific task and compare it to the desired results. Such process continues until the achieved task matches the ideal to a certain extent. This automation simulates the genetically breeding process based on Darwin's theory. The "Genetic Programming" repetitively transforms a group of programs into a new group of programs based on the same concept of biological operations. The operations cover mutation, crossover, reproduction, and the change of structure caused by the gene duplication or deletion [11, 12].

14.6 Administering Genetic Programming Scheme

The genetic programming requires a group of parameters to be defined to the process and produce an output. To address parameter requirements, the following parameters must be defined:

1. The set of terminals and the set of primitive functions for each branch of the to-be-evolved program. The set of terminals includes independent variables of the problem, zero argument functions, and random constants.
2. The fitness measure
3. Parameters that control the run
4. The criteria for termination, and
5. Technique for designating result of the run [10, 13].

14.7 Genetic Programming In Details

The "Genetic Programming" (GP) starts by randomly generating a group of programs/functions based on the introduced parameters (inputs). As mentioned above, the GP repetitively transforms group of programs/functions into a new group of programs by applying mutation, crossover, and reproduction. In GP the selected programs, to continue to work on, are probabilistically selected to participate in the genetic operation based on their performance with regard to their outcomes. Only the programs that have high percentage of similarity to the desired outcome are selected and this factor is called "fitness".

The fitness is a scale provided by the user to measure the suitability of an outcome. The operation of running the programs, getting results, measuring the fitness, and selecting the best programs continues until the desired result is produced.

The genetic programming starts the following steps:

1. Create the first generation (a group) of programs with the given function and terminals.
2. Repetitively perform the following operations:
   a. Run each program in the group and determine its fitness by applying the fitness measure established.
   b. Select one or more program(s) from the group based on the probability for the fittest programs in the group; these are used to participate in the following genetic operations.
   c. Create new group of programs from the filtered programs by implementing the operation of reproduction, crossover, mutation, and change of structure on them. See the explanation below:
      i. "Reproduction": Copy the selected individual program to the new population.
      ii. "Crossover": Create new offspring program(s) for the new population. This is done by recombining randomly chosen parts from two selected programs.
      iii. "Mutation": Create one new offspring program for the new population. It is done by randomly mutating a randomly chosen part of one selected program.
      iv. "Change of structure": Choose a change of structure operation from the available repertoire of such operations. This operation also creates one new offspring program for the new population upon applying a chosen architecture-altering operation to one selected program.
3. Create group of programs until termination criteria is satisfied.
4. Process of GP is stopped once the termination criterion is satisfied. The result of such termination is a program that is capable of predicting a desired outcome exactly or within an acceptable tolerance limit [13].

14.8 Genetic Expression Programming

Genetic expression programming (GEP) is a "population-based evolutionary algorithm" developed by Ferreira and it is directed descendent of GP [14, 15].

The individuals in GEP are encoded here as linear strings of fixed length, and are transformed upon processing into expression that are no-linear entities with different lengths and shapes. The produced entities are known as expression trees (ET). As is apparent, these ET are produced by one chromosome, which ultimately might have one or more genes.

The ETs are expression of chromosome and they are subjected to the selection process which is guided by the fitness value. The selected group becomes a candidate for reproduction and during this, the chromosomes are modified by the genetic operator and not the ET.

"In genetic modification, population of ETs is capable of discovering traits and therefore adapting to particular problem they are employed to solve. This means that, within enough time and setting the stage correctly, a good solution to the problem will be discovered" (Ferreira, 2001a, b) [14].

14.9 Developing Model With Genexpo Software

GeneXPro is software that uses the GEP as a learning algorithm for the introduced variable to produce the prediction model.

This research introduces eleven parameters as terminals (inputs); the same was observed in the case of Neural Networks. The same training and testing files were used to build the GEP model. Graph 14.2 shows the comparison of test results conducted in the laboratory to the results of current

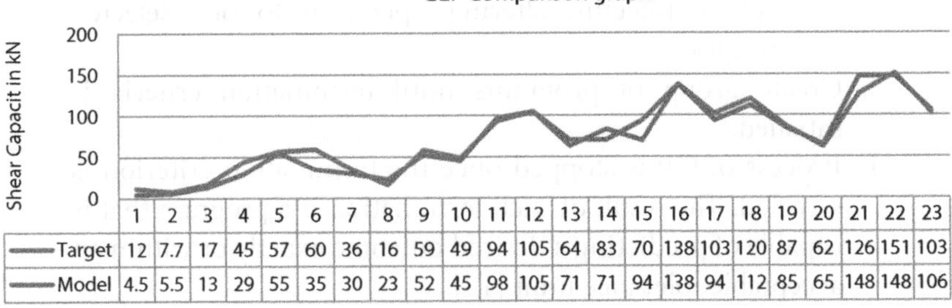

	1	2	3	4	5	6	7	8	9	10	11	12	13	14	15	16	17	18	19	20	21	22	23
Target	12	7.7	17	45	57	60	36	16	59	49	94	105	64	83	70	138	103	120	87	62	126	151	103
Model	4.5	5.5	13	29	55	35	30	23	52	45	98	105	71	71	94	138	94	112	85	65	148	148	106

Graph 14.2 The comparison of testing result for Genetic Programming.

study's model which was built by Genetic Expression Programing. There is a good fit between the laboratory results and theoretical GEP model results.

Table 14.3 shows statistical data that are extracted from the comparison between the laboratory results and the current study's model.

The following table (Table 14.4) lists the details of the parameters of the model developed.

Table 14.3 Statistical data for the selected model.

Statistical tool	GEP
MSE	57.59
MAPE	0.09517
MAE	1.58
Correlation Coefficient " r " for testing	0.97195
Coefficient of Determination R-Sqr	0.94469

Table 14.4 List of the parameters of the model developed.

Parameters	Details
Number of generation	4771
Number of chromosomes	30
Head size	11
Number of genes	5
Linking function	+
Mutation rate	0.044
One-point recombination rate	0.3
Two-point recombination rate	0.3
Gene recombination rate	0.1
Gene transportation rate	0.1
Function sets and weights*	+ (2) ; - (2); * (2); / (2); Sqrt (1); Exp (1); Ln (1); x^2 (1); x^3 (1); 3rd root (1); Sin (1); Cos (1); Atan (1)

The concluded model's mathematical equation is:

Val_01 = (−9.771 – ((fc – AG) / (Cos((Pow(2.01, (1.0 / 3.0)) * UP)) * −9.771)))

Val_02 = (diameter + Sqrt(((E – (Cos(C) –)) – (Exp(2.62) – (R + R)))))

Val_03 = (Cos((Atan(Log((fc + E))) * Log(Pow(8.509, 2)))) * C)

Val_04 = ((Exp(Exp((UP * R))) + Pow((C – −8.237), (1.0 / 3.0))) * Cos(((−2.285 * diameter) – −8.237)))

Val_05 = ((diameter / ((UP + (ED + 1.370)) – (Sqrt(AG) * (TH * ED)))) – fc)

Predicted Shear Value = [Val_01] + [Val_02] + [Val_03] + [Val_04] + [Val_05]

The above formula has been integrated into a program that takes the input parameters and in a click of a mouse calculates the predicted shear capacity of the anchor (see Figure 14.1). The program may be downloaded from www.clickme.com.

The results are very accurate in predicting the shear capacity of the adhesive anchors where the Correlation Coefficient (r) value is 0.971 for

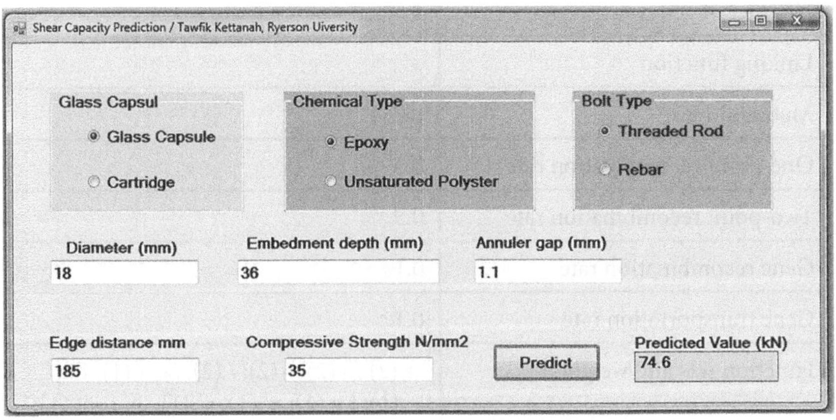

Figure 14.1 Screenshot for the shear capacity prediction program.

Table 14.5 Model comparison.

Statistical tool	GEP	ANN
MSE	57.59	94.80
MAPE	0.09517	0.12211
MAE	1.58	2.03
Correlation Coefficient " r " for testing	0.97195	0.97258
Coefficient of Determination R-Sqr	0.94469	0.94590

testing, Mean Absolute Error (MSE) 1.58%, Mean Absolute Percent Error (MAPE) is 0.09517, and best fitness 827 (see Table 14.5 for comparison).

14.10 Comparing NN and GEP Results

The results of NN and GEP were very satisfactory for the use of a prediction model and the difference between them is negligible enough to make it acceptable to use either as a model. The two developed models were tested for performance using the same test dataset that was taken from the 96 records dataset. The laboratory results were compared to the results predicted by each developed model separately and Graphs 14.3 and 14.4 were plotted thusly. We used Graphs 14.3 and 14.4 to record the performance of the models using statistical tools such as MSE, MAPE, MAE, coefficient of Correlation, and Coefficient of Determination R-Square as shown in Table 14.5 [5].

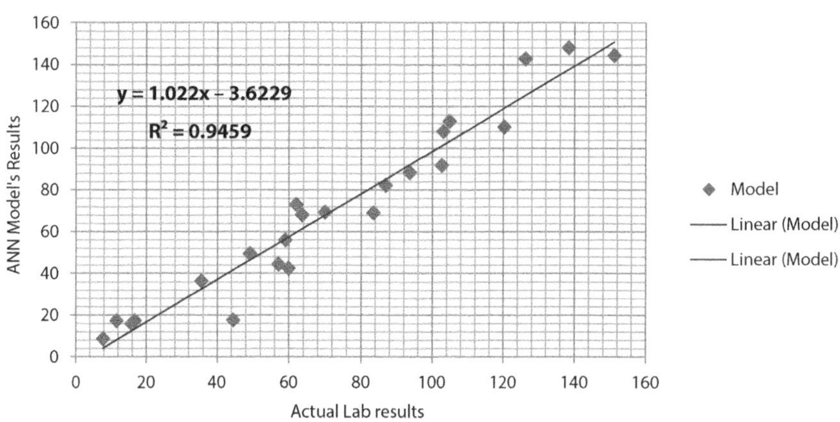

Graph 14.3 Coefficient of determination (the results of ANN).

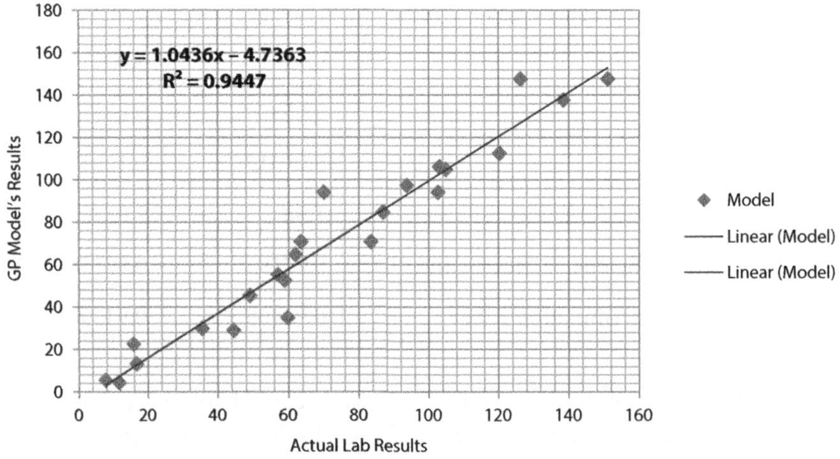

Graph 14.4 Coefficient of determination (the results for GEP).

The above graphs and table are used to record the following observations:

- The values of Coefficient of Correlation (r) and the Coefficient of determination (R-Square) for both models are very close and show high accuracy.
- The GEP model demonstrates a better performance when MSE, MAPE, and MAE were compared as shown in Table 14.4.
- This research favours the GEP model for adaption and the formula is tailored for the ease of use. This is achieved by having the formula written into the "Visual Basic .NET" program and a small (in size) executable file is attached to achieve the calculation for the introduced parameters without the need to manually calculate the predicted shear capacity of the anchor (see Figure 14.1 screenshot of the program).

14.11 Conclusions

This research work has been successful in developing a model to predict the shear capacity for adhesive anchors in uncracked concrete to an accuracy of 93%. Such results are very good when the fuzzy nature of the parameters involved, and the variance of the accuracy of laboratory tests is taken into account.

This research renders the decision of selecting an anchor for certain criterion possible without the need to conduct an actual test.

It is important to conduct experiments to evaluate the suggested anchor for use as this has long been the most common recommended practice in the industry.

The formula extracted from the model is an excellent prediction tool to further evaluate the suggested anchors.

References

1. Rolf Eligehausen, Ronald A. Cook, and Jörg Appl, Behavior and Design of Adhesive Bonded Anchors, *ACI Structural Journal*, Vol. 103, No. 6, November-December 2006.
2. Building Code Requirements for Structural Concrete (ACI 318M-08) and Commentary.
3. Mehmet Gesoglu, Erhan Guneyisi, Prediction of load-carrying capacity of adhesive anchors by soft computing techniques, *Materials and Structures* (2007) 40:939-951.
4. Sanad A, Saka MP (2001) Prediction of ultimate shear strength of reinforced concrete deep beams using neural networks. *J Struct Eng–ASCE* 127(7):818–828.
5. Baykasoglu A, Dereli T, Tanıs S (2004) Prediction of cement concrete using soft computing techniques. *Cement Concrete Res* 34:2083–2090.
6. Goh ATC (1995) Prediction of ultimate shear strength of deep beams using neural networks. *ACI Struct J* 92(1):28–32.
7. Seung-Jun Kwon, Ha-Won Song, Analysis of carbonation behavior in concrete using neural network algorithm and carbonation modeling, *Cement and Concrete Research*, Vol. 40, Issue 1, January 2010, pp. 119-127.
8. Sherief S. S. Sakla, Ashraf F. Ashour, Prediction of tensile capacity of single adhesive anchors using neural networks, *Computers & Structures* 83 (2005) 1792–1803.
9. Jerzy Hola, Krzysztof Schabowicz , Application of Artificial Neural Network to determine concrete compressive strength based on Non-destructive tests, *Journal of Civil Engineering and Management*, 2005, Vol XI, No 1, 23-32.
10. Cook RA, Kunz J, Fuchs W, Konz RC (1998) Behavior and design of single adhesive anchors under tensile load in uncracked concrete. *ACI Struct J* 95(1):9–26.
11. Linge Bai, Manolya Eyiyurekli, David E. Breen , Automated Shape Composition Based on Cell Biology and Distributed Genetic Programming, Department of Computer Science, Drexel University.
12. Qualification of Post-Installed Mechanical Anchors in Concrete (ACI 355.2-07) and Commentary.

13. Ferreira, Cândida, Gene Expression Programming: A New Adaptive Algorithm for Solving Problems, *Complex Systems*, Vol. 13, Issue 2, pp. 87-129, 2001.
14. Roland A. Cook, Jacob Kunz, Werner Fuchs, and Robert C. Konz; Behavior and Design of Single Adhesive Anchors under Tensile Load in uncracked Concrete; *ACI Structural Journal*, Vol. 95, No. 1, January-February 1998.
15. Mustafa Sarıdemir, Genetic programming approach for prediction of compressive strength of concretes containing rice husk ash, *Construction and Building Materials*, Vol. 24, Issue 10, October 2010.

15

Adaptive Heuristic - Genetic Algorithms

R. Anandan

Department of Computer Science and Engineering, Vels Institute of Science, Technology & Advanced Studies, PV Vaithiyalingam Rd, Pallavaram, Chennai, Tamil Nadu, India

Abstract

Genetic Algorithm (GA) is a search-based optimization technique that is based on genetics and natural selection principles. It's routinely used to find optimal or near-optimal solutions to problems that would take an eternity to solve otherwise. It's commonly utilized to tackle optimization problems in research and machine learning.

John Holland first brought GAs to the public, and produced a theoretical analysis based on a schema, which became known as the schema theorem. He was trying to figure out how likely it is for a schema to survive from one generation to the next, as well as how many schema are likely to be present in the next.

Holland's Schema Theorem is a step in the right direction for researchers trying to find out the mathematics behind how genetic algorithms function. The Schema Theorem has undergone several adjustments and suggestions over the past year in order to make it more generic. We won't go into the mathematics of the Schema Theorem in this section; instead, we'll aim to acquire a basic understanding of what the Schema Theorem is. According to the schema theorem, a schema with above-average fitness, a short defining length, and a lower order is more likely to withstand crossover and mutation.

***Keywords*:** Genetics and natural selection principles, optimization technique, schema, crossover and mutation

15.1 Introduction

Genetic Algorithms (GENETIC ALGORITHMs) is the method of Fraser and Bremermann, considered to be an effectual usual algorithm method projected to do algorithms that simulate genetic systems and much seminal work. In the 1960s Holland with colleagues and students practiced this method and

Email: anandan.se@velsuniv.ac.in

Ramakant Bhardwaj, Jyoti Mishra, Satyendra Narayan and Gopalakrishnan Suseendran (eds.) Mathematics in Computational Science and Engineering, (329–342) © 2022 Scrivener Publishing LLC

reprinted it in 1992. Holland's book of 1975 (Holland, 1992 - reprinted) is recognized as one of the seminal works for GENETIC ALGORITHMs.

15.2 Genetic Algorithm

Genetic Algorithms is an easy form of learning and improvising design of Algorithm. The foundation of the principle is Darwin's Natural selection, which states survival of the fittest. It offers solutions to various problems and helps to evaluate the value and also permits to combine with each other [1].

This method gives confidence to reproduce good solutions; it will steadily produce enhanced solutions. It proves that any range of problem can be supported and solved.

As we trace the origins in genetics, it can be carried through into the computer counterparts. A gene is generally made up of a cell which represents individual problems. It represents individual solution of the problem. The group is known as populace.

In travelling salesman problem (TSP) the individual gene represents individual cites. The complete tour and its number are represented as gene and populace, respectively. Finally we found that GENETIC ALGORITHM is the best to work with present populace.

Let us take a hundred as a populace and work the values and results.

Generally selected genes are checked at random. Different problems are represented as bit strings, but it is not necessary to represent individuals and later the same would be evaluated. There are some functions set down to identify individual genes to find valuable response or answer to the problem. So genes are permitted to combine with one another.

Initially an appropriate "parents of the fittest" should be selected to breed. It should not be determined wrongly. When an individual has a low evaluation, like low contribution, a parent should be selected by proportion to their evaluation rating. It is an opportunity to allow weaker members to take part in Evolution.

It is not mandatory to repeat many times; there should be a possibility with each breeding pair as they yield Child or not. Usually set 60% and attempt other values for experiment.

Sometimes in life, a mutation can occur and GENETIC ALGORITHMs imitate it. Mutation happens with less likelihood and occurs with reference to the coding [2]. There is a fair and simple apply of mutation by bit strings. It is easy to identify that individual bit of the gene decide and tried at random.

As conclusion, GENETIC ALGORITHM has knowledge to solve various problems and also teaches domain knowledge on evaluation function. This rating method breeds mechanism, which just manipulating bit strings.

15.3 The Genetic Algorithm

The various implementation of a Genetic Algorithm:

1. Initialize a group of genes
2. To estimate individual gene in the group.
   2.1 To generate latest genes by joint genes at mutation.
   2.2 Removes old member to replace new members.
   2.3 To estimate new members to add in group.
3. It is recommended to Repeat 2nd stage to termination condition.
4. Obtain the best gene as result.

This is the fundamental GENETIC ALGORITHM.

15.4 Evaluation Module

In order to evaluate gene the recommended method is evaluation module. It is important to know that the module helps to solve the problem by the part of the Genetic Algorithm. The remaining modules operate on bit strings.

To find the result, it is essential to identify the problem and the appropriate module should be used.

For example, Genetic Algorithm tries to find solution to evaluate travelling salesman problem (TSP) by using individual gene on calculating the distance travelled.

A salesman can visit many cities. It is instructed to start and complete at a particular city only once, so that we can find the shortest distance as possible.

15.5 Populace Module

15.5.1 Introduction

We can maintain populace by using the following techniques of populace module.

15.5.2 Initialisation Technique

The method calculates the initial populace by varying the case and values. In the condition of a binary coded gene, a single bit is initialized to arbitrary zero or one.

The module can be initializing the populace with some known good solutions. It can be applied anywhere to improve. We can get better results by seeding the populace with the good solutions. But the seed would disappear from the populace. On the other hand, the combination of seed populace and elitism would give an assured result of good solutions not poor result.

15.5.3 Deletion Technique

This technique describes an idea to find the deleted populace of individual GENETIC ALGORITHM.

The following are the three general deletion techniques:

Delete-All: By using the technique we delete the present populace and replace the same count of genes which has created recently.

Steady-State: By using the technique we delete n ancient members and replace by n fresh members. The same count should be deleting and replaced once by detection technique. If you want to remove the poor, do it random.

Steady-State-No-Duplicates: It is like the previous but the referred algorithm verifies that no replica genes are included to the populace.

15.5.4 Parent Selection Procedure

In reproduction decision of selecting parent is based on user perspective. The fittest individuals from the populace as well as less fit individuals explore and increase the probability of giving valuable offspring. There is a danger in small number of genes, which converges the individuals to an inferior solution.

Two communal parent selection procedures are desc ribed below.

Roulette Wheel Selection: In this an individual has been given an opportunity to play a role of parent in percentage to good report; it is called roulette wheel selection, whose parent could be identified as spinning a roulette wheel to fit to proportional parent.

The method can be applied for the following:

1. TF (total fitness) is the average of all the populace members.
2. Create a haphazard no. m, among 0 and TF.
3. Fitness of first populace member is combined to the existing members to give a result which is >= m.

It is known that the difficulty in this method is an individual member can dominate others to select an elevated quantity of times. The overturn is also true. There is a good chance for close members to get selected. To understand the problems of different, normalization techniques can be used.

Tournament Selection: The other successful method is tournament selection because potential parents are selected whereas tournament will choose individual's parent. This can be done in different ways and the two suggestions are:

1. Choose a couple of individuals at haphazard. Create a haphazard no., S, among 0 and 1. If $S < r$ use the primary as a parent. If the $S >= r$ then use the secondary as the parent. It should be repeated to choose the secondary parent "r" represents parameter.
2. Choose two individuals at haphazard. The parent will be the highest individual evaluation. Do a Genetic Algorithm in to locate a second parent.

15.5.5 Fitness Technique

By using the evaluation method we can get a parent which can lead to a problem, as an example of an individual will dominate the whole when it is higher than other members [3]. In the same way, there may be a chance of choosing equal that will lead to a random search.

To resolve the problem, individual gene will be given two values, an evaluation and fitness. A good parent would be selected by normalizing the fitness evaluation. Some common methods used for calculating fitness are detailed below.

Fitness by Assessment: The fitness of the gene should be equal to the evaluation.

Windowing: This technique consumes the least evaluation and allots a fitness to individual gene to attain minimum.

Linear Normalization: Here evaluation values are decreased to sort gene value and it is assigned to the fitness value which begins at a constant value and decreases linearly.

15.5.6 Populace Size

It calculates the number of genes to be in the populace at once.

15.5.7 Elitism

Sometimes in Genetic Algorithm progresses, an excellent solution might be found early in the Genetic Algorithm run however the deleted from the populace. The best solution would be one solution. This technique is known as elitism.

It assures the best member to the next generation and also predicts the percentage of the populace.

15.6 Reproduction Module

15.6.1 Introduction

This module is conscientious for the populace Genetic Algorithm of genes. Normally, it will work for two parents and it will breed and the Child module would be added to the populace.

This module is mostly prepared by operators with mutation playing a lesser, but it plays an imperative role.

15.6.2 Operators

Genetic Algorithm operator has developed One Point Crossover. There are some specific operators for exact problems and be considered as ideal.

One-Point Crossover

It is done with two parents and breeds two Child. The function as follows

P1	***1***	0	1	1	1	0	1
P2	1	1	0	0	1	1	***0***

C1	1	0	0	0	1	1	0
C2	1	1	1	1	1	0	1

(Take no. notice of the emphasized bits)

- Select two parents.
- A *Crossover Point* is selected haphazard (Ref: the dotted line).
- Child 1 consumes genes from their respective parents at left crossover point.
- Child 2 consumes genes from their respective parents at right crossover point.

Two-Point Crossover

Both two-point and one-point crossover works in the same way but two crossover points has chosen. *n*-point crossover also is chosen. The advantage is Two-point crossover because the highlighted bits (shown in the column) should give a better result, since they are close to each other.

Uniform Crossover

At random we choose two Child we decide, which parent would give bit value to the child. The implementation of the algorithm as follows.

P1	1	0	1	1	1	0	1
P2	1	1	0	0	1	1	0
T	0	1	1	0	0	1	0

C1	1	0	1	0	1	0	0
C2	1	1	0	1	1	1	1

- Select two parents.
- Random bits template has been created.
- 1 and 0 indicate the Child 1 and 2 get bits from parent 1 and parent 2 respectively.
- 1 and 0 indicate the Child 1 and 2 get bits from parent 1.

Order-Based Crossover

One of the problems of crossover operators led to Genetic Algorithm solutions. The TSP gene is identified as a listing of towns. We produce one Genetic Algorithm solution by allowing one-point crossover operator. Meanwhile we identify duplicate cities and delete odd. To deal the problem two approaches can be dealt. It is recommended to use a repair function to get a potential solution for examination. If the result is a Genetic Algorithm solution then it is allowed to carry repair to ensure its validity. Evidently the function which is carried out is problem reliant. The other developing method of this approach produces Genetic Algorithm solutions. It can be completed through domain information and also to work on a whole range of problems. One of the operators is Order-based crossover which assures the travelling salesman case identifies no cities are duplicate or lost. The procedure is described as

P1	A	B	C	D	E	F	G
P2	E	B	D	C	F	G	A
T	0	1	1	0	0	1	0

C1	E	B	C	D	G	F	A
C2	A	B	D	C	E	G	F

Select two parents

- Random bits template has been created.
- 1 and 1 indicate the Child 1 and 2 get bits from parent 2 and parent 2 respectively.
- 1 and 1 indicate the Child 1 and 2 get bits from parent 1 and 2.
- A prototype T is created which consists of random bits
- variety of genes look in identical group as in parent 2
- Replace by child 1 in the given order.
- Child 2 will be created in the same order.

Moderately Matched Crossover (MMX)

MMX is related order-based crossover used to discuss travelling salesman problem. It is a best way to find the operations of MMX. It is a Goldberg's example, 1989.

Parents, A and B are chosen as two Crossover Points.

Parent A	9	8	4	5	6	7	1	3	2	10
Parent B	8	7	1	2	3	10	9	5	4	6

With reference to B the genes the Crossover site. In initial 2 to 5 steps in map are parent A. so exchange genes 2 to 5 in Parent B. In similar way exchange 3 and 6, 10 and 7.

The same steps should be used for parent A (exchanging 5 and 2, 6 and 3, 7 and 10). The results are given in the following.

Parent A	9	8	4	2	3	10	1	6	5	7
Parent B	8	10	1	5	6	7	9	2	4	3

As result we have got duplicate to order one gene in the other gene in reverse.

One of the advantages is MMX helps to conserve city positions to take order from another parent. The reason to select the operator is to identify a two-dimensional stock cutting problem. It is already discussed in the previous discussion.

Cycle Crossover

Another crossover operator is TSP cycle operator. It is an example of Goldberg, 1989. The concept of this crossover of each city should from one parent from the other. Parents A and B work as follows. Select the first city from A.

Parent A	9	8	2	1	7	4	5	10	6	3
Parent B	1	2	3	4	5	6	7	8	9	10

Since 9 selected from A, likely choose 1 in given in table on top of the similar position in B as in A. City in the same location occupies in A.

Parent A	9			1						

City 1 in A occupies to city 4 in B, so city 4 in A occupies in the same location.

Parent A	9			1		4				

We prolong this process a Genetic Algorithm to arrive at

Parent A	9			1		4			6	

Here we should choose city 9 and put it in A to complete a cycle; which derives the name of the operator. B gives the details of missing cities,

Parent A	9	2	3	1	5	4	7	8	6	10

Execute same Crossover operation with B

Parent B	1	8	2	4	7	6	5	10	9	3

15.6.3 Mutation

This crossover operator is used to get better genes in case of the two parents as the similar value in a convinced location then the other won't change that.

It is intended to conquer this problem and insert some diversity to the populace.

One of the general ways, mutation is used to choose a bit at random to change to zero or one. It is possible to change the bit to the similar value which has previously existed.

Other operator may exchange parts of the gene or build up with precise mutation operator's problem.

15.6.4 Mutation Rate

It is a parameter of GENETIC ALGORITHM explains the work of mutation. A characteristic value is 0.008. So the gene mutation blows down the bit string and each bit has one chance in 8000 of being mutated.

15.6.5 Crossover Rate

It explains how frequently crossover should be implemented. 'A' value is 0.6, when open with two parents and the chance of 60% for breeding.

15.6.6 Dynamic Mutation and Crossover Rates

During Genetic Algorithm research, crossover is more significant in initial phases of the Genetic Algorithm, even as populace pursuit's large areas of the search space. It explores a smaller area has found.

In the same way we do appropriate change in the mutation and crossover rates to progress so that there is a small chance of mutation in initial phases and advanced likelihood next phases. In similar way the crossover rate begins with a high likelihood and drops.

15.7 Example

To understand the concept of Genetic Algorithm operation a simple example of Reeves, 1995, has been given [4]. But we use roulette wheel selection to describe Genetic Algorithm operation.

In order to display the Genetic Algorithm we use a Crossover likelihood, $P_C = 1.0$ and a mutation likelihood, $P_M = 0.0$, which is functional Crossover, but by no means of mutation.

Guess we have a function

$$f(y) = y^3 - 6 * y^2 + 9 * y + 10 \quad - \; 64-96+36+10$$

and we wish to make the most of it.

In additional we set the limit t40.

We set populace size to four and produce the subsequent random genes.

Gene	Binary	y	f(y)	% of RW
P_1	1110	14	1704	2.67
P_2	0111	7	122	0.19
P_3	1011	11	714	1.11
P_4	0010	2	12	0.01
	TOTAL		**2552**	**3.98**
	AVG		**638**	

The subsequent stage is to choose parents to breed. Choose two parents and allow creating two fresh individuals in simplified GENETIC ALGORITHM. We repeat this to create a new populace.

The initial breeding can be done as following.

Roulette Wheel at f(y)	Parent Chosen	Crossover Point
714	P_3	N/A
122	P_2	1

The random number of Roulette Wheel is between 0 and 2552.

We choose Parent by this method.

Applying single point crossover between P3 and P2 we reach our destination at two Child.

$$C_1 = 1011$$

$$C_2 = 0111$$

We bring out the similar process a Genetic Algorithm to generate C3 and C4.

Roulette Wheel at f(y)	Parent Chosen	Crossover Point
12	P_4	N/A
122	P_2	2

$$C_3 = 0011$$

$$C_4 = 0100$$

The new solutions should be reevaluated to form a new populace (Described in P1..P4).

Gene	Binary	y	f(y)
P_1	1011	11	714
P_2	0111	7	122
P_3	0011	3	10
P_4	0100	4	14
	TOTAL	**25**	**860**
	AVG		**215**

Even though it is an easy method, it has a few important things.

- A rise in the average evaluation.
- The strongest individual in the initial populace is P_2 has chosen twice but it has lost in the populace.
- y=3 in the populace, so the near value may be10.

We consider the first populace was 11, 7, 3 and 4. We imply GENETIC ALGORITHM methods for above ($P_C = 1.0$, $P_M = 0.0$), so a probability of finding the global optimum.

15.8 Schema Theorem

15.8.1 Introduction

Schema is a theoretical analysis introduced by John Holland, based on a schema theorem. He tries to explain a schema is to live from one generation to the next generation.

Work with the following

0	0	0	1	0	1	1
0	1	0	0	0	0	1

The best examples of schema

*	*	0	*	0	*	1

'*' is a trump card can be replaced by a 0 or 1. The other way thinks about schema's subset of all possible genes (for example the left most bits are the same in both gene).

Gene is of length *n* then it contains 3^n schemata (as each position can have the value 0, 1 or *).

We estimate the fitness of an individual to get information about every schema which exists in the gene. *M* means that for a populace of individuals to evaluate M3n schemata. Though, it must be borne in mind that a few schemata will not be accessible and others will partly cover with additional schemata. This is precisely what we look into. We ultimately generate populace with best fit in schemata.

It is the fact that we are changing *M* individuals but $M3^n$ schemata that provides genetic algorithms is known as *implicit parallelism.*

We require describing three terms to continue the discussion.

Length: The distance between the start and end of the schema. It is used to calculate the length of a schema as defined by (Goldberg, 1989).The referred schema has a length of 4.

Order: The number of definite positions. The schema of the order is 3.

FR: The ratio of the fitness of a schema to the average fitness of the populace.

15.9 Conclusion

The genetic algorithm is very effective when we have a large number of populace. In travelling salesman problem when the cities to be visited is more and the solution is very trivial then genetic algorithm is the only optimal solution.

15.10 Future Scope

Since the algorithm is very effective for unknown solution, it can be used widely in Pharmaceutical Industry for drug discovery and analysis. It may also lead to a solution when the data is enormous.

References

1. Goldberg and David E, *Genetic Algorithms in Search. Optimization and Machine Learning*, Pearson Education, New Delhi, 2006.
2. Kalyamoy Deb, *Multiobjective Optimization using Evolutionary Algorithms*, John Wiley & Sons, First Edition, USA, 2003.
3. Koza, John, Wolfgang Banzhaf, Kumar Chellapilla, Kalyanmoy Deb, Marco Dorigo, David Fogel, Max Garzon, David Goldberg, Hitoshi Iba, and Rick Riolo (Eds.), *Genetic Programming*, Academic Press. Morgan Kaufmann, USA, 1998.
4. John R.Koza, Forrest H Bennett III, David Andre, Martin A Keane, *Genetic Programming III: Darwinian Invention and Problem Solving*, Morgan Kaufmann, USA, 1999.

16

Mathematically Enhanced Corrosion Detection

SeyedBijan Mahbaz[1,2], Giovanni Cascante[3], Satyendra Narayan[4*], Maurice B. Dusseault[1] and Philippe Vanheeghe[5]

[1]Research and Development Department, InspecTerra Inc., Waterloo, Ontario, Canada
[2]Department of Earth and Environmental Sciences, University of Waterloo, Waterloo, Ontario, Canada
[3]Department of Civil and Environmental Engineering, University of Waterloo, Waterloo, Ontario, Canada
[4]Department of Applied Computing, Faculty of Applied Science & Technology, Sheridan College, Ontario, Canada
[5]LAGIS UMR CNRS 8219, École Centrale de Lille, Villeneuve d'Ascq, France

Abstract

Defect detection is a prime goal of Non-Destructive Testing (NDT) methods; detection accuracy and quantification are important secondary goals. Intelligent detection methods using "smart" digital processing techniques are important to increase the probability of detection and the accuracy and reliability of test results in many NDT applications. Passive Magnetic Inspection (PMI), an innovative NDT method, is used to inspect a reinforced concrete sample with three holes, in three different positions and locations of steel reinforcement. Simultaneously, a sound rebar is tested as a base case. Principal Component Analysis (PCA) is used as the basis of a "smart" signal processing method for automatically locating possible defects in a digital data array (a three-component scan line) and increasing the accuracy of detection location. The PCA analysis approach has the capability to be used as an automatic signal processing method for analysing PMI data to accurately locate defects and to statistically "learn" to reduce errors and uncertainty.

Keywords: Non-destructive testing (NDT), reinforced concrete, passive magnetic inspection (PMI), intelligent signal processing, principal component analysis (PCA), reinforced concrete

**Corresponding author*: narayan.satyendra@gmail.com

Ramakant Bhardwaj, Jyoti Mishra, Satyendra Narayan and Gopalakrishnan Suseendran (eds.) Mathematics in Computational Science and Engineering, (343–358) © 2022 Scrivener Publishing LLC

16.1 Introduction

One of the most essential construction materials in civil structures (e.g., buildings, bridges, platforms), and also underground structures (e.g., roadbeds, concrete pipelines and tunnels), is reinforced concrete (RC) [1]. External environmental conditions such as exposure to corrosive industrial fluids, solutions of road salt, and service temperature, are known to considerably affect the durability and longevity of RC structures. Defects (corrosion and cracks) in steel reinforcement result from these conditions, and decrease the strength of RC structures, thus increasing their failure risk.

Defects play a key role in the mechanical behavior of reinforced concrete structures during loading [2]. These defects are mainly cracks or small holes, both the results of steel rebar corrosion, caused by electrochemical and chemical processes. Fine cracks in the concrete let chloride ions penetrate and reach the rebar. Another key process in the deterioration of RC structures is the development of carbonation depth over the concrete cover and reach to the steel rebar [3]. Although there is a specified service life for RC structures based on standards [4], corrosion processes may cause steel reactions, cracking, and premature loss of structural integrity [5]. Steel corrosion is almost certainly the primary reason for prematurely reduced lifespans and service levels in RC structures [6].

Locating a deteriorated section of a rebar and recognizing the deteriorations severity is essential to defining a mandatory repair or replacement plan. The main advantage of non-destructive testing methods is their quick assessment of the problem without damage of structural elements. Passive Magnetic Inspection method (PMI), a recently invented NDT method and applied in oil and gas pipeline area, works based on the residual magnetic field of ferromagnetic materials. Therefore, this method is suitable for inspecting the soundness of steel reinforcements. Steel reinforcement gains natural magnetic memory during production and installation under the effect of the earth's magnetic field. This natural magnetic property of ferromagnetic reinforcement generates Self Magnetic Flux Leakage (SMFL), which is the basis of the PMI method [2, 4, 7].

In addition to the earth's magnetic field, mechanical force is another reason for changes in the magnetic response of steel rebars. Mechanical force affects the alignment of electron dipoles in magnetic domains, which causes changes in the self-magnetic response of steel reinforcement [8]. Deterioration, corrosion, and cracks also affect the magnetic domain configuration, and direction of electron dipoles in the domains, which in turn causes changes in the magnetic response of reinforcement [16]. Previous applications of PMI method were mostly for non-destructive testing of exposed industrial structures, such as metal reservoir tanks and aboveground/underground pipelines [4, 5, 7]. Few studies have focused on using this method to detect the location of damage and deterioration in RC structures [8, 9].

The non-destructive testing (NDT) methods most commonly used to detect corrosion in engineering applications are visual identification, ultrasonic, radiographic, and magnetic flux leakage approaches. According to the literature, visual inspection is perhaps the most common method, followed by ultrasonic techniques. The three "active" NDT techniques listed require an emitter and a receiver; for example, radiographic methods require a gamma ray source and a method of recording the gamma ray amplitudes after they have passed through the corrosion site. Magnetic flux methods require active excitation and measurement of the resulting change rate (decay) and location of the induced field. Ultrasonic devices have the emitters and detectors often in the same probe, and they generally require direct contact with the steel. These methods take time and require suitable conditions. For example, if ultrasonic methods are used to detect steel corrosion under 10 cm of concrete cover, the porosity, moisture condition, and nature of the aggregate will greatly impact the results, so quantitative interpretation becomes challenging, and at best a qualitative assessment ensues. In all electrical methods, moisture and salinity present complications, and in radiographic methods, certified technologists are needed, and transmission mode measurements are necessary.

There are several risk-based inspection approaches and assessment techniques, sometimes referred to as "fitness-for-service" methods, which help provide some insights to condition assessment over time (e.g., annual corrosion site inspections). However, corrosion in real situations is highly complex; rarely is it consistent and uniform corrosion of steel plates such as the chloride corrosion of ships' hulls from seawater below the water line. More often, it is localized and variable, such as galvanic corrosion associated with stray currents and unknown current pathways (e.g., pinholes in sheathing), or crevice corrosion from accumulation of agents such as salt and moisture in the base of a crack. Other evidence of corrosion is in the form of highly localized pitting (mm to cm scale), often invisible if it is on the inside of a pipe or on the backside of a structural steel element. Intergranular corrosion, stress-induced corrosion, leaching-induced corrosion, and erosion-accelerated corrosion are some other examples of processes that may be local to more general, and may give different responses to the various NDT approaches used to locate and quantify defects in steel goods.

Steel corrosion may be avoided, prevented or slowed. The most common methods are using corrosion-resistant materials (aluminum, stainless steel, carbon fibre/epoxy elements...), applying preventive coatings (sheathing, epoxy paints, asphaltic compounds...), implementing environmental measures (avoiding use of NaCl de-icing agents), design modification (no crevices or pooling sites), corrosion inhibitors (protective chemicals), sacrificial coatings, cathodic protection, etc. Despite these options, steel use will continue largely unabated because of its strength and versatility. Thus, corrosion will also continue, but steel corrosion and defects such as fractures also are costly: derailments, unexpected loss-of-service events, tank rupture, pipeline accidents, and unexpectedly short life spans are but a few examples.

Some of the mathematics associated with NDT analysis and interpretation are discussed in this article, including useful signal processing methods used in NDT, and application of a principal component analysis (PCA) method for detection of defects in a digital data set. The novel contribution is the development and application of PCA as an advanced mathematical signal-processing tool for defect detection of reinforcement using the digital data base provided by the Passive Magnetic Inspection (PMI) method.

16.1.1 Mathematics in NDT

Non-destructive testing (NDT) is a generic name for a vast range of indirect and non-invasive methods to evaluate the properties of a material, component or structure (the "specimen") for quantification of response level and characterization of response differences or discontinuities that may be associated with flaws and defects, including internal manufacturing defects (e.g., in welds using radiography), without causing damage to the original specimen. Physical information from the specimen at different times needs to be measured and analyzed, involving the real-life application of physics and mathematics methods.

The measured information is called a "signal"; these are different for various NDT methods. In ultrasonic methods, the information is a displacement-time signal for wave propagation through the specimen; this signal contains information related to the physical properties of the material and the modification of that information by defects or other anomalies. Specific mathematical approaches are needed to extract useful features from the signals, and these are referred to as "signal processing". This signal processing may be done in real time (e.g., a fast Fourier transform or use of a pre-set Wiener filter), yielding a graphical or digital output that is recorded and can be further interpreted. In most cases, a series of signal processing steps must be applied after the data acquisition phase because there is a need for operator guidance, because different processing pathways are available, and because a real-time processing option is not feasible. The goal is always to take a raw data stream, properly acquired, and analyze it for the specific features of interest, which may be quite different for various cases. An ancillary goal is to render the outcomes of signal processing into 2-D and 3-D images of various types so that the results can be carefully examined to guide practical decisions; for example, redo the NDT examination for greater detail and resolution, replace or repair the element, repeat the assessment at a later date, or certify the element's integrity.

16.1.2 Principal Component Analysis (PCA)

There are several signal processing techniques for corrosion detection, some of them based on statistical analyses. One of these that has proven particularly useful is based on the Principal Component Analysis (PCA) method [10]. Interpretation of raw data, pattern recognition, and selective filtering and elimination of unnecessary data to reduce analysis size are some PCA capabilities, making it a powerful signal processing method suitable for NDT application. PCA is widely used in other industries such as petroleum and electrical engineering [11–13] and, taking advantage of high-speed computations, it has become a useful tool for pattern recognition and artificial intelligence applications [14].

In this paper, a PCA-based signal processing method for analyzing and interpretation of data from the passive magnetic inspection (PMI) method is applied to detect defects in rebar (reinforcing steel) embedded within a concrete specimen.

16.2 Case Study: PCA Applied to PMI Data for Defect Detection

Reinforced concrete (RC) is the most common construction material in civil infrastructure (e.g., buildings, bridges, platforms, concrete pipes, highways, tanks...) and is widely used in mining (e.g., shaft lining, tunnel support) [15]. Exposure to corrosive industrial fluids, aqueous solutions of road salt, and severe service temperature fluctuations affect the service level and longevity of RC structures. Defects (corrosion and cracks) in the embedded steel reinforcement arise from these processes, decreasing the load capacities of the RC structures, and increasing the probability of sudden loss-of-service events.

Defects in steel and concrete play a key role in the load response behavior of reinforced concrete structures [16]. These defects are mainly cracks or small holes, both the result of steel corrosion caused by electrochemical and chemical processes. Fine cracks in the concrete allow water with chloride ions to penetrate and reach the rebar, and the ensuing corrosion products generate expansion, accelerating the process. Another key process in the deterioration of RC structures is the development of a carbonation layer in the concrete cover, reaching the steel rebar, and triggering or enhancing corrosion [17]. Although there are specified service life guidelines for RC structures [18], steel corrosion may lead to loss of cross-sectional area, development of local weak spots (most common in highly stressed sections), and premature deterioration of structural integrity [19]. Steel corrosion is the primary reason for prematurely reduced lifespans and service levels in RC structures [20].

Locating a deteriorated rebar section and quantifying the deterioration severity are essential to defining a mandatory repair or replacement plan (asset management). The main advantage of NDT methods is their quick assessment of the problem without damage to structural elements (coring, drilling, cover removal). The Passive Magnetic Inspection method (PMI), a recently developed NDT approach, is based on corrosion-induced detectable alterations in the residual magnetic field of ferromagnetic materials. PMI is suitable for quantifying the condition (soundness) of steel reinforcement under concrete cover. Steel reinforcement bars (and cable tendons, and other steel elements) have an "imprinted" magnetic memory developed during production and installation under the effect of the earth's magnetic field, and this natural magnetic property of ferromagnetic reinforcement generates Self Magnetic Flux Leakage (SMFL), the basis of the PMI method [21–23].

In addition to the earth's magnetic field, mechanical stress affects changes in the magnetic response of steel rebars; if the stress is large enough, it alters the alignment of electron dipoles in magnetic domains, changing the steel's self-magnetic response [24]. Corrosion and cracks also impact the magnetic domain configuration, causing changes in the steel's magnetic response [25]; more specifically, corrosion destroys the ferromagnetic properties, and cracks create anomalous variations in the surrounding magnetic field. Previous PMI method applications were mostly used for non-destructive testing of exposed industrial structures, such as metal reservoir tanks and aboveground/underground pipelines [22, 23, 26]. Some studies have focused on using PMI to detect and locate damage and deterioration in RC structures [27, 28].

In RC structure condition assessment, quantitative steel condition state is achieved through "smart" signal processing methods based on various mathematical approaches (signal conditioning, filtering, anomaly detection, defect classification, gradient analysis...). Smart processing gives consistent outcomes, and anomalies may be quantified and ranked for further, more detailed PMI and other NDT assessments to reduce uncertainty. The quantification of the mathematical relations between defects and magnetic fields is the basis of smart processing. This study presents a "proof-of-concept" application of Principal Component Analysis (PCA) to PMI signal data to locate defects, using the results of experimental tests using the PMI approach on RC specimens. This simple proof-of-concept study is intended only to demonstrate the PCA method's ability to locate the sites of deterioration in steel reinforcement covered by concrete.

16.3 PCA Feature Extraction for PMI Method

A feature of PMI arising from its physical basis using the self-magnetic potential of steel rebar is that PMI is a passive NDT method; there is no need to apply a strong magnetic field and then measure the response. The self-magnetic potential occurs because of magnetic domain boundaries and accumulations of electron dipoles in an aligned direction under the earth's magnetic field. Changes in the metal structure that destroy the self-magnetic properties (corrosion) or radically alter the alignment of these domains (plastic strain, cracking) change the magnetic response of ferromagnetic specimens [21–28], generating anomalies in the surrounding magnetic field that are detectable. These anomalies are identified through feature extraction mathematical methods. The feature extraction technique applied in this study is Principal Component Analysis (PCA). The PMI data are collected in the time domain, and PCA anomaly identification allows this NDT method to quickly identify corrosion sites in a consistent manner [29].

PCA is a statistical method to extract the principal components of a relevant feature, and the raw magnetic field data around the reinforced concrete is a suitable multivariate data source available in the time-domain for PCA analysis. To explain how to achieve unique and valuable scalar measures in different directions using PCA analysis, the concept of eigenvalues is introduced. More specifically, "eigensignals", *Hn*, are defined as a column vector with n variables. The PMI magnetic field data is collected in orthogonal axes (x, y, and z), giving an array matrix *[H]* of size $3 \times n$:

$$H = [Hx, Hy, Hz]_n \tag{16.1}$$

The average signal Ψ is defined a direct numerical average:

$$\psi = \frac{1}{3}\sum_{n=1}^{3} H_n \tag{16.2}$$

By subtracting the average signal from each signal column, "difference signals" that deviate substantially from the average can be identified and subjected to principal component analysis. The difference signals, $\phi_{x,y,z}$, are:

$$\phi_{x,y,z} = H_{x,y,z} - \psi_{x,y,z} \tag{16.3}$$

A covariance matrix C is calculated in order to find the orthogonal eigenvectors:

$$\mathrm{C} = \frac{1}{3}\sum_{\mathrm{n}=1}^{3} \phi_{\mathrm{n}} \times \phi_{\mathrm{n}}^{\mathrm{T}} = \frac{1}{3}\mathrm{A} \times \mathrm{A}^{\mathrm{T}} \tag{16.4}$$

Where $\mathrm{A}=[\phi_x, \phi_y, \phi_z]$.
If we consider V_{i} as the eigenvectors of $\mathrm{A} \times \mathrm{A}^{\mathrm{T}}$, then:

$$\mathrm{A} \times \mathrm{A}^{\mathrm{T}} v_i = u_i v_i, i = x, y, z \tag{16.5}$$

Therefore;

$$u_i = \mathrm{A} v_i \tag{16.6}$$

These u_i values are referred to as "Eigensignals", and the principal components for any signal H are defined to be:

$$w_i = u_i^T \left(H - \psi\right) \tag{16.7}$$

The values of w_i are new features extracted from the PMI data to be correlated with the properties of the defects to provide more consistent and reliable anomaly detection and defect condition quantification.

16.4 Experimental Setup and Test

Figure 16.1a shows the reinforced concrete specimen used for this study. The steel rebar placed within the concrete specimen has three small holes (defects) in different positions and orientations (Figure 16.1b). The rebar type is A572-G60, based on ASTM standards, and its relative magnetic permeability is around 75 [30, 31]. Table 16.1 presents information about the geometry of the three holes, as part of the assessment is to evaluate the detectability of flaws in different orientations.

Figure 16.2 shows raw magnetic field data obtained directly above the steel rebar within the concrete specimen that provided 5.4 cm thickness of cured and aged concrete above the top of the rebar. These raw magnetic field data are measured in the x, y, and z directions with respect to the top of the concrete specimen. These experimental data make it clear that directly locating the three holes from the raw signals is impossible (Table 16.1), and that sophisticated signal processing is necessary to extract useful (quantitative) data from the PMI traces.

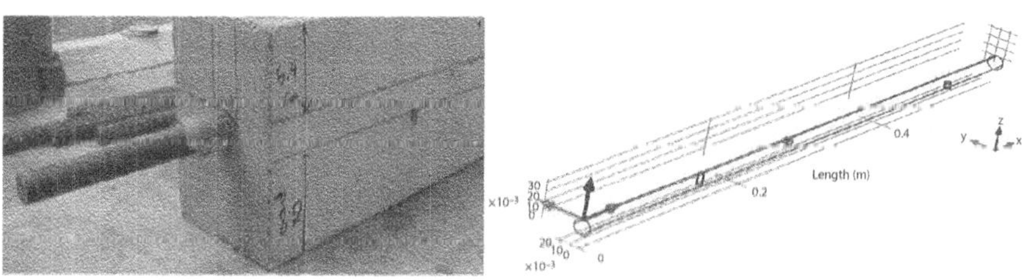

(a) (b)

Figure 16.1 (a) Concrete beam specimen. (b) Three small holes in sound steel reinforcement (rebar).

Table 16.1 Geometry information for the three holes in the steel reinforcement.

Holes	Radius (cm)	Depth (cm)	Hole location on rebar (m)	Position
H1	0.29	1.24	0.14	Top
H2	0.34	0.57	0.27	left side
H3	0.33	0.67	0.49	Bottom

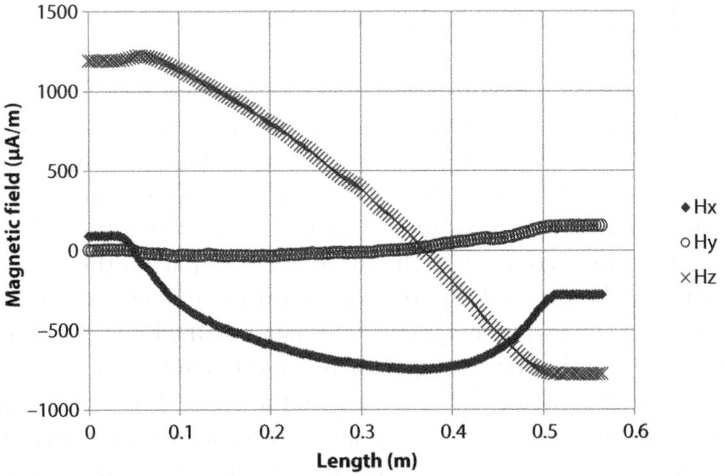

Figure 16.2 Raw magnetic field data from the surface of the drilled reinforced concrete beam.

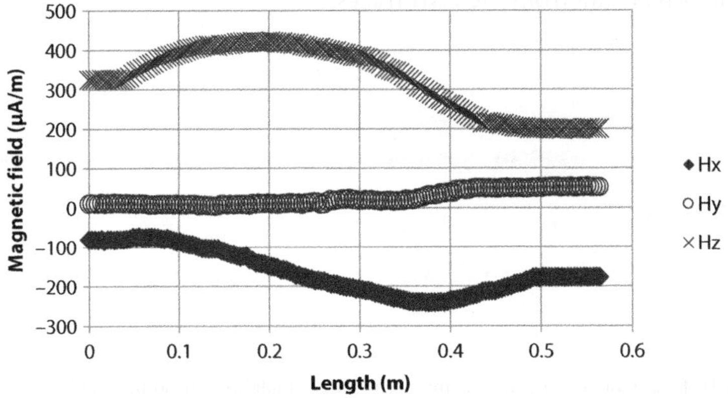

Figure 16.3 Raw magnetic field data from the surface of sound reinforced concrete beam.

As mentioned in Eq.16.1, these data are input as column vectors. The same experimental test results using a fully intact (sound) steel reinforcement bar were used as threshold data (Figure 16.3).

16.5 Results

Based on applying equations 6 and 7 to the experimental test data scan, a set of Eigensignals is directly calculated from the scan line data. The first Eigensignals, shown in Figure 16.4, represent the PCA analysis results for

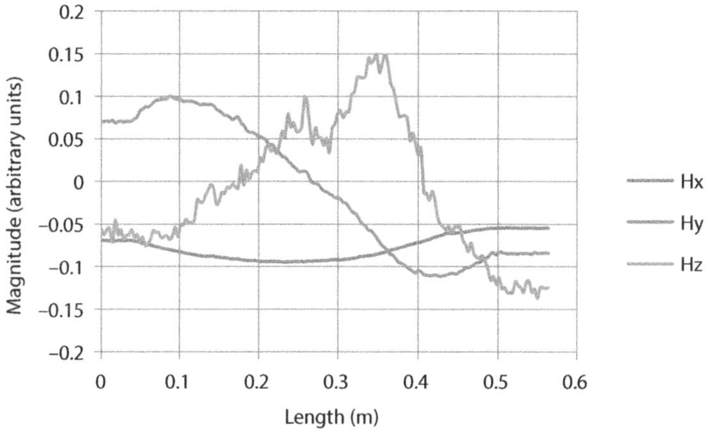

Figure 16.4 Eigensignals of sound steel reinforcement.

the sound steel reinforcement. These results were used as a threshold [21] to distinguish between the sound and deteriorated reinforcement sections.

Figure 16.5 shows the Eigensignals derived from the drilled (i.e., flawed) steel reinforcement experimental scan. Again, the results from the PCA analysis of the sound and drilled reinforcements fail to show significant changes that could be classified as anomalies arising from the presence of the flaws. Some relatively smooth changes occur in the pattern of magnetic field changes in the x and y directions, but the important (diagnostic) changes are to be found in the z-direction magnetic field.

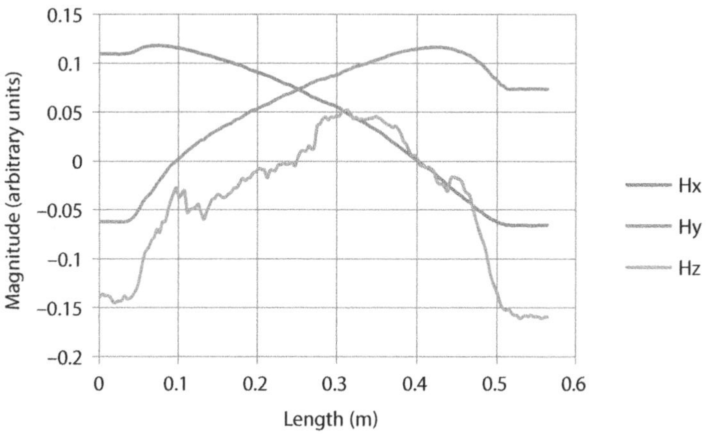

Figure 16.5 Eigensignals of drilled steel reinforcement.

Figure 16.6 shows the results of subtracting the drilled bar Eigensignals from the sound bar Eigensignals. The graphs from the x, and y axes do not present meaningful results, but magnetic data from the z axis show changes from negative to positive values that appear to be directly related to the locations of the drilled holes (Table 16.1). In the field magnetic analysis carried out, the area above the zero line [22] is the diagnostic metric, and the zero line is shown in red in Figure 16.6.

Red box No.1 in Figure 16.6 shows the top hole on the rebar. In this location the Eigensignal of the Z axis data crosses the zero line and, exactly at the start of the hole, it dips to below the zero line, stays below zero until the end of hole, returns to above it after and then drops below zero again. The second hole's anomalous response is different because it is at the side of the rebar, and has a different depth (Figure 16.6). In this case (red box No.2), the Eigensignal increased to a value just above the zero line and then returned to values below it. For the hole at the bottom of the rebar (hole-3), the Eigensignal shows a different pattern, but it still passes significantly above the zero line before the hole and returns to a negative value after it. The anomaly width for this hole is wider than its diameter, perhaps because this hole's signal is a little bit wider than the other two. Because the Eigensignal graph shows different patterns for the three holes (top, right side, and bottom), this method may also be suitable for detecting the clock position of defects in reinforcement.

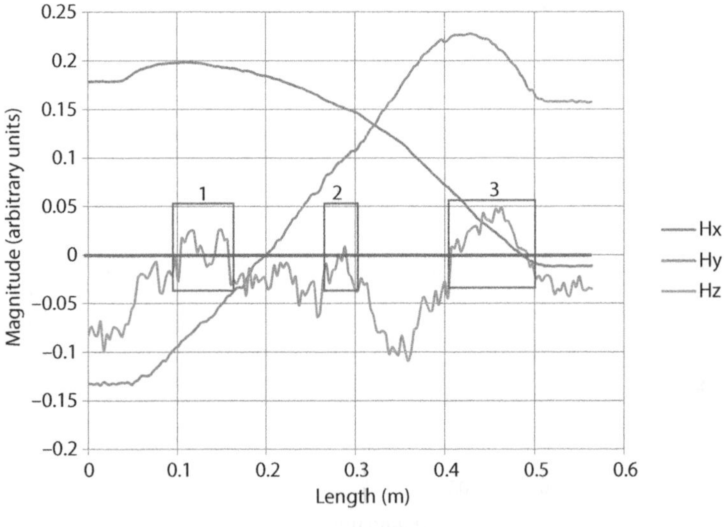

Figure 16.6 Subtraction of Eigensignal from threshold Eigensignal. Red line shows the zero line. Three red boxes show the locations of holes.

16.6 Conclusions

This brief article describes a PCA-based feature extraction method applied to a new, non-destructive, passive magnetic anomaly detection method used to detect defects in steel rebar under concrete cover. In this proof-of-concept article, the method is shown to be capable of detecting the location of holes in three different positions in a rebar. Furthermore, for each defect's position and depth, the Eigensignals show different scan line response patterns, and these different patterns may be candidates for determination of more geometrical features of the defects. The PCA features can be calculated directly from the digital PMI data set, and also compared with other approaches such as wavelet methods and gradient methods.

The results show that small flaws (3-4 mm diameter, 5-6 mm long) give detectable magnetic field anomalies under substantial concrete cover thickness. Calibration of corrosion defects in rebar in active infrastructure through various methods allows the condition of each corrosion site to be quantified, and this is important for management of repairs and making decisions about safety and replacement scheduling, or whether large-scale refurbishing is suitable for sustaining service life.

References

1. Bohni, H., (Ed.). (2005). *Corrosion in reinforced concrete structures* (First ed.). England and USA: Woodhead Publishing and CRC Press.
2. Wilson, J. W., Tian, G. Y., & Barrans, S. (2007). Residual magnetic field sensing for stress measurement. *Sensors & Actuators: A. Physical*, 135(2), 381-387.
3. Khan, A., & Teja, T. S. (2010). An experimental study on prevention of reinforcement corrosion in concrete structures. *International Journal of Research & Reviews in Applied Sciences*, 5(2), 190.
4. Doubov, A. A. (1998). Screening of weld quality using the metal magnetic memory. *Weld World*, 41, 196-199.
5. Vlasov V.T., Dubov A.A. (2004). *Physical bases of the metal magnetic memory method*. Moscow: Tisso Co,. 389 p.
6. Hubert, A., & Schäfer, R. (1998). *Magnetic Domains: The Analysis of Magnetic Microstructures*. Springer, Berlin, New York.
7. Doubov, A. A. (2000). The express-technique of welded joints examination with use of metal magnetic memory. *NDT and E International*, 33(6), 351-362.
8. Wang, Z. D., Yao, K., Deng, B., & Ding, K. Q. (2010). Quantitative study of metal magnetic memory signal versus local stress concentration. *NDT & E International*, 43(6), 513-518.

9. Gallo, G. E., (2010). Investigation of Magnetic Sensing System for In-Place Corrosion Characterization in Metals.
10. Turk MA, Pentland AP. Face recognition using eigenfaces. *J CognitNeurosci* 1991;(3):71-86.
11. Kim M, Lee YH, Han C. Real-time classification of petroleumproducts using near-infrared. *Comput Chemical Engineering* 2000;24:513-7.
12. Gilbert Saporta, Ndeye Niang Keita. Principal Component Analysis: Application to Statistical Process Control. Gérard Govaert. Data Analysis, ISTE, pp. 1-23, 2009, 1-84821-098-1. ff10.1002/9780470611777.ch1ff. ffhal-01125713f.
13. B. Moore, "Principal component analysis in linear systems: Controllability, observability, and model reduction," in *IEEE Transactions on Automatic Control*, vol. 26, no. 1, pp. 17-32, February 1981, doi: 10.1109/TAC.1981.1102568.
14. Finlayson GD, Tian GY. Colour normalisation for colour object recognition. *Int J Pattern Recog Artif Intell* 1999;13(8):1271-85.A. Sophian *et al.* / NDT&E International 36 (2003) 37-4141.
15. Bohni, H. (Ed.). (2005) *Corrosion in reinforced concrete structures* (First Ed.). England and USA: Woodhead Publishing and CRC Press.
16. Oliver, J., Linero, D.L., Huespe, A.E. & Manzoli, O.L. 2008, "Two-dimensional modeling of material failure in reinforced concrete by means of a continuum strong discontinuity approach", *Computer Methods in Applied Mechanics and Engineering*, vol. 197, no. 5, pp. 332-348.
17. Khan., A., and Teja., T.S., 2010, An Experimental Study on Prevention of Reinforcement Corrosion in Concrete Structures, *Ijrras* 5 (2), November 2010.
18. Saraswathy, V. & Song, H.W. (2007) Improving the durability of concrete by using inhibitors. *Building and Environment*, 42(1), 464-472.
19. Abdulrahman, A. S., Ismail, M., & Hussain, M. S. (2011). Corrosion inhibitors for steel reinforcement in concrete: A review. *Scientific Research and Essays*, 6(20), 4152-4162.
20. Ahmad, S. (2003). Reinforcement corrosion in concrete structures, its monitoring and service life prediction--a review. *Cement and Concrete Composites*, 25(4-5), 459-471.
21. Mahbaz S.B., Dusseault M.B., Cascante G., Vanheeghe Ph Detecting defects in steel reinforcement using the passive magnetic inspection method. *J Environ Eng Geophys* 2017 ;22 (2):153–166. doi: 2113 /JEEG22.2.153.
22. M.B. Dusseault, S.B. Mahbaz. System and method for detecting irregularities in rebar in reinforced concrete. United States Patent. Patent No: US 10,533,970 B2. https://patents.google.com/patent/US10533970B2/en, 2020.
23. Mahbaz S.B. Non-destructive passive magnetic and ultrasonic inspection methods for condition assessment of reinforced concrete Ph.D. thesis. Canada: Department of Civil and Environmental Engineering, University of Waterloo; 2016. http://hdl.handle.net/10012/10547.

24. Hubert, A., & Schäfer, R. (1998) *Magnetic Domains: The Analysis of Magnetic Microstructures*. Springer, Berlin, New York.
25. Hironaka, K. & Uedaira, S. 1990, "Soft magnetic properties of Co-Fe-P and Co-Fe-Sn-P amorphous films formed by electroplating", Magnetics, IEEE Transactions on, vol. 26, no. 5, pp. 2421-2424.
26. M. Mosharafi, S.B. Mahbaz, M.B. Dusseault, Bridge deck assessment using infrastructure corrosion assessment magnetic method (iCAMM™) technology, a case study of a culvert in Markham city, Ontario, Canada, *NDT & E International*, Volume 116, 2020, 102356, ISSN 0963-8695, https://doi.org/10.1016/j.ndteint.2020.102356.
27. Mosharafi M., Mahbaz S., Dusseault M.B. Simulation of real defect geometry and its detection using passive magnetic inspection (PMI) method. *Appl Sci* 2018;8(7):1147. doi:10.3390/app8071147.
28. Mosharafi M., Mahbaz S.B., Dusseault M.B., Vanheeghe Ph Magnetic detection of corroded steel rebar: reality and simulations. *NDT E Int* 2020; 110:102225. doi: 10.1016/j.ndteint.2020.102225.
29. Mosharafi M., Mahbaz S.B., Dusseault M.B. Size and location detection of transverse cracks using a passive magnetic method. *Measurement* 2020; 154:107485. doi: 10.1016/j.measurement.2020.107485.
30. Chen, T., Tian, G.Y., Sophian, A. & Que, P.W. 2008, "Feature extraction and selection for defect classification of pulsed eddy current NDT", *NDT & E International*, vol. 41, no. 6, pp. 467-476.
31. Murthy, I.V.R. & Rao, D.B. 1978, "The zero-line method of interpreting total field magnetic anomalies of spherical ore-bodies", *Pure and Applied Geophysics*, vol. 116, no. 6, pp. 1191-1199.

17

Dynamics of Malaria Parasite with Effective Control Analysis

Nagadevi Bala Nagaram[1*] and Suresh Rasappan[2]

[1]*Department of Mathematics, Vel Tech Rangarajan Dr. Sagunthala R&D Institute of Science and Technology, Chennai, Tamil Nadu, India*
[2]*Department of Mathematics, University of Technology and Applied Sciences-Ibri, Ibri, Sultanate of Oman*

Abstract
In this paper, a mathematical model for plasmodium life cycle is designed with parameters to describe the states. Through this construction one can forecast the rate of infection. The deterministic model is designed for plasmodium life cycle if exponential growth occurs at each state. Further some basic qualitative properties are investigated in and around equilibrium point. The local stability is done here through Lasalle's invariance principle. The global stability analysis is done by defining suitable Lyapunov function for the deterministic model. The control strategy is used to control the infected states. Through this control technique the recovery of the disease is possible mathematically. So far back propagation technique is used for the described model to investigate the stability analysis of plasmodium life cycle. And also the analysis is done for stochastic perturbation for the described model. The determination of stability via Lyapunov functions are described here. In numerical simulation, the diagrams are presented which support the theory part. The simulation work is done here through MATLAB.

Keywords: Exponential growth, Lyapunov function, stochastic perturbation, Equilibrium point, plasmodium

17.1 Introduction

At the end of the 19th century, it was believed that people were infected by the malaria parasite by inhalation of dirty water. According to Manson's

**Corresponding author*: nagadevibalaarun@gmail.com

Ramakant Bhardwaj, Jyoti Mishra, Satyendra Narayan and Gopalakrishnan Suseendran (eds.)
Mathematics in Computational Science and Engineering, (359–382) © 2022 Scrivener Publishing LLC

theory, malaria was caused by the bite of mosquito vector. While it drew blood the parasite was absorbed into the mosquito. This was spread by biting the human or through water by leaving the mosquito parasites [1]. In 1898 Ronald Ross described his investigation that when the parasites got maturity in the mosquitoes midgut the disease was spread by biting the human through salivary gland of the mosquito. Also he felt that Manson's theory was correct.

Many researchers have done analysis about the dynamics of Plasmodium [2]. Several experts illustrated the pictorial representation of plasmodium life cycle. It contains several stages between mosquito and human host. The transmitter has sporozoites in its saliva [3, 4]. This sporozoite infected a human through the bite of a mosquito and traveled into the human (liver) hepatocyte. At this stage some species of plasmodium go to sleep. It is so far named as hypnozoite, which can stay in the hepatocyte for several years. The next stage of plasmodium life cycle was merozoite and these merozoite were mixed with blood to infect the red blood cells [5]. Then further typical stages are ring-shaped form of merozoites, trophozoite and schizont. At this stage new merozoite was produced and infected red blood cells, allowing the new merozoite to travel within the bloodstream to infect new red blood cells. The above event repeats several times [6, 7]. Thereafter asexual creatures of male and female gametocytes are produced. This gametocyte freely moves through in the blood, until the mosquito take blood from the infected human host [8].

After entering into the midgut of mosquito, zygote is produced by sexual reproduction. Then the stages were ookinete, oocyst. In this stage, sporozoite are produced by oocyst. The sporozoite migrate to the mosquito's saliva; when the mosquito bites a subsequent human, it injects the parasites into the human blood, repeating the entire life cycle.

The characteristic analysis for exponential growth plasmodium life cycle (EGPLC) is described. Stochastic perturbation is introduced and analysis done in this range over area.

The stability analysis is done by using various techniques, which have been done by researchers in recent years. Among them, one of the valuable techniques to stabilize is backstepping control [9–11]. In this technique, by introducing the parametric control in the states which are required. To obtain the stability, the opted Lyapunov functions are defined.

The structure of the article is in sequence. The mathematical representation of exponential growth plasmodium life cycle has been described in

Sections 17.2 and 17.3. The analysis of local and global stability of the equilibrium points are discussed in Sections 17.4 and 17.5. Section 17.6 describes controlling the system by back propagation technique. Stochastic perturbation were introduced and analyzed stability in section 17.7. Section 17.8 presented the diagram which supports the theory part and section 17.9 is supported to the conclusion.

17.2 The Mathematical Structure of EGPLC

The representation of the EGPLC model is described below:

- The stages of EGPLC model are classified as 10 states, such as first state is populace of Anopheles mosquito, after inject the parasite, name as sporozoite is second state. Followed by this the developed further states are hyponozoite, infected hepatocyte, hepatic schizont, ruptured schizont, merozoite, first trophozoite, new trophozoite and gametocyte.
- At every stage, exponential growth are taken as initial population.

Figure 17.1 depicts the flow diagram of EGPLC model.

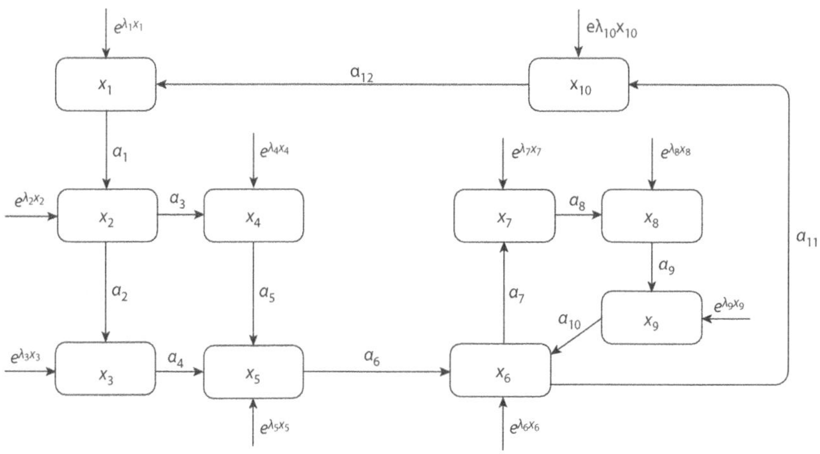

Figure 17.1 Flow diagram of EGPLC model.

The Mathematical structure of EGPLC are given below:

$$\frac{dx_1}{dt} = x_1 e^{\lambda_1 x_1} + \alpha_{12} x_{10} - \alpha_1 x_1$$
$$\frac{dx_2}{dt} = x_2 e^{\lambda_2 x_2} + \alpha_1 x_1 - (\alpha_2 + \alpha_3) x_2$$
$$\frac{dx_3}{dt} = x_3 e^{\lambda_3 x_3} + \alpha_2 x_2 - \alpha_4 x_3$$
$$\frac{dx_4}{dt} = x_4 e^{\lambda_4 x_4} + \alpha_3 x_2 - \alpha_5 x_4$$
$$\frac{dx_5}{dt} = x_5 e^{\lambda_5 x_6} + \alpha_4 x_3 + \alpha_5 x_4 - \alpha_6 x_5$$
$$\frac{dx_6}{dt} = x_6 e^{\lambda_6 x_6} + \alpha_6 x_5 - \alpha_{10} x_9 - (\alpha_7 + \alpha_{11}) x_6$$
$$\frac{dx_7}{dt} = x_7 e^{\lambda_7 x_7} + \alpha_7 x_6 - \alpha_8 x_7$$
$$\frac{dx_8}{dt} = x_8 e^{\lambda_8 x_8} + \alpha_8 x_7 - \alpha_9 x_8$$
$$\frac{dx_9}{dt} = x_9 e^{\lambda_9 x_9} + \alpha_9 x_8 - \alpha_{10} x_9$$
$$\frac{dx_{10}}{dt} = x_{10} e^{\lambda_{10} x_{10}} + \alpha_{11} x_6 - \alpha_{12} x_{10} \quad (17.1)$$

where x_1 is the populace of adult mosquitoes, x_2 is the populace of sporozoite x_3 is the populace of Hypnozoite, x_4 is the populace of Infected hepatocyte, x_5 is the populace of Hepatic schizont, x_6 is the populace of Ruptured schizont, x_7 is the number of Merozoites, x_8 is the populace of Early trophozoite(Ring form), x_9 is the populace of late trophozoite, x_{10} is the populace of gametocyte. b is the natural birth rate at initial stage, α_1 is the rate of sporozoite from adult mosquito, α_2 is the rate of sporozoite splited and one amount of them push to hypnozoite, α_3 is the rate of growth of infected hepatocyte from sporozoite, α_4 is the rate of hypnozoite push to hepatic schizont, α_5 is the rate of growth of hepatic schizont from infected hepatocyte, α_6 is the rate of growth of ruptured schizont from hepatic schizont, α_7 is the rate of ruptured schizont turn to merozoite, α_8 is the rate of growth of ring form trophozoite from merozoite, α_9 is the rate of growth of new trophozoite from early trophozoite, α_{10} is the rate of late trophozoite to ruptured schizont, α_{11} is the rate of ruptured schizont to gametocyte, α_{12} is the rate of gametocyte enter into anopheles mosquito through blood meal and the exponential growth populace is taken as $e^{\lambda_i x_i}$, for i=1 to 10.

17.3 The Modified EGPLC Model

The modified EGPLC system is given below:

$$\begin{aligned}
\frac{dx_1}{dt} &= \alpha_{12}x_{10} - \alpha x_1 \\
\frac{dx_2}{dt} &= \alpha_1 x_1 - Px_2 \\
\frac{dx_3}{dt} &= \alpha_2 x_2 - Nx_3 \\
\frac{dx_4}{dt} &= \alpha_3 x_2 - Tx_4 \\
\frac{dx_5}{dt} &= \alpha_4 x_3 + \alpha_5 x_4 - Ix_5 \\
\frac{dx_6}{dt} &= \alpha_6 x_5 + \alpha_{10} x_9 - Ax_6 \\
\frac{dx_7}{dt} &= \alpha_7 x_6 - Lx_7 \\
\frac{dx_8}{dt} &= \alpha_8 x_7 - Gx_8 \\
\frac{dx_9}{dt} &= \alpha_9 x_8 - Rx_9 \\
\frac{dx_{10}}{dt} &= \alpha_{11} x_6 - Hx_{10}
\end{aligned} \tag{17.2}$$

where,

$$E=\alpha_1-e^{\lambda_1 x_1}, P=\alpha_2+\alpha_3-e^{\lambda_2 x_2}, N=\alpha_4-e^{\lambda_3 x_3}, T=\alpha_5-e^{\lambda_4 x_4}, I=\alpha_6-e^{\lambda_5 x_5},$$
$$A=\alpha_7+\alpha_{11}-e^{\lambda_6 x_6}, L=\alpha_8-e^{\lambda_7 x_7}, G=\alpha_9-e^{\lambda_8 x_8}, R=\alpha_{10}-e^{\lambda_9 x_9}, H=\alpha_{12}-e^{\lambda_{10} x_{10}}.$$

17.4 Equilibria and Local Stability Analysis

The equilibrium and interior equilibrium point are discussed for the EGPLC model. The equilibrium points are got by equating the system(2) to zero.

$$E_1=(\frac{\alpha_{12}}{E}x_{10},0,0,0,0,0,0,0,0,0),\ E_2=(0,\frac{\alpha_1}{P}x_1,0,0,0,0,0,0,0,0),$$
$$E_3=(0,0,\frac{\alpha_2}{N}x_2,0,0,0,0,0,0,0),\ E_4=(0,0,0,\frac{\alpha_3}{T}x_2,0,0,0,0,0,0),$$
$$E_5=(0,0,0,0,\frac{\alpha_4 x_3+\alpha_5 x_4}{I},0,0,0,0,0),\ E_6=(0,0,0,0,0,\frac{\alpha_6}{A}x_5,0,0,0,0),$$
$$E_7=(0,0,0,0,0,0,\frac{\alpha_7}{L}x_6,0,0,0),\ E_8=(0,0,0,0,0,0,0,\frac{\alpha_8}{G}x_7,0,0),$$
$$E_9=(0,0,0,0,0,0,0,0,\frac{\alpha_9}{R}x_8,0),\ E_{10}=(0,0,0,0,0,0,0,0,0,\frac{\alpha_{11}}{H}x_6)$$

and the interior exponential growth equilibrium points are

$$E^*=(x_1^*,x_2^*,x_3^*,x_4^*,x_5^*,x_6^*,x_7^*,x_8^*,x_9^*,x_{10}^*)$$

where,

$$x_1^*=\frac{\alpha_{12}}{E}x_{10}^*, x_2^*=\frac{\alpha_1}{P}x_1^*, x_3^*=\frac{\alpha_2}{N}x_2^*, x_4^*=\frac{\alpha_3}{T}x_2^*, x_5^*=\frac{\alpha_4 x_3^*+\alpha_5 x_4^*}{I},$$
$$x_6^*=\frac{\alpha_6 x_5^*+\alpha_{10} x_9^*}{A}, x_7^*=\frac{\alpha_7}{L}x_6^*, x_8^*=\frac{\alpha_8}{G}x_7^*, x_9^*=\frac{\alpha_9}{R}x_8^*, x_{10}^*=\frac{\alpha_{11}}{H}x_6^* \tag{17.3}$$

17.5 Analysis of Global Stability

Theorem 1: The point ***E**** is globally asymptotically stable if

$$\alpha_1 = (\frac{-x_2(x_2 - x_2^*) + Px_2}{x_1}), \qquad \alpha_2 = (\frac{-x_3(x_3 - x_3^*) + Nx_3}{x_2}),$$

$$\alpha_3 = (\frac{-x_4(x_4 - x_4^*) + Tx_4}{x_2}), \qquad \alpha_4 = (\frac{-x_5(x_5 - x_5^*) - \alpha_5 x_4 + Ix_5}{x_3}),$$

$$\alpha_6 = (\frac{-x_6(x_6 - x_6^*) - \alpha_{10} x_9 + Nx_6}{x_5}), \quad \alpha_7 = (\frac{-x_7(x_7 - x_7^*) + Gx_7}{x_6}),$$

$$\alpha_8 = (\frac{-x_8(x_8 - x_8^*) + Cx_8}{x_7}), \qquad \alpha_9 = (\frac{-x_9(x_9 - x_9^*) + Tx_9}{x_8}),$$

$$\alpha_{11} = (\frac{-x_{10}(x_{10} - x_{10}^*) + Hx_{10}}{x_6}), \qquad \alpha_{12} = (\frac{-x_1(x_1 - x_1^*) + Ex_1}{x_{10}}) \tag{17.4}$$

Proof: The Lyapunov function is taken as,

$$V_g(x_i) = \sum_{i=1}^{10} \beta_i[(x_i - x_i^*) - x_i^* \ln(\frac{x_i}{x_i^*})], for\ i = 1,2,...,10 \tag{17.5}$$

In the region, the positive definite function V_g is obtained except at E^*. The rate of change of V_g is,

$$\dot{V}_g = \sum_{i=1}^{10} \beta_i \frac{x_i}{x_i^*}(x_i - x_i^*) \tag{17.6}$$

Solving equation (17.6) along (17.3) we get,

$$\begin{aligned}
\dot{V}_g &= \beta_1(x_1 - x_1^*)\left(\frac{\alpha_{12}x_{10} - Ex_1}{x_1}\right) + \beta_2(x_2 - x_2^*)\left(\frac{\alpha_1 x_1 - Px_2}{x_2}\right) \\
&+ \beta_3(x_3 - x_3^*)\left(\frac{\alpha_2 x_2 - Nx_3}{x_3}\right) + \beta_4(x_4 - x_4^*)\left(\frac{\alpha_3 x_2 - Tx_4}{x_4}\right) \\
&+ \beta_5(x_5 - x_5^*)\left(\frac{\alpha_4 x_3 + \alpha_5 x_4 - Ix_5}{x_5}\right) + \beta_6(x_6 - x_6^*)\left(\frac{\alpha_6 x_5 + \alpha_{10} x_9 - Ax_6}{x_6}\right) \\
&+ \beta_7(x_7 - x_7^*)\left(\frac{\alpha_7 x_6 - Lx_7}{x_7}\right) + \beta_8(x_8 - x_8^*)\left(\frac{\alpha_8 x_7 - Gx_8}{x_8}\right) \\
&+ \beta_9(x_9 - x_9^*)\left(\frac{\alpha_9 x_8 - Rx_9}{x_9}\right) + \beta_{10}(x_{10} - x_{10}^*)\left(\frac{\alpha_{11} x_6 - Hx_{10}}{x_{10}}\right) \qquad (17.7)
\end{aligned}$$

Now choosing (17.4) to (17.7), we get

$$\begin{aligned}
\dot{V}_g = &-\beta_1(x_1 - x_1^*)^2 - \beta_2(x_2 - x_2^*)^2 - \beta_3(x_3 - x_3^*)^2 \\
&- \beta_4(x_4 - x_4^*)^2 - \beta_5(x_5 - x_5^*)^2 - \beta_6(x_6 - x_6^*)^2 \\
&- \beta_7(x_7 - x_7^*)^2 - \beta_8(x_8 - x_8^*)^2 - \beta_9(x_9 - x_9^*)^2 \\
&- \beta_{10}(x_{10} - x_{10}^*)^2 \qquad (17.8)
\end{aligned}$$

and hence $\dot{V}_g$ is negative definite. Since by Lasalle's invariance principle, $\dot{V}_g$ is globally asymptotically stable.

17.6 Global Stability Analysis with Back Propagation

$$\frac{dx_1}{dt} = (e^{\lambda_1 x_1} - \alpha_1)x_1 + \alpha_{12}x_{10}$$

$$\frac{dx_2}{dt} = (e^{\lambda_2 x_2} - \alpha_1 - \alpha_3)x_2 + \alpha_1 x_1$$

$$\frac{dx_3}{dt} = (e^{\lambda_3 x_3} - \alpha_4)x_3 + \alpha_2 x_2$$

$$\frac{dx_4}{dt} = (e^{\lambda_4 x_4} - \alpha_5)x_4 + \alpha_3 x_2$$

$$\frac{dx_5}{dt} = (e^{\lambda_5 x_6} - \alpha_6)x_5 + \alpha_4 x_3 + \alpha_5 x_4$$

$$\frac{dx_6}{dt} = (e^{\lambda_6 x_6} - \alpha_7 - \alpha_{11})x_6 + \alpha_6 x_5 - \alpha_{10}x_9$$

$$\frac{dx_7}{dt} = (e^{\lambda_7 x_7} - \alpha_8)x_7 + \alpha_7 x_6$$

$$\frac{dx_8}{dt} = (e^{\lambda_8 x_8} - \alpha_9)x_8 + \alpha_8 x_7$$

$$\frac{dx_9}{dt} = (e^{\lambda_9 x_9} - \alpha_{10})x_9 + \alpha_9 x_8$$

$$\frac{dx_{10}}{dt} = (e^{\lambda_{10} x_{10}} - \alpha_{12})x_{10} + \alpha_{11}x_6$$

Let us consider the first state in EGPLC model,

$$\dot{x}_1 = (e^{\lambda_1} - \alpha_1)x_1 + \alpha_{12}x_{10} + u_1$$

where x_3 is a mosq- pseudo controller. By defining Lyapunov function as,

$$V_1 = \frac{x_1^2}{2}$$

Assume that $x_2 = \vartheta_1$ as a pseudo controller and if $\vartheta_1 = 0$ then

$$\dot{V}_1 = -(e^{\lambda_1} - \alpha_1)x_1^2$$

Thus obtained negative definite. The mosq- state is asymptotically stable.

The function ϑ_1 is on estimative when x_2 as a ruler, the occurance of error in the system be,

$$w_2 = x_2 - \vartheta_1 = x_2$$

consider the (x_1, w_2) subsystem defined by

$$w_2 = (e^{\lambda_2} - (\alpha_2 + \alpha_3))x_2 + \alpha_1 x_1 + u_2$$

where x_3 is a sporo- pseudo controller. By defining Lyapunov function as,

$$V_2 = V_1 + \frac{x_2^2}{2}$$

Assume that $x_3 = \vartheta_2$ as a pseudo controller and if $\vartheta_2 = 0$ then

$$\dot{V}_2 = -(e^{\lambda_1} - \alpha_1)x_1^2 - (e^{\lambda_2} - \alpha_2 - \alpha_3)w_2^2$$

Thus obtained negative definite. The sporo-state is asymptotically stable.

The function ϑ_2 is on estimative when x_3 as a ruler, the occurance of error in the system be,

$$w_3 = \mathrm{x}_3 - \vartheta_2 = x_3$$

consider the (x_1, w_2, w_3) subsystem defined by

$$\dot{w}_3 = (e^{\lambda_3} - \alpha_4)x_3 + \alpha_2 x_2 + u_3$$

where x_4 is a hypno-pseudo controller. By defining Lyapunov function as,

$$V_3 = V_2 + \frac{x_3^2}{2}$$

Assume that $x_4 = \vartheta_3$ as a pseudo controller and if $\vartheta_3 = 0$ then

$$\dot{V}_3 = -(e^{\lambda_1} - \alpha_1)x_1^2 - (e^{\lambda_2} - \alpha_2 - \alpha_3)w_2^2 - (e^{\lambda_3} - \alpha_4)w_3^2$$

Thus obtained negative definite. The hypno-state is asymptotically stable.

The function ϑ_3 is on estimative when x_4 as a ruler, the occurance of error in the system be,

$$w_4 = x_4 - \vartheta_3 = x_4$$

consider the (x_1, w_2, w_3, w_4) subsystem defined by

$$\dot{w}_4 = (e^{\lambda_4} - \alpha_5)x_4 + \alpha_3 x_2 + u_4$$

where x_5 is a hepato-pseudo controller. By defining Lyapunov function as,

$$V_4 = V_3 + \frac{x_4^2}{2}$$

Assume that $x_5 = \vartheta_4$ as a pseudo controller and if $\vartheta_4 = 0$ then

$$\dot{V}_4 = -(e^{\lambda_1} - \alpha_1)x_1^2 - (e^{\lambda_2} - \alpha_2 - \alpha_3)w_2^2 - (e^{\lambda_3} - \alpha_4)w_3^2 - (e^{\lambda_4} - \alpha_5)w_4^2$$

Thus obtained negative definite. The hepato-state is asymptotically stable.

The function ϑ_4 is on estimative when x_5 as a ruler, the occurance of error in the system be,

$$w_5 = x_5 - \vartheta_4 = x_5$$

consider the $(x_1, w_2, w_3, w_4, w_5)$ subsystem defined by

$$\dot{w}_5 = (e^{\lambda_5} - \alpha_6)x_5 + \alpha_4 x_3 + \alpha_5 x_4 + u_5$$

where x_6 is a hepatic-pseudo controller. By defining Lyapunov function as,

$$V_5 = V_4 + \frac{x_5^2}{2}$$

Assume that $x_6 = \vartheta_5$ as a pseudo controller and if $\vartheta_5 = 0$ then

$$\dot{V}_5 = -(e^{\lambda_1} - \alpha_1)x_1^2 - (e^{\lambda_2} - \alpha_2 - \alpha_3)w_2^2 - (e^{\lambda_3} - \alpha_4)w_3^2 \\ -(e^{\lambda_4} - \alpha_5)w_4^2 - (e^{\lambda_5} - \alpha_6)w_5^2$$

Thus obtained negative definite. The hepatic-state is asymptotically stable.

The function ϑ_5 is on estimative when x_6 as a ruler, the occurance of error in the system be,

$$w_6 = x_6 - \vartheta_5 = x_6$$

consider the $(x_1, w_2, w_3, w_4, w_5, w_6)$ subsystem defined by

$$\dot{w}_6 = (e^{\lambda_6} - \alpha_7 + \alpha_{11})x_6 + \alpha_6 x_5 + \alpha_{10} x_9 + u_6$$

where x_7 is a schizo-pseudo controller. By defining Lyapunov function as,

$$V_6 = V_5 + \frac{x_6^2}{2}$$

Assume that $x_7 = \vartheta_6$ as a pseudo controller and if $\vartheta_6 = 0$ then

$$\dot{V}_6 = -(e^{\lambda_1} - \alpha_1)x_1^2 - (e^{\lambda_2} - \alpha_2 - \alpha_3)w_2^2 - (e^{\lambda_3} - \alpha_4)w_3^2 \\ -(e^{\lambda_4} - \alpha_5)w_4^2 - (e^{\lambda_5} - \alpha_6)w_5^2 - (e^{\lambda_6} - \alpha_7 + \alpha_{11})w_6^2$$

Thus obtained negative definite. The schizo-state is asymptotically stable.

The function ϑ_6 is on estimative when x_7 as a ruler, the occurance of error in the system be,

$$w_7 = x_7 - \vartheta_6 = x_7$$

consider the $(x_1, w_2, w_3, w_4, w_5, w_6, w_7)$ subsystem defined by

$$\dot{w}_7 = (e^{\lambda_7} - \alpha_8)x_7 + \alpha_7 x_6 + u_7$$

where x_8 is a merozoi-pseudo controller. By defining Lyapunov function as,

$$V_7 = V_6 + \frac{x_7^2}{2}$$

Assume that $x_8 = \vartheta_7$ as a pseudo controller and if $\vartheta_7 = 0$ then

$$\begin{aligned}\dot{V}_7 = &-(e^{\lambda_1} - \alpha_1)x_1^2 - (e^{\lambda_2} - \alpha_2 - \alpha_3)w_2^2 - (e^{\lambda_3} - \alpha_4)w_3^2 \\ &- (e^{\lambda_4} - \alpha_5)w_4^2 - (e^{\lambda_5} - \alpha_6)w_5^2 - (e^{\lambda_6} - \alpha_7 + \alpha_{11})w_6^2 \\ &- (e^{\lambda_7} - \alpha_8)w_7^2\end{aligned}$$

Thus obtained negative definite. The merozoi-state is asymptotically stable.

The function ϑ_7 is on estimative when x_8 as a ruler, the occurance of error in the system be,

$$w_8 = x_8 - \vartheta_7 = x_8$$

consider the $(x_1, w_2, w_3, w_4, w_5, w_6, w_7, w_8)$ subsystem defined by

$$\dot{w}_8 = (e^{\lambda_8} - \alpha_9)x_8 + \alpha_8 x_7 + u_8$$

where x_9 is a tropho-pseudo controller. By defining Lyapunov function as,

$$V_8 = V_7 + \frac{x_8^2}{2}$$

Assume that $x_8 = \vartheta_7$ as a pseudo controller and if $\vartheta_7 = 0$ then

$$\begin{aligned}\dot{V}_8 = &-(e^{\lambda_1} - \alpha_1)x_1^2 - (e^{\lambda_2} - \alpha_2 - \alpha_3)w_2^2 - (e^{\lambda_3} - \alpha_4)w_3^2 \\ &- (e^{\lambda_4} - \alpha_5)w_4^2 - (e^{\lambda_5} - \alpha_6)w_5^2 - (e^{\lambda_6} - \alpha_7 + \alpha_{11})w_6^2 \\ &- (e^{\lambda_7} - \alpha_8)w_7^2 - (e^{\lambda_8} - \alpha_9)w_8^2\end{aligned}$$

Thus obtained negative definite. The tropho-state is asymptotically stable.

The function ϑ_8 is on estimative when x_9 as a ruler, the occurance of error in the system be,

$$w_9 = x_9 - \vartheta_8 = x_9$$

consider the $(x_1, w_2, w_3, w_4, w_5, w_6, w_7, w_8, w_9)$ subsystem defined by

$$\dot{w}_9 = (e^{\lambda_9} - \alpha_{10})x_9 + \alpha_9 x_8 + u_9$$

where x_{10} is a newtropho-pseudo controller. By defining Lyapunov function as,

$$V_9 = V_8 + \frac{x_9^2}{2}$$

Assume that $x_9 = \vartheta_8$ as a pseudo controller and if $\vartheta_8 = 0$ then

$$\begin{aligned}\dot{V}_9 = &-(e^{\lambda_1} - \alpha_1)x_1^2 - (e^{\lambda_2} - \alpha_2 - \alpha_3)w_2^2 - (e^{\lambda_3} - \alpha_4)w_3^2 \\ &- (e^{\lambda_4} - \alpha_5)w_4^2 - (e^{\lambda_5} - \alpha_6)w_5^2 - (e^{\lambda_6} - \alpha_7 + \alpha_{11})w_6^2 \\ &- (e^{\lambda_7} - \alpha_8)w_7^2 - (e^{\lambda_8} - \alpha_9)w_8^2 - (e^{\lambda_9} - \alpha_{10})w_9^2\end{aligned}$$

Thus obtained negative definite. The newtropho-state is asymptotically stable.

The function ϑ_9 is on estimative when x_{10} as a ruler, the occurance of error in the system be,

$$w_{10} = x_{10} - \vartheta_9 = x_{10}$$

consider the $(x_1, w_2, w_3, w_4, w_5, w_6, w_7, w_8, w_9, w_{10})$ subsystem defined by

$$\dot{w}_{10} = (e^{\lambda_{10}} - \alpha_{12})x_{10} + \alpha_{11} x_6 + u_{10}$$

By defining Lyapunov function as,

$$V_{10} = V_9 + \frac{x_{10}^2}{2}$$

Assume that $x_{10} = \vartheta_9$ as a pseudo controller and if $\vartheta_9 = 0$ then

$$\begin{aligned}\dot{V}_{10} = &-(e^{\lambda_1} - \alpha_1)x_1^2 - (e^{\lambda_2} - \alpha_2 - \alpha_3)w_2^2 - (e^{\lambda_3} - \alpha_4)w_3^2 \\ &- (e^{\lambda_4} - \alpha_5)w_4^2 - (e^{\lambda_5} - \alpha_6)w_5^2 - (e^{\lambda_6} - \alpha_7 + \alpha_{11})w_6^2 \\ &- (e^{\lambda_7} - \alpha_8)w_7^2 - (e^{\lambda_8} - \alpha_9)w_8^2 - (e^{\lambda_9} - \alpha_{10})w_9^2 - (e^{\lambda_{10}} - \alpha_{12})w_{10}^2\end{aligned} \tag{17.9}$$

Thus obtained negative definite from equation (17.9). The above state is asymptotically stable.

Thus the infected states are controlled by given the required controller in needed states. This technique is more powerful than the other control stratergy.

17.7 Stability Analysis of Non-Deterministic EGPLC Model

In some cases the vertebrate host may be affected by some other disease; also then the infection rate due to new disturbance is occurred in the described system. So far stochastic analysis is also done for the model.

The perturbation is in the non-deterministic model is described. The white noise is allowed in all the parameter around E^*, which are proportion to $(x_i - x_i^*), i = 1,2,...,10$.

$$\begin{aligned}dx_1 &= [\alpha_{12}x_{10} - Ex_1]dt + \sigma_1[x_1 - x_1^*]dwt_1 \\ dx_2 &= [\alpha_1 x_1 - Px_2]dt + \sigma_2[x_2 - x_2^*]dwt_2 \\ dx_3 &= [\alpha_2 x_2 - Nx_3]dt + \sigma_3[x_3 - x_3^*]dwt_3 \\ dx_4 &= [\alpha_3 x_2 - Tx_4]dt + \sigma_4[x_4 - x_4^*]dwt_4 \\ dx_5 &= [\alpha_4 x_3 + \alpha_5 x_4 - Ix_5]dt + \sigma_5[x_5 - x_5^*]dwt_5 \\ dx_6 &= [\alpha_6 x_5 + \alpha_{10}x_9 - Ax_6]dt + \sigma_6[x_6 - x_6^*]dwt_6 \\ dx_7 &= [\alpha_7 x_6 - Lx_7]dt + \sigma_7[x_7 - x_7^*]dwt_7 \\ dx_8 &= [\alpha_8 x_7 - Gx_8]dt + \sigma_8[x_8 - x_8^*]dwt_8 \\ dx_9 &= [\alpha_9 x_8 - Rx_9]dt + \sigma_9[x_9 - x_9^*]dwt_9 \\ dx_{10} &= [\alpha_{11}x_6 - Hx_{10}]dt + \sigma_{10}[x_{10} - x_{10}^*]dwt_{10}\end{aligned} \tag{17.10}$$

The real constants σ_i and wt_i, *for* $i = 1,2,\ldots,10$ are independent to each other. The dynamical behavior of the system is investigated around E^*. Therefore, to linearize the system using stochastic differential equation's (SDE) about E^* given below:

Let $u_{gi} = (x_i - x_i^*), \, for \; i = 1,2,\ldots,10$

$$du_g(t) = f(u_g(t))dt + g(u_g(t))dw(t) \tag{17.11}$$

where, $u_g(t) = [u_{g1}(t)\; u_{g2}(t)\; u_{g3}(t)\; u_{g4}(t)\; u_{g5}(t)\; u_{g6}(t)\; u_{g7}(t)\; u_{g8}(t)\; u_{g9}(t)\; u_{g10}(t)]^T$ and

$$f(u_g(t)) = \begin{bmatrix} -E & 0 & 0 & 0 & 0 & 0 & 0 & 0 & 0 & \alpha_{12} \\ \alpha_1 & -P & 0 & 0 & 0 & 0 & 0 & 0 & 0 & 0 \\ 0 & \alpha_2 & -N & 0 & 0 & 0 & 0 & 0 & 0 & 0 \\ 0 & \alpha_3 & 0 & -T & 0 & 0 & 0 & 0 & 0 & 0 \\ 0 & 0 & \alpha_4 & \alpha_5 & -I & 0 & 0 & 0 & 0 & 0 \\ 0 & 0 & 0 & 0 & \alpha_6 & -A & 0 & 0 & \alpha_{10} & 0 \\ 0 & 0 & 0 & 0 & 0 & \alpha_7 & -L & 0 & 0 & 0 \\ 0 & 0 & 0 & 0 & 0 & 0 & \alpha_8 & -G & 0 & 0 \\ 0 & 0 & 0 & 0 & 0 & 0 & 0 & \alpha_9 & -R & 0 \\ 0 & 0 & 0 & 0 & 0 & \alpha_{11} & 0 & 0 & 0 & -H \end{bmatrix} u_g(t) \tag{17.12}$$

$$g(u_g(t)) =$$

$$\begin{bmatrix} \sigma_1 u_{g1} & 0 & 0 & 0 & 0 & 0 & 0 & 0 & 0 & 0 \\ 0 & \sigma_2 u_{g2} & 0 & 0 & 0 & 0 & 0 & 0 & 0 & 0 \\ 0 & 0 & \sigma_3 u_{g3} & 0 & 0 & 0 & 0 & 0 & 0 & 0 \\ 0 & 0 & 0 & \sigma_4 u_{g4} & 0 & 0 & 0 & 0 & 0 & 0 \\ 0 & 0 & 0 & 0 & \sigma_5 u_{g5} & 0 & 0 & 0 & 0 & 0 \\ 0 & 0 & 0 & 0 & 0 & \sigma_6 u_{g6} & 0 & 0 & 0 & 0 \\ 0 & 0 & 0 & 0 & 0 & 0 & \sigma_7 u_{g7} & 0 & 0 & 0 \\ 0 & 0 & 0 & 0 & 0 & 0 & 0 & \sigma_8 u_{g8} & 0 & 0 \\ 0 & 0 & 0 & 0 & 0 & 0 & 0 & 0 & \sigma_9 u_{g9} & 0 \\ 0 & 0 & 0 & 0 & 0 & 0 & 0 & 0 & 0 & \sigma_{10} u_{g10} \end{bmatrix} \tag{17.13}$$

$g(t) = 0$ obtained about E^*. Let U be the set

$$U = (t \geq t_0) \times R^+.$$

Hence $V \in C_2^U(U)$ is twice continuously differentiable function with respect to u and a continuous functions with respect to t. The $It\hat{o}'$ SDE is given below:

$$\frac{\partial V_g(t,u_g)}{\partial t} + f^T u_g(t) \frac{\partial V_g(t,u_g)}{\partial u_g} + \frac{1}{2} trace[g^T(u_g(t)) \frac{\partial^2 V_g(t,u_g)}{\partial u_g^2} g(u_g(t))] \tag{17.14}$$

$$where \frac{\partial V_g}{\partial u_g} = [\frac{\partial V_g}{\partial u_{g1}} \frac{\partial V_g}{\partial u_{g2}} \frac{\partial V_g}{\partial u_{g3}} \frac{\partial V_g}{\partial u_{g4}} \frac{\partial V_g}{\partial u_{g5}} \frac{\partial V_g}{\partial u_{g6}} \frac{\partial V_g}{\partial u_{g7}} \frac{\partial V_g}{\partial u_{g8}} \frac{\partial V_g}{\partial u_{g9}} \frac{\partial V_g}{\partial u_{g10}}]^T$$

$$\frac{\partial^2 V_g(t,u_g)}{\partial u_g^2} = col(\frac{\partial^2 V_g}{\partial w_j \partial u_{gi}}), i, j = 1,2,...,10.$$

Remark: If

$$V_g(t,u_g) \leq K_2 |u_g|^\rho \tag{17.15}$$

and

$$LV_g(t,u_g) \leq K_3 |u_g|^\rho, K_i > 0, \rho > 0. \tag{17.16}$$

Then the system of equation (17.10) is globally asymptotically mean square stable in probability (17.14).

Theorem 2: If

$$\begin{aligned}&\alpha_{12}u_{g10}u_{g1}w_1 + \alpha_1 u_{g1}u_{g2}w_2 + \alpha_2 u_{g2}u_{g3}w_3 + \alpha_3 u_{g2}u_{g4}w_4\\&+w_5(\alpha_4 u_{g5}u_{g3} + \alpha_5 u_{g5}u_{g4}) + w_6(\alpha_6 u_{g5}u_{g6} + \alpha_{10}u_{g9}u_{g6})\\&+\alpha_7 u_{g6}u_{g7}w_7 + \alpha_8 u_{g8}u_{g7}w_8 + \alpha_9 u_{g8}u_{g9}w_9 + \alpha_{11}u_{g10}u_{g6}w_{10} = 0\end{aligned} \tag{17.17}$$

then the population free solution (17.10) is asymptotically mean square stable.

Proof: The inequalities (17.16) hold with $\rho = 2$.
Now the *Itô* SDE (17.14) becomes

$$\begin{aligned}LV_g(t,u_g) = &-w_1 E u_{g1}^2 - w_2 P u_{g2}^2 - w_3 N u_{g3}^2 - w_4 T u_{g4}^2 - w_5 I u_{g5}^2 - w_6 A u_{g6}^2\\&- w_7 L u_{g7}^2 - w_8 G u_{g8}^2 - w_9 R u_{g9}^2 - w_{10} H u_{g10}^2 + \alpha_{12}u_{g10}u_{g1}w_1\\&+ \alpha_1 u_{g1}u_{g2}w_2 + \alpha_2 u_{g2}u_{g3}w_3 + \alpha_3 u_{g2}u_{g4}w_4\\&+ w_5(\alpha_4 u_{g5}u_{g3} + \alpha_5 u_{g5}u_{g4}) + w_6(\alpha_6 u_{g5}u_{g6} + \alpha_{10}u_{g9}u_{g6})\\&+ \alpha_7 u_{g6}u_{g7}w_7 + \alpha_8 u_{g8}u_{g7}w_8 + \alpha_9 u_{g8}u_{g9}w_9 + \alpha_{11}u_{g10}u_{g6}w_{10}\\&+ \frac{1}{2} trace[g^T(u_g(t)) \frac{\partial^2 V_g(t,u_g)}{\partial u_g^2} g(u_g(t))]\end{aligned} \tag{17.18}$$

here

$$\frac{\partial^2 V_g}{\partial u_g^2} = \begin{bmatrix} w_1 & 0 & 0 & 0 & 0 & 0 & 0 & 0 & 0 & 0 \\ 0 & w_2 & 0 & 0 & 0 & 0 & 0 & 0 & 0 & 0 \\ 0 & 0 & w_3 & 0 & 0 & 0 & 0 & 0 & 0 & 0 \\ 0 & 0 & 0 & w_4 & 0 & 0 & 0 & 0 & 0 & 0 \\ 0 & 0 & 0 & 0 & w_5 & 0 & 0 & 0 & 0 & 0 \\ 0 & 0 & 0 & 0 & 0 & w_6 & 0 & 0 & 0 & 0 \\ 0 & 0 & 0 & 0 & 0 & 0 & w_7 & 0 & 0 & 0 \\ 0 & 0 & 0 & 0 & 0 & 0 & 0 & w_8 & 0 & 0 \\ 0 & 0 & 0 & 0 & 0 & 0 & 0 & 0 & w_9 & 0 \\ 0 & 0 & 0 & 0 & 0 & 0 & 0 & 0 & 0 & w_{10} \end{bmatrix} \tag{17.19}$$

and

$$g^T(u_g(t)) \frac{\partial^2 V_g}{\partial u_g^2} g(u_g(t)) = \begin{bmatrix} w_1\sigma_1^2 u_{g1}^2 & 0 & 0 & 0 & 0 & 0 & 0 & 0 & 0 & 0 \\ 0 & w_2\sigma_2^2 u_{g2}^2 & 0 & 0 & 0 & 0 & 0 & 0 & 0 & 0 \\ 0 & 0 & w_3\sigma_3^2 u_{g3}^2 & 0 & 0 & 0 & 0 & 0 & 0 & 0 \\ 0 & 0 & 0 & w_4\sigma_4^2 u_{g4}^2 & 0 & 0 & 0 & 0 & 0 & 0 \\ 0 & 0 & 0 & 0 & w_5\sigma_5^2 u_{g5}^2 & 0 & 0 & 0 & 0 & 0 \\ 0 & 0 & 0 & 0 & 0 & w_6\sigma_6^2 u_{g6}^2 & 0 & 0 & 0 & 0 \\ 0 & 0 & 0 & 0 & 0 & 0 & w_7\sigma_7^2 u_{g7}^2 & 0 & 0 & 0 \\ 0 & 0 & 0 & 0 & 0 & 0 & 0 & w_8\sigma_8^2 u_{g8}^2 & 0 & 0 \\ 0 & 0 & 0 & 0 & 0 & 0 & 0 & 0 & w_9\sigma_9^2 u_{g9}^2 & 0 \\ 0 & 0 & 0 & 0 & 0 & 0 & 0 & 0 & 0 & w_{10}\sigma_{10}^2 u_{g10}^2 \end{bmatrix} \tag{17.20}$$

with

$$\frac{1}{2}trace[g^T(u_g(t))\frac{\partial^2 V_g(t,u_g)}{\partial u_g^2}g(u_g(t))] = \frac{1}{2}[w_1\sigma_1^2u_{g1}^2 + w_2\sigma_2^2u_{g2}^2 + w_3\sigma_3^2u_{g3}^2 + w_4\sigma_4^2u_{g4}^2 + w_5\sigma_5^2u_{g5}^2 + w_6\sigma_6^2u_{g6}^2 + w_7\sigma_7^2u_{g7}^2 + w_8\sigma_8^2u_{g8}^2 + w_9\sigma_9^2u_{g9}^2 + w_{10}\sigma_{10}^2u_{g10}^2] \tag{17.21}$$

Use (17.21) along (17.18) we get

$$\begin{aligned}LV_g(t,u_g) = &-w_1[E-\frac{1}{2}\sigma_1^2]u_{g1}^2 - w_2[P-\frac{1}{2}\sigma_2^2]u_{g2}^2 - w_3[N-\frac{1}{2}\sigma_3^2]u_{g3}^2\\ &-w_4[T-\frac{1}{2}\sigma_4^2]u_{g4}^2 - w_5[I-\frac{1}{2}\sigma_5^2]u_{g5}^2 - w_6[A-\frac{1}{2}\sigma_6^2]u_{g6}^2\\ &-w_7[L-\frac{1}{2}\sigma_7^2]u_{g7}^2 - w_8[G-\frac{1}{2}\sigma_8^2]u_{g8}^2\\ &-w_9[R-\frac{1}{2}\sigma_9^2]u_{g9}^2 - w_{10}[H-\frac{1}{2}\sigma_{10}^2]u_{g10}^2\end{aligned}$$

Thus obtained negative definite. The above process is asymptotically mean square stable. So when the system is perturbed the stability is achieved through suitable Lyapunov function.

17.8 Discussion on Numerical Simulation

The present article designed the EGPLC mathematical model. Around equilibrium point the feasibility obtained. The system is stabilized which converges to E. And also obtained asymptotically mean square stable when E^* increases. The numerical part is done here by using fourth order Runge-kutta method to solve the system of EGPLC model. For the values of α_i for $= 1$ *to* 12 are taken as $c = 0.5$, $e = 0.904$, $f = 0.432$, $g = 0.169$, $i = 0.753$, $j = 0.896$, $h = 0.228$, $b = 0.567$, $a = 0.999$, $k = 0.732$, $l = 0.907$, $d = 0.741$, $\sigma = 9$ and the initial densities are $x_1 = 0.124$, $x_2 = 0.324$, $x_3 = 0.654$, $x_4 = 0.342$. $x_5 = -0.43919$, $x_6 = -0.36817$, $x_7 = 0.28$, $x_8 = 0.555$, $x_9 = 0.444$, $x_{10} = 0.456$ taken.

While the increasing of time span all the stages are stable for all values of initial population.

We observed that the more randomness occur in adult mosquito stages only. That is, the more randomness occur in x_1 stage. Compare to x_1 stage, the stages of sporozoite, hepatic schizont and hypnozoite are disturbed but not much and maintain the interior equilibrium state. Remaining stages are in zero equilibrium state.

Figure 17.2 depicts the deterministic of EGPLC model.

Figure 17.3 depicts the stability of deterministic of EGPLC model. When some random noise occur at all stages around interior equilibrium point then it stabilized at equilibrium point.

Figure 17.4 depicts the stochastic of EGPLC model.

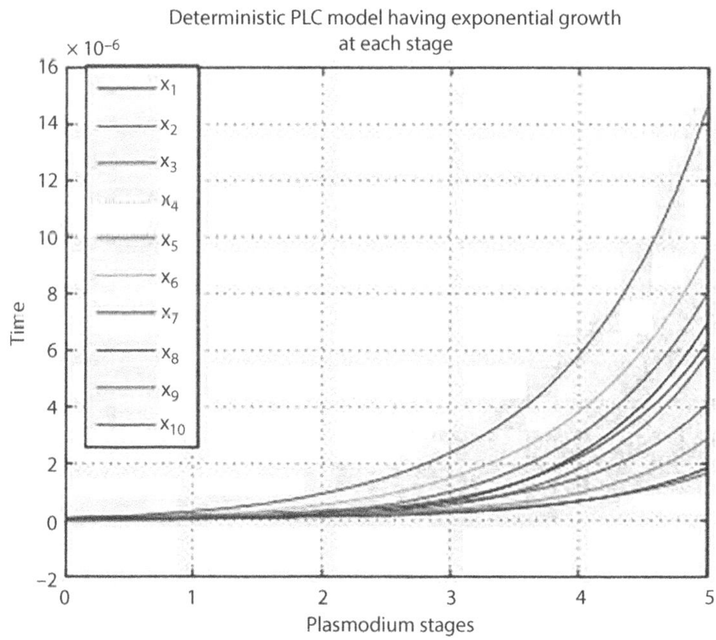

Figure 17.2 Pictorial representation of EGPLC.

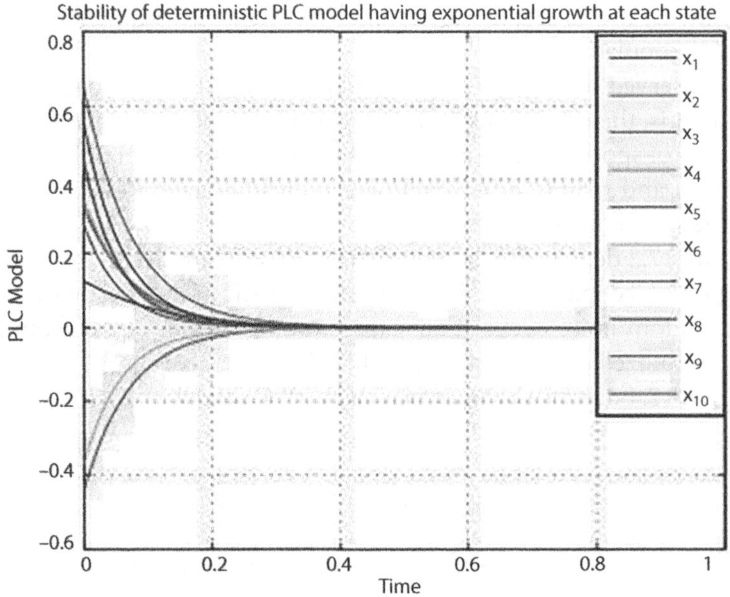

Figure 17.3 Stability of deterministic EGPLC model.

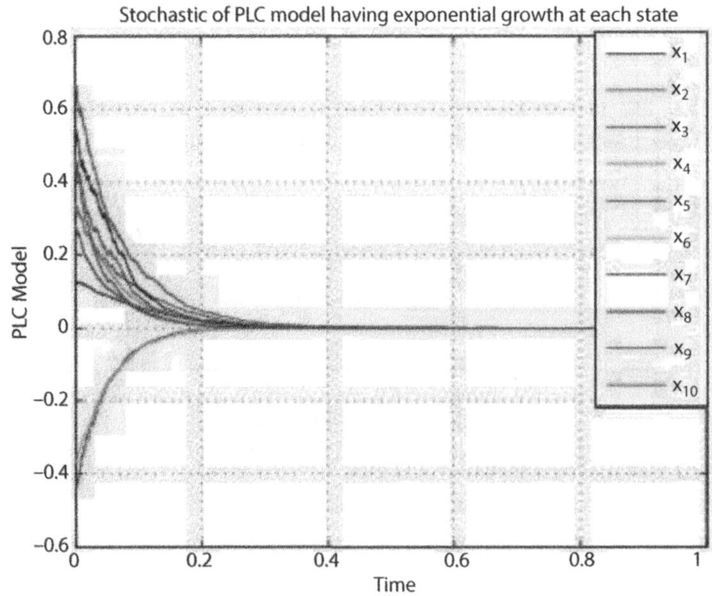

Figure 17.4 Stochastic of EGPLC model.

17.9 Conclusion

In this article, the complexity of EGPLC model is investigated. The E and E^* of the system have been found. The characterization of the model is investigated around E and obtained stability. The local stability is obtained by the criterion of Lasalle's invariance principle. The global stability is achieved by choosing suitable Lyapunov function. Around equilibrium point the feasibility obtained. The system is stabilized which converges to E. And also obtained asymptotically mean square stable when E^* increases.

To control the infection rate at each state the back propagation is applied and the system is controlled. This article described mathematically about the stability from the infection states. If the system is disturbed by some factors, like some other disease which may infect the vertebrate host's blood cells, also investigate through Ito process. The adult mosquito stage x_1 is more disturbed and also stability is achieved. And also conclude that the sporozoite, hepatic schizont and hypnozoite are disturbed but not much compared with adult mosquito state x_1 and maintain the interior equilibrium state. Remaining stages are in zero equilibrium state.

Thus stochastic perturbation is analyzed and obtained stability for EGPLC model. By constructing suitable Lyapunov function, the global asymptotically stable is achieved for the EGPLC model. Finally numerical examples are given and respective diagrams are presented which support the result. The simulation work is done here through MATLAB.

17.10 Future Scope of the Work

The mathematical model on dynamics of plasmodium life cycle is a nascent field of research. There are some strategies to control the infection rate at required states. One of the techniques is sterile Insect technology. In this paper, we proposed back propagation technique to control the infection rate in required state. Also we can analyze the bifurcation which is enormous growth by small disturbance in the system.

References

1. Chitnis, N., Smith, T. and Steketee, R., 2008. A mathematical model for the dynamics of malaria in mosquitoes feeding on a heterogeneous host population. *Journal of Biological Dynamics*, 2(3), pp. 259–285.

2. Gebremeskel, A.A. and Krogstad, H.E., 2015. Mathematical modelling of endemic malaria transmission. *American Journal of Applied Mathematics*, 3(2), pp. 36–46.
3. Kwiatkowski, D.P., 2005. How malaria has affected the human genome and what human genetics can teach us about malaria. *The American Journal of Human Genetics*, 77(2), pp. 171–192
4. Alano, P., Roca, L., Smith, D., Read, D., Carter, R. and Day, K., 1995. Plasmodium falciparum: parasites defective in early stages of gametocytogenesis. *Experimental parasitology*, 81(2), pp. 227–235
5. Tsukamoto, M., 1977. An imported human malarial case characterized by severe multiple infections of the red blood cells. *Tropical Medicine*, 19(2), pp. 95–104.
6. Sinden, R.E., 2009. Malaria, sexual development and transmission: retrospect and prospect. *Parasitology*, 136(12), pp. 1427–1434.
7. Tabo, Z., Luboobi, L.S. and Ssebuliba, J., 2017. Mathematical modelling of the in-host dynamics of malaria and the effects of treatment. *Journal of Mathematics and Computer Science*, 17(1), pp. 1–21.
8. Annan, K. and Mukinay, C.D., 2017. Stability and time-scale analysis of malaria transmission in human-mosquito population. *International Journal of Systems Science and Applied Mathematics*, 2(1), pp. 1–9.
9. Yau, M.A., 2011. A Mathematical model to break the life cycle of Anopheles mosquito. *Shiraz E-Medical Journal*, 12, pp. 120–128.
10. Rasappan, S., Jothi, N.K. and Subramani, J., 2014. Dynamics of anopheles mosquito life cycle break-up using backstepping control. In *International Conference on Mathematical Sciences*, Elsevier Proceedings.
11. Nagaram, N.B. and Rasapan, S., 2019. Plasmodium Life Cycle in Hepatocyte with Varying Population. *Indian Journal of Public Health Research & Development*, 10(6).

18

Dynamics, Control, Stability, Diffusion and Synchronization of Modified Chaotic Colpitts Oscillator with Triangular Wave Non-Linearity Depending on the States

Suresh Rasappan[1*] **and Niranjan Kumar K.A.**[2]

[1]*Department of Mathematics, University of Technology and Applied Sciences-Ibri, Ibri, Sultanate of Oman*
[2]*Department of Mathematics, Vel Tech Rangarajan Dr. Sagunthala R&D Institute of Science and Technology, Avadi, Chennai, Tamil Nadu, India*

Abstract
The aim of this article is to present a new chaotic oscillator. Although several chaotic systems have been formulated, despite that a few chaotic systems exhibit chaotic behavior. A new chaotic system with chaotic attractor is introduced for the nonlinearity of triangular waves. It is interesting to note that this striking phenomenon occurs rarely compared to chaotic systems. The results from the numerical simulation indicate the feasibility of the proposed chaotic system. In addition, the chaos control, stability, diffusion and synchronization of such a system were discussed.

***Keywords*:** Chaos, colpitts oscillator, lyapunov exponent, diffusion, stability, synchronization, triangular wave non-linearity

**Corresponding author*: mrpsuresh83@gmail.com

Ramakant Bhardwaj, Jyoti Mishra, Satyendra Narayan and Gopalakrishnan Suseendran (eds.)
Mathematics in Computational Science and Engineering, (383–414) © 2022 Scrivener Publishing LLC

18.1 Introduction

The study of chaotic dynamical systems has drawn the attention of researchers in recent times. Research on a chaotic system with chaotic attractor is posing several challenges thereby making the study quite interesting.

A non-linear dynamical system exhibiting complex and unpredictable behavior is called a chaotic system [1]. The parameter values are varying with range, and the sensitivity depends on initial conditions. These are the remarkable properties [2] of chaotic systems. Sometimes, the chaotic systems are deterministic [3, 4] and they have long-term unpredictable behavior [5, 6].

While chaotic systems are highly sensitive, their sensitivity depends on their initial conditions. The chaotic nature is one of the qualitative [7, 8] properties of a dynamical system [9, 10].

The controlling of the chaotic systems may be accomplished in three ways, such as stabilization [11, 12] of unstable periodic motion "contained" in the chaotic set, suppression of chaotic behavior by external forcing like periodic noise, periodic parametric perturbation and algorithm of various automatic control like feedback [13, 14], backstepping [15], sample feedback, time delay feedback, etc.

There exist two ways for the application of controls in a chaotic system. The first one is the change of attractor of the system. The second one is the change in the point position of the phase space for the system which is a constant value in its parameter.

A continuous, repeated and alternating wave production without any input is an oscillator. Converting power supply to an alternating current signal is one of the primary properties of oscillators. The signal of feedback containing a pair of coils and an inductive divider in the server is called Colpitts oscillator [16, 17]. Due to some parametric change and the variation of input, the chaotic nature may occur in Colpitts oscillators.

In this paper, a new chaotic Colpitts oscillator is proposed. It is a modified form of the earlier version of Colpitts oscillators. In section 18.2, the modified form of Colpitts oscillator [18, 19] is presented with the formulation of the mathematical model. In addition, invariant property, equilibrium point and Lyapunov exponents [20–23] are investigated. In section 18.3, adaptive backstepping technique [24] is explained for the proposed system. In section 18.4, a non-linear feedback system is established. The control strategy of backstepping is employed to analyze the non-linear feedback system in section 18.5. Finally, the numerical simulation [25–28] is upheld for the hypothetical outcomes.

18.2 The Mathematical Model of Chaotic Colpitts Oscillator

The depiction of a simplified illustrative diagram for modified Colpitts oscillator is undertaken in Figure 18.1. In addition to Electronic devices, communication systems also have wide usage of the Colpitts oscillator. It is a single-transistor implementation of a sinusoidal oscillator.

The following are the hypotheses for simplifying the extensive simulation of the complete circuit model.

- The base-emitter(B-E) driving point(V-I) characteristic of the R_E with triangular Wave function is

$$I_E = f(V_{BE}) = I_S\left[\frac{2a}{\pi}\sin^{-1}\left(\sin\left(\frac{2\pi}{p}(x_3)\right)\right)\right]$$

$$\text{and } IE = f(V_{BE}) = I_S\left[\frac{2a}{\pi}\sin-1\left(\sin\left(\frac{2\pi}{p}(x_1)\right)\right)\right]$$

where I_s is the emitter current (inverse saturation current), a is amplitude and p is period of the B-E junction.

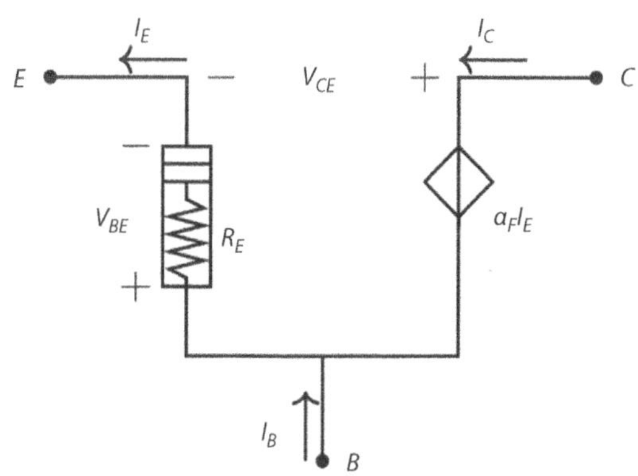

Figure 18.1 The circuit diagram.

- The state space is schematically represented in Figure 18.1.

$$R_C C_1 \frac{dV_{C_1}}{dt} = V_0 - V_{C_1} - V_{C_2} + R_C I_L - R_C f(V_{BE})$$
$$R_C C_2 \frac{dV_{C_2}}{dt} = V_0 - V_{C_1} - V_{C_2} - R_C I_0 + R_C I_L$$
$$C_3 \frac{dV_{C3}}{dt} = I_L - (1-\alpha) f(V_{BE})$$
$$L \frac{dI_L}{dt} = -R_b I_L - V_{C_1} - V_{C2} - V_{C_3}$$

The following is the proposed new system with Colpitts oscillator:

$$\begin{aligned} \dot{x}_1 &= \sigma_1(-x_1 - x_2) + x_4 - \gamma\phi_1(x_3) \\ \dot{x}_2 &= \varepsilon_1\sigma_1(-x_1 - x_2) + \varepsilon_1 x_4 \\ \dot{x}_3 &= \varepsilon_2(-x4 - (1-\alpha)\gamma\phi_2(x_1)) \\ \dot{x}_4 &= -x_1 - x_2 - x_3 - \sigma_2 x_4 \end{aligned} \tag{18.1}$$

where $\phi_1(x_3) = \frac{2a}{\pi}\sin^{-1}\left(\sin\left(\frac{2\pi}{p}(x_3)\right)\right)$, $\phi_2(x_1) = \frac{2a}{\pi}\sin^{-1}\left(\sin\left(\frac{2\pi}{p}(x_1)\right)\right)$.

In system (1), the state variables are assumed as x_1, x_2, x_3 and x_4 along with six positive parameters, σ_1, γ, ε_1, ε_2, σ_2 and α. The system (1) is an autonomous system to which a triangular wave expression is associated.

With the modification of coordinates provided by the scheme $(x_1, x_2, x_3, x_4,) \mapsto (-x_1, -x_2, -x_3, -x_4)$, the system (1) is found to be invariant.

The mathematical system of the Colpitts oscillator mathematical system when equated to zero gives the eqilibrium points of the system as specified below:

$$\begin{aligned} \sigma_1(-x_1 - x_2) + x_4 - \gamma\phi_1(x_3) &= 0 \\ \varepsilon_1\sigma_1(-x_1 - x_2) + \varepsilon_1 x_4 &= 0 \\ \varepsilon_2(x_4 - (1-\alpha)\gamma\phi_2(x_1)) &= 0 \\ -x_1 - x_2 - x_3 - \sigma_2 x_4 &= 0 \end{aligned} \tag{18.2}$$

Solving the system (2), it is seen that the new chaotic system (2) has a unique equilibrium at the origin.

The Jocobian matrix of the system (1) at the equilibrium point E is given by

$$J_E = \begin{bmatrix} -\sigma_1 & -\sigma_1 & -4\gamma\alpha/p & 1 \\ -\varepsilon_1\sigma_1 & -\varepsilon_1\sigma_1 & 0 & \varepsilon_1 \\ -\varepsilon_2(1-\alpha)4\gamma\alpha/p & 0 & 0 & \varepsilon_2 \\ -1 & -1 & -1 & -\sigma_2 \end{bmatrix} \tag{18.3}$$

The corresponding characteristic equation of Colpitts oscillator system (1) with respect to E is given by the relation

$$\Delta_1\lambda^4 + \Delta_2\lambda^3 + \Delta_3\lambda^2 + \Delta_4\lambda + \Delta_5 = 0 \tag{18.4}$$

where

$$\Delta_1 = 1$$

$$\Delta_2 = \varepsilon_1\sigma_1 + \sigma_1 + \sigma_2$$

$$\Delta_3 = \frac{[16\alpha\varepsilon_2\gamma^2 a^2 + \varepsilon_1\sigma_1\sigma_2 p^2 + \varepsilon_1 p^2 - 16\varepsilon_2\gamma^2 a^2 + \varepsilon_2 p^2 + \sigma_1\sigma_2 p^2 + p^2]}{p^2}$$

$$\Delta_4 = \frac{[16\alpha\varepsilon_1\varepsilon_2\gamma^2\sigma_1 a^2 + 16a\varepsilon_2\gamma^2\sigma_2 a^2 + 4\alpha\varepsilon_2\gamma ap - 16\varepsilon_1\varepsilon_2\gamma^2\sigma_1 a^2 + \varepsilon_1\varepsilon_2\sigma_1 p^2 - 16\varepsilon_2\gamma^2\sigma_2 a^2 - 8\varepsilon_2\gamma ap + \varepsilon_2\sigma_1 p^2]}{p^2}$$

$$\Delta_5 = \frac{[16\alpha\varepsilon_1\varepsilon_2\gamma^2\sigma_1\sigma_2 a^2 + 16a\varepsilon_1\varepsilon_2\gamma^2 a^2 - 16\varepsilon_1\varepsilon_2\gamma^2\sigma_1\sigma_2 a^2 - 16\varepsilon_1\varepsilon_2\gamma^2 a^2]}{p^2}$$

Applying Routh-Hurwitz stability criterion [29] to the characteristic equation, we conclude that the system is unstable for all values of the parameters at the equilibrium position E.

From the Jacobian matrix (18.3), among the states x_1, x_2, x_3 and x_4, if x_1 and x_3 are both positive or negative or of opposite signs, it implies "*Hopf bifurcation*". This phenomenon is also known as "*Poincaré–Andronov–Hopf bifurcation*". This bifurcation leads a local birth of "*chaos*" nature in modified Colpitts oscillator (18.1).

Interestingly, the system (1) is chaotic for the parameters

$$\varepsilon_1 = 1, \quad \varepsilon_2 = 20, \quad \sigma_1 = 1.49, \quad \sigma_2 = 0.872, \quad \gamma = 1.475, 32.90, \quad \alpha = \frac{255}{256}$$

Lyapunov exponents may be considered as one of the keys to differentiate between chaotic, hyperchaotic, stable and periodic nature of the systems.

Table 18.1 gives the details of the chaotic and hyperchaotic nature of the system. For this calculation, the observation time (T) is considered as 500 and the sampling time (Δt) is taken as 0.5. For various initial conditions, the system (1) exhibits chaotic and hyperchaotic nature.

By applying Wolf algorithm [30], the Lyapunov exponents corresponding to the new chaotic system (1) are obtained as follows:

From Table 18.1, the Lyapunov exponential dimension is calculated. The attractor of the new system is observed to be a strange attractor with fractal dimensions.

Through numerical simulation, the chaotic attractor of the system (1) is obtained as shown in Figure 18.3.

Figure 18.2 depicts the Lyapunov exponents of the modified Colpitts oscillator and Figure 18.3 shows the chaotic nature of the modified Colpitts oscillator and Poincaré Map of the modified Colpitts oscillator.

Table 18.1 LEs of system (1) for observation time (T) = 500, sampling time (Δt) = 0.5, $\varepsilon_1 = 1$, $\varepsilon_2 = 20$, $s_1 = 1.49$, $s_2 = 0.872$, $\alpha = \frac{255}{256}$, $\gamma = 1.475, 32.90, 32.95$ with various sampling and observation times using Wolf algorithm.

Sl. no.	Parameter, a, p	Initial condition	LEs	Sign of the LEs	Nature
1	$\gamma = 1.475$, $a = 1$, $p = 1$	0.00001, 0.00001, 0.00001, 0.00001	2.024442, −0.093200, −2.892090, −02.891145	[+, ≈ 0, −, −]	Chaotic
2	$\gamma = 32.90$, $a = 1$, $p = 1$	0.00001, 0.00001, 0.00001, 0.00001	36.698109, −0.630107, +0.005158, +01.187488	[+, −, ≈ 0, +]	Hyperbolic
3	$\gamma = 32.90$, $a = 1$, $p = 2$	0.00001, 0.00001, 0.00001, 0.00001	18.790350, −0.402598, −6.563959, −15.673989	[+, ≈ 0, −, −]	Chaotic
4	$\gamma = 32.95$, $a = 1$, $p = 1$	0.00001, 0.00001, 0.00001, 0.00001	36.753129, −0.630701, +0.157932, +01.523079	[+, −, ≈ 0, +]	Hyperbolic

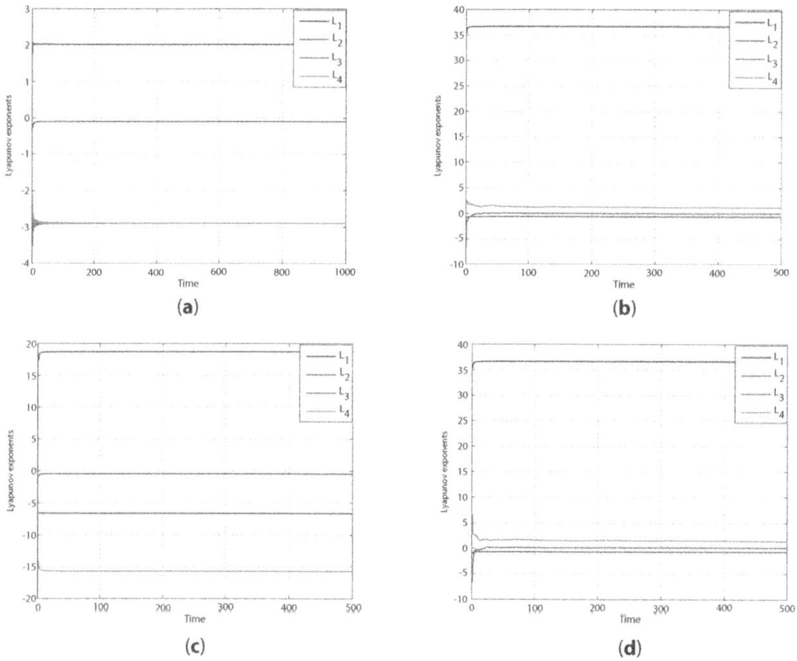

Figure 18.2 Lyapunov exponents of the Modified Colpitts oscillator. (a) The Lyapunov exponent for Modified Colpitts oscillator with $\varepsilon_1 = 1, \varepsilon_2 = 20, \sigma_1 = 1.49, \sigma_2 = 0.872, \alpha = \frac{255}{256}$ $\gamma = 1.475, a = 1, p = 1$ with initial condition $(x_1, x_2, x_3, x_4) = (.00001, .00001, .00001, .00001)$. (b) The Lyapunov exponent for Modified Colpitts oscillator with $\varepsilon_1 = 1, \varepsilon_2 = 20, \sigma_1 = 1.49, \sigma_2 = 0.872, \alpha = \frac{255}{256}$ $\gamma = 32.90, a = 1, p = 1$ with initial condition $(x_1, x_2, x_3, x_4) = (.00001, .00001, .00001, .00001)$. (c) The Lyapunov exponent for Modified Colpitts oscillator with $\varepsilon_1 = 1, \varepsilon_2 = 20, \sigma_1 = 1.49, \sigma_2 = 0.872, \alpha = \frac{255}{256}$ $\gamma = 32.90, a = 1, p = 2$ with initial condition $(x_1, x_2, x_3, x_4) = (.00001, .00001, .00001, .00001)$. (d) The Lyapunov exponent for Modified Colpitts oscillator with $\varepsilon_1 = 1, \varepsilon_2 = 20, \sigma_1 = 1.49, \sigma_2 = 0.872, \alpha = \frac{255}{256}$ $\gamma = 32.95, a = 1, p = 1$ with initial condition $(x_1, x_2, x_3, x_4) = (.00001, .00001, .00001, .00001)$.

The study of qualitative properties is one of the utilities of this paradigm. The stability control, limit cycle, periodicity and chaos are some notable qualitative properties. The following theorems bring out the local stability properties of the modified Colpitts oscillator.

Theorem 1. The interior equilibrium point E is locally asymptotically stable in the positive octant.

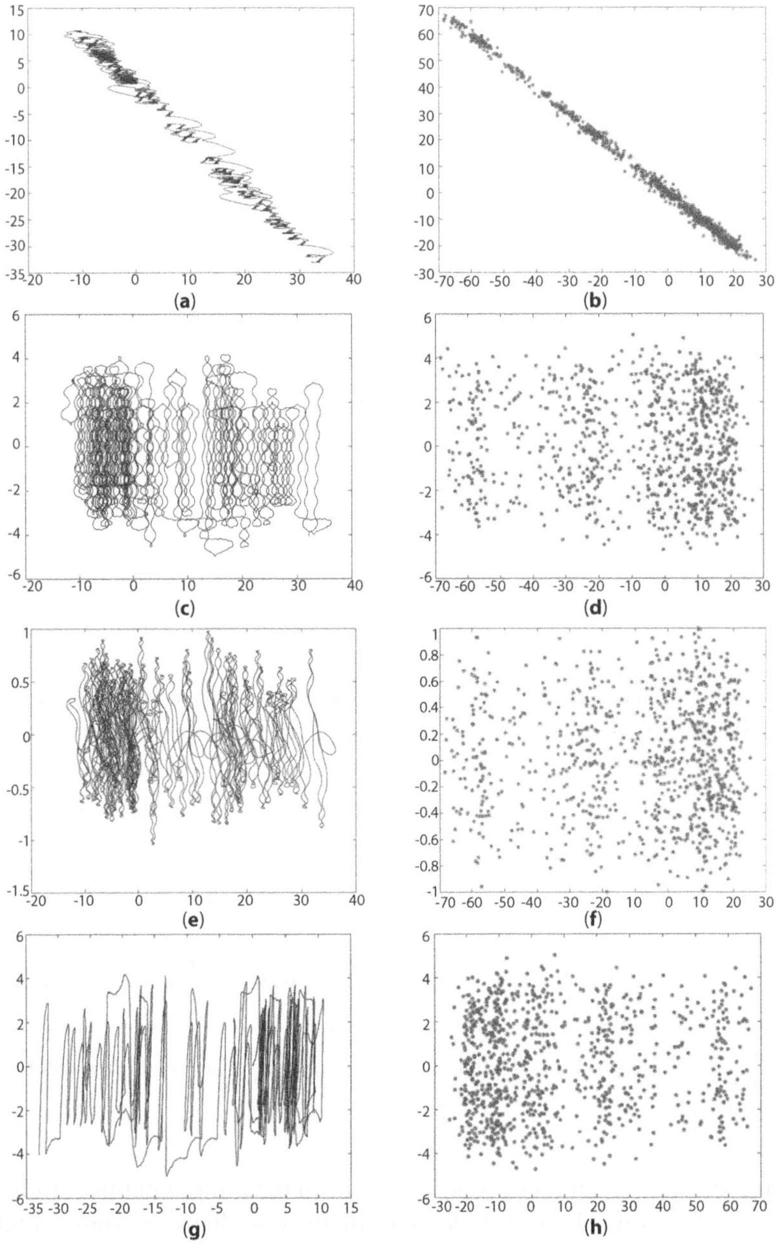

Figure 18.3 Portrait of Colpitts. (a) Chaotic nature between x_1 and x_2. (b) Poincaré Map between x_1 and x_2. (c) Chaotic nature between x_1 and x_3. (d) Poincaré Map between x_1 and x_3. (e) Chaotic nature between x_1 and x_4. (f) Poincaré Map between x_1 and x_4. (g) Chaotic nature between x_2 and x_3. (h) Poincaré Map between x_2 and x_3. *(Continued)*

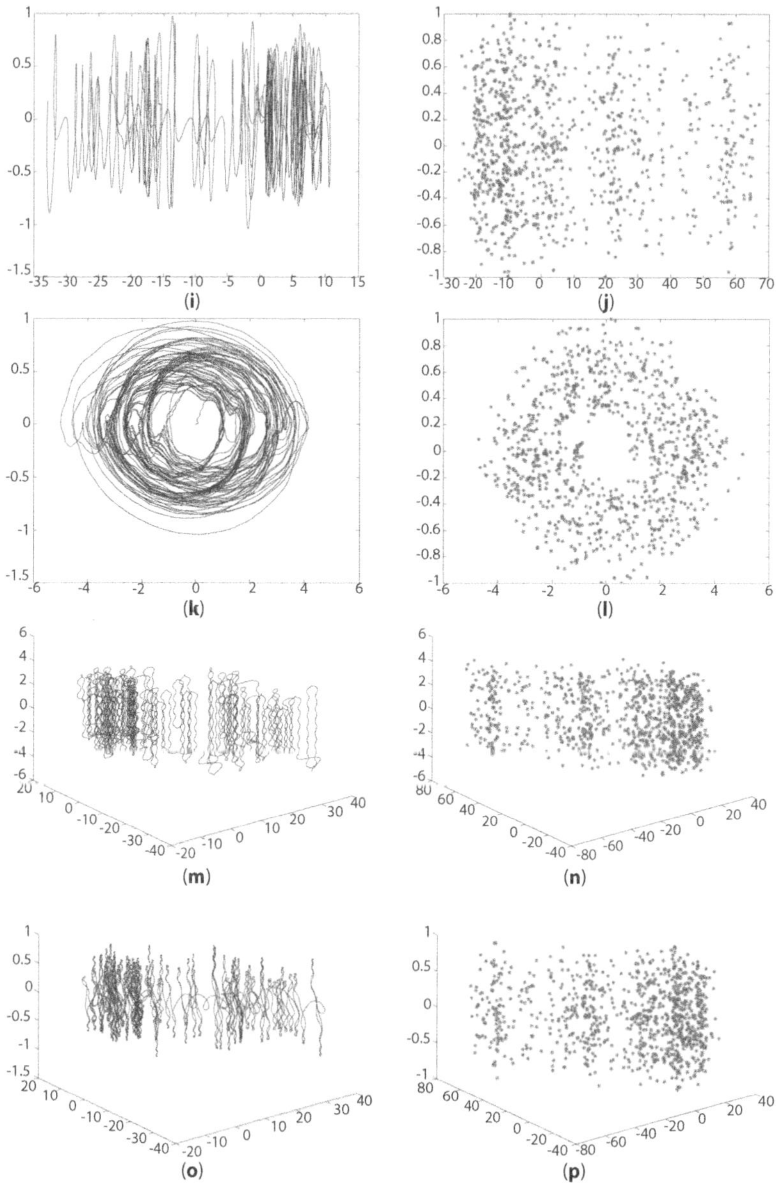

Figure 18.3 (**Continued**) Portrait of Colpitts. (i) Chaotic nature between x_2 and x_4. (j) Poincaré Map between x_2 and x_4. (k) Chaotic nature between x_3 and x_4. (l) Poincaré Map between x_3 and x_4. (m) Chaotic nature between x_1, x_2 and x_3. (n) Poincaré Map between x_1, x_2 and x_3. (o) Chaotic nature between x_1, x_2 and x_4. (p) Poincaré Map between x_1, x_2 and x_4.

(Continued)

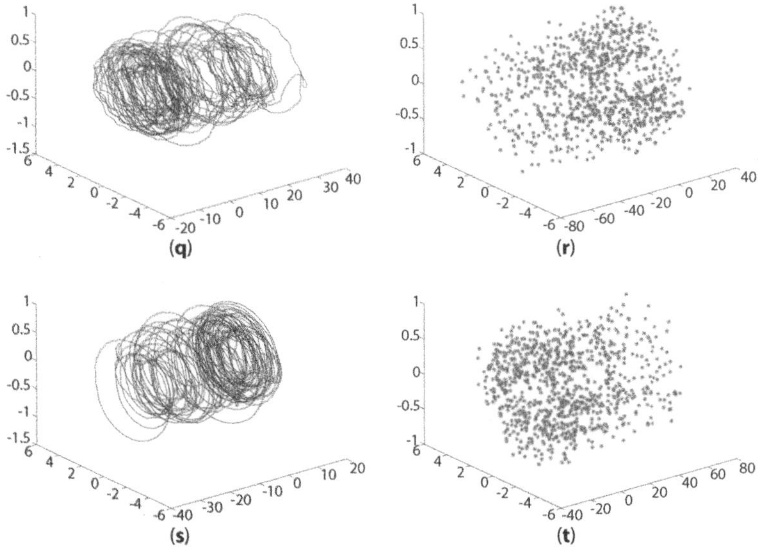

Figure 18.3 (Continued) Portrait of Colpitts. (q) Chaotic nature between x_1, x_3 and x_4. (r) Poincaré Map between x_1, x_3 and x_4. (s) Chaotic nature between x_2, x_3 and x_4. (t) Poincaré Map between x_2, x_3 and x_4.

Proof. By *divergence criterion* theorem, assume

$$\theta(x_1,x_2,x_3,x_4)=\frac{1}{x_1x_2x_3x_4} \tag{18.5}$$

where $\theta(x_i, i = 1,2,3,4) > 0$ if $x_i > 0, i = 1,2,3,4$.

Now consider

$$\begin{aligned} p_1 &= \sigma_1(-x_1-x_2)+x_4-\gamma\phi_1(x_3) \\ p_2 &= \varepsilon_1\sigma_1(-x_1-x_2)+\varepsilon_1 x_4 \\ p_3 &= \varepsilon_2(x_4-(1-\alpha)\gamma\phi_2(x_1)) \\ p_4 &= -x_1-x_2-x_3-\sigma_2 x_4 \end{aligned} \tag{18.6}$$

where $\phi_1(x_3)=\frac{2a}{\neq}\sin^{-1}\left(\sin\left(\frac{2\neq}{p}(x_3)\right)\right), \phi_2(x_1)=\frac{2a}{\neq}\sin^{-1}\left(\sin\left(\frac{2\neq}{p}(x_1)\right)\right).$

Define

$$\nabla = \frac{\partial}{\partial x_1}(p_1\theta) + \frac{\partial}{\partial x_2}(p_2\theta) + \frac{\partial}{\partial x_3}(p_3\theta) + \frac{\partial}{\partial x_4}(p_4\theta) \tag{18.7}$$

We have to determine ∇ given by Eq. (18.7) along with the trajectories provided by Equations (18.5) and Eq. (18.6). We obtain

$$\begin{aligned}\nabla = &-\frac{[\sigma_1]x_1x_2x_3x_4 + [\sigma_1(-x_1 - x_2) + x_4 - \gamma\phi_1(x_3)]x_2x_3x_4}{x_1^2x_2^2x_3^2x_4^2} \\ &-\frac{\varepsilon_1\sigma_1x_1x_2x_3x_4 + [\varepsilon_1\sigma_1(-x_1 - x_2) + \varepsilon_1x_4]x_1x_3x_4}{x_1^2x_2^2x_3^2x_4^2} \\ &-\frac{\varepsilon_2[x_4 - (1-\alpha)\gamma\phi_2(x_1)]x_1x_2x_4}{x_1^2x_2^2x_3^2x_4^2} \\ &-\frac{\sigma_2x_1x_2x_3x_4 + (-x_1 - x_2 - x_3 - \sigma_2x_4)x_1x_2x_3}{x_1^2x_2^2x_3^2x_4^2}\end{aligned}$$

which is less than zero.

From *Benedixon-Dulac criterion*, it it clear that the first octant does not contain any limit cycle.

Consequently, the equilibrium provided by E is found to be locally asymptotically stable.

The relation between the limit cycle and closed trajectories exhibits the local asymptotic stability. The following theorem is concerned with the stability under closed trajectory using Bendixson's criteria theorem.

Theorem 2. There is no closed trajectory for the interior equilibrium point.

Proof. Define

$$\psi(x_i, i = 1,2,3,4) = \frac{\partial p_1}{\partial x_1} + \ldots + \frac{\partial p_4}{\partial x_4} \tag{18.8}$$

Find Ψ along with the trajectories associated with Eq. (18.8). It follows that

$$\Psi = -s_1 - \varepsilon_1 s_1 - s_2 \neq 0 \tag{18.9}$$

Hence, by applying *Bendixson's criteria theorem* to Eq. (18.9), it is seen that there is no closed trajectory surrounding the point E.

Hence, limit cycle does not exist emcompassing *E*.

Therefore, the point *E* is evidential to be locally asymptotically stable. In oscillator, exhibiting stable periodic orbit and it corresponds to a special type of solution for a oscillator. The following theorem focuses attention on the nontrivial periodic solution.

Theorem 3. The modified Colpitts oscillator given by Eq. (18.1) has a nontrivial periodic solution.

Proof. Define

$$\Phi = \frac{d}{dt}\left(\frac{x_1^2+x_2^2+x_3^2+x_4^2}{2}\right) = x_1\frac{dx_1}{dt}+x_2\frac{dx_2}{dt}+x_3\frac{dx_3}{dt}+x_4\frac{dx_4}{dt}$$

$$= x_1\dot{x}_1 + x_2\dot{x}_2 + x_3\dot{x}_3 + x_4\dot{x}_4 = \textstyle\sum_{i=1}^{4} x_i\frac{dx_i}{dt} \tag{18.10}$$

Find ϕ from Eq. (18.10) along the trajectories Eq. (18.1). We see that

$$\begin{aligned}
\Phi &= x_1[\sigma_1(-x_1-x_2)+x_4-\gamma\phi_1(x_3)] \\
&\quad + x_2[\varepsilon_1\sigma_1(-x_1-x_2)+\varepsilon_1 x_4] \\
&\quad + x_3[\varepsilon_2(x_4-(1-\alpha)\gamma\phi_2(x_1))] \\
&\quad + x_4[-x_1-x_2-x_3-\sigma_2 x_4] \\
&= -\sigma_1 x_1^2 - \sigma_1 x_1 x_2 + x_1 x_4 - \gamma x_1\phi_1(x_3) \\
&\quad - \varepsilon_1\sigma_1 x_1 x_2 - \varepsilon_1\sigma_1 x_2^2 + \varepsilon_1 x_2 x_4 \\
&\quad + \varepsilon_2 x_3 x_4 - \varepsilon_2(1-\alpha)x_3\gamma\phi_2(x_1) \\
&\quad - x_1x_4 - x_2x_4 - x_3x_4 - \sigma_2 x_4^2 \\
&= -(\sigma_1 x_1^2 + \varepsilon_1\sigma_1 x_2^2 + \sigma_2 x_4^2) \\
&\quad - \sigma_1 x_1 x_2(1+\varepsilon_1) - (1-\varepsilon_1)x_2x_4 - (1-\varepsilon_2)x_3x_4 - x_1\gamma\phi_1(x_3) \\
&\quad - x_3\varepsilon_2(1-\alpha)\gamma\phi_2(x_1) \\
&= -(\nabla_1 + \nabla_2)
\end{aligned} \tag{18.11}$$

where $\nabla_1 = \sigma_1 x_1^2 + \varepsilon_1\sigma_1 x_2^2 + \sigma_2 x_4^2$

$\nabla_2 = \sigma_1 x_1 x_2(1+\varepsilon_1) + (1-\varepsilon_1)x_2x_4 + (1-\varepsilon_2)x_3x_4 + x_1\gamma\phi_1(x_3) + x_3\varepsilon_2(1-\alpha)$
$\gamma\phi_2(x_1)$

It is observed that $\nabla_1 + \nabla_2$ is positive for $x_1^2 + x_2^2 + x_3^2 + x_4^2 < a$ and negative for $x_1^2 + x_2^2 + x_3^2 + x_4^2 < b$, where *a*, *b* are positive constants.

This implies that any solution $x_i(t)$ of (1) which starts in the annulus $a < \sum_{i=1}^{4} x_i^2 < b$.

Hence, by *Poincaré-Bendixson* theorem, there exists atleast one periodic solution $x_i(t)$, $i = 1,2,3,4$ of Eq. (18.1) lying in this annulus.

Hence, the modified Colpitts oscillator Eq. (18.1) has a nontrivial periodic solution.

The study of control refers to the process of influencing the behaviour of an oscillator to achieve a desired goal, primarily through the use of feedback control. The following section describes the backstepping control when the parameter values are unknown.

18.3 Adaptive Backstepping Control of the Modified Colpitts Oscillator with Unknown Parameters

18.3.1 Proposed System

The modified Colpitts oscillator system is given by the dynamics with controllers

$$
\begin{aligned}
\dot{x}_1 &= \sigma_1(-x_1 - x_2) + x_4 - \gamma\phi_1(x_3) + u_1 \\
\dot{x}_2 &= \varepsilon_1\sigma_1(-x_1 - x_2) + \varepsilon_1 x_4 + u_2 \\
\dot{x}_3 &= \varepsilon_2(x_4 - (1-\alpha)\gamma\phi_2(x_1)) + u_3 \\
\dot{x}_4 &= -x_1 - x_2 - x_3 - \sigma_2 x_4 + u_4
\end{aligned}
\tag{18.12}
$$

where $\phi_1(x_3) = \frac{2a}{\neq}\sin^{-1}\left(\sin\left(\frac{2\neq}{p}(x_3)\right)\right)$, $\phi_2(x_1) = \frac{2a}{\neq}\sin^{-1}\left(\sin\left(\frac{2\neq}{p}(x_1)\right)\right)$.

In system (12), x_1, x_2, x_3 and x_3 are state variables and u_1, u_2, u_3 and u_4 are adaptive controllers.

The synchronization error is defined as $e_i = y_i - x_i$, $i = 1,2,3,4$.

The unknown parameters are updated by

$$
\begin{aligned}
e_{\sigma_1} &= \sigma_1 - \hat{\sigma}_1(t), & e_{\sigma_2} &= \sigma_2 - \hat{\sigma}_2(t) \\
e_{\varepsilon_1} &= \varepsilon_1 - \hat{\varepsilon}_1(t), & e_{\varepsilon_2} &= \varepsilon_2 - \hat{\varepsilon}_2(t) \\
e_{\alpha} &= \alpha - \hat{\alpha}_1(t), & e_{\gamma} &= \gamma - \hat{y}(t)
\end{aligned}
\tag{18.13}
$$

By differentiating (18.13) with respect to 't', one obtains

$$\dot{e}_{\sigma 1} = -\dot{\hat{\sigma}}_1(t), \quad \dot{e}_{\sigma 2} = -\dot{\hat{\sigma}}_2(t)$$
$$\dot{e}_{\varepsilon 1} = -\dot{\hat{\varepsilon}}_1(t), \quad \dot{e}_{\varepsilon 2} = -\dot{\hat{\varepsilon}}_2(t)$$
$$\dot{e}_{\alpha} = -\dot{\hat{\alpha}}_1(t), \quad \dot{e}_{\gamma} = -\dot{\hat{y}}(t)$$

At this stage, the state of the system is considered as

$$\dot{x}_1 = s_1(-x_1 - x_2) + x_4 - \gamma\phi_1(x_3) + u_1 \tag{18.14}$$

where x_2 is regarded as virtual controller.

In order to stabilize the system, the suitable Lyapunov function is defined as

$$V_1(x_1) = \frac{1}{2}x_1^2 + \frac{1}{2}e_{\sigma_1}^2 + \frac{1}{2}e_\gamma^2$$

By differentiating V_1 with respect to t,

$$\begin{aligned}\dot{V}_1 &= x_1\dot{x}_1 + e_{\sigma_1}\dot{e}_{\sigma_1} + e_\gamma\dot{e}_\gamma \\ &= x_1[\sigma_1(-x_1 - x_2) + x_4 - \gamma\phi_1(x_3) + u_1] + e_{\sigma_1}(-\dot{\hat{\sigma}}_1) + e_\gamma(-\dot{\hat{\gamma}})\end{aligned} \tag{18.15}$$

where x_2 is regarded as virtual controller and is defined as

$$x_2 = \beta_1(x_1) \text{ and } \beta_1(x_1) = 0.$$

The controller u_1 is assumed as

$$u_1 = -x_1 + \hat{\sigma}_1 x_1 - x_4 + \hat{\gamma}\phi_1(x_3) \tag{18.16}$$

and the unknown parameters $\hat{\sigma}_1$ and $\hat{\gamma}$ are updated by

$$\begin{aligned}\dot{\hat{\sigma}}_1 &= -x_1^2 + e_{\sigma_1} \\ \dot{\hat{\gamma}} &= -x_1\phi_1(x_3) + e_\gamma\end{aligned} \tag{18.17}$$

On substitution of (18.16) and (18.17) into (18.15), we get

$$\dot{V}_1 = -x_1^2 - e_{\sigma_1}^2 - e_y^2$$

which is found to be a negative definite function.

Hence by Lyapunov stability theory, the system is globally asymptotically stable.

Now define the relation between β_1 and x_2 by

$$\omega_2 = x_2 - \beta_1$$

Consider the subsystem (x_1, ω_2). We have

$$\dot{x}_1 = -e_{\sigma_1} x_1 - \sigma_1 \omega_2 - e_\gamma \phi_1(x_3) - x_1$$
$$\dot{\omega}_2 = -\varepsilon_1 \sigma_1 x_1 - \varepsilon_1 \sigma_1 \omega_2 + \varepsilon_1 x_4 + u_2$$

Define V_2 by the Lyapunov function as

$$V_2 = V_1 + \frac{1}{2}\omega_2^2 + \frac{1}{2}e_{\varepsilon_1}^2$$

On differentiating V_2 with respect to t, we get

$$\dot{V}_2 = x_1 \dot{x}_1 + e_{\sigma_1}(-\dot{\hat{\sigma}}_1) + e_\gamma(-\dot{\hat{\gamma}}) + e_{\varepsilon_1}(-\dot{\hat{\varepsilon}}_1) + \omega_2 \dot{\omega}_2 \tag{18.18}$$

The controller u_2 is assumed as

$$u_2 = \sigma_1 x_1 + \hat{\varepsilon}_1(\sigma_1 x_1 + \sigma_1 \omega_2 - x_4) + x_3 - \omega_2 \tag{18.19}$$

Let x_3 be the virtual controller. It is defined as $x_3 = \beta_2(x_1, \omega_2)$ with the assumption that $\beta_2(x_1, \omega_2) = 0$.

The parameter ε_1 is estimated as

$$\dot{\hat{\varepsilon}}_1 = -\omega_2(\sigma_1 x_1 + \sigma_1 \omega_2 - x_4) + e_{\varepsilon_1} \tag{18.20}$$

Substituting (18.19) and (18.20) into (18.18), we get

$$\dot{V}_2 = -x_1^2 - e_{\sigma_1}^2 - e_\gamma^2 - w_2^2 - e_{\varepsilon_1}^2$$

which is a negative definite function.

Hence by Lyapunov stability theory, the system is globally asymptotically stable.

The relation between x_3 and β_2 is defined by

$$\omega_3 = x_3 - \beta_2$$

Consider the subsystem $(x_1, \omega_2, \omega_3)$. We have

$$\begin{aligned}
\dot{x}_1 &= -e_{\sigma_1} x_1 - \sigma_1 \omega_2 - e_\gamma \phi_1(x_3) - x_1 \\
\dot{\omega}_2 &= -e_{\varepsilon_1}(\sigma_1 x_1 + \sigma_1 \omega_2 - x_4) - \omega_2 + \sigma_1 x_1 + \omega_3 \\
\dot{\omega}_3 &= \varepsilon_2 \left(x_4 - (1-\alpha)\gamma\phi_2(x_1)\right) + u_3
\end{aligned}$$

Now consider the Lyapunov function

$$V_3 = V_2 + \frac{1}{2}\omega_3^2 + \frac{1}{2}e_{\varepsilon_2}^2 + \frac{1}{2}e_\alpha^2$$

The derivative V_3 of with respect to is obtained as

$$\dot{V}_3 = \dot{V}_2 + \omega_3 \dot{\omega}_3 + e_{\varepsilon_2}\dot{e}_{\varepsilon_2} + e_\alpha \dot{e}_\alpha \tag{18.21}$$

where

$$u_3 = -\omega_2 - \omega_3 + \hat{\varepsilon}_2 \gamma \phi_2(x_1) - \varepsilon_2 \hat{\alpha} \gamma \phi_2(x_1) \tag{18.22}$$

Let us denote the virtual controller by x_4. It is defined as $x_4 = \beta_3(x_1, \omega_2, \omega_3)$ and we assume that $\beta_3(x_1, \omega_2, \omega_3) = 0$.

The parameters are estimated as

$$\begin{aligned}
\dot{\hat{\varepsilon}}_2 &= -\omega_3 \gamma \phi_2(x_1) + e_{\varepsilon_2} \\
\dot{\hat{\alpha}}_2 &= -\omega_3 \varepsilon_2 \gamma \phi_2(x_1) + e_\alpha
\end{aligned} \tag{18.23}$$

Substitute (18.22) and (18.23) into (18.21). Then we get

$$\dot{V}_3 = -x_1^2 - e_{\sigma_1}^2 - e_\gamma^2 - w_2^2 - e_{\varepsilon_1}^2 - w_3^2 - e_{\varepsilon_2}^2 - e_\alpha^2$$

which is a negative definite function.

Hence by the theory of Lyapunov, it follows that the system provided by Eq. (18.12) is stable.

Now the relation between x_4 and β_3 is defined by

$$\omega_4 = x_4 - \beta_3$$

Consider the subsystem $(x_1, \omega_2, \omega_3, \omega_4)$ provided by

$$\begin{aligned}
\dot{x}_1 &= -e_{\sigma_1}x_1 - \sigma_1\omega_2 - e_\gamma\phi_1(x_3) - x_1 \\
\dot{\omega}_2 &= -e_{\varepsilon_1}(\sigma_1 x_1 + \sigma_1\omega_2 - x_4) - \omega_2 + \omega_3 + \sigma_1 x_1 \\
\dot{\omega}_3 &= \varepsilon_2\omega_4 - e_{\varepsilon_2}\gamma\phi_2(x_1) + e_\alpha\varepsilon_2\gamma\phi_2(x_1) - \omega_2 - \omega_3 \\
\dot{\omega}_4 &= -x_1 - x_2 - x_3 - \sigma_2\omega_4 + u_4
\end{aligned}$$

Now consider the Lyapunov function

$$V_4 = V_3 + \frac{1}{2}\omega_4^2 + \frac{1}{2}e_{\sigma_2}^2$$

The derivative of with respect to is obtained as

$$\dot{V}_4 = \dot{V}_3 + \omega_4\dot{\omega}_4 + e_{\sigma_2}\dot{e}_{\sigma_2} \tag{18.24}$$

where

$$u_4 = -\varepsilon_2\omega_3 + x_1 + x_2 + x_3 + \hat{\sigma}_2\omega_4 - \omega_4 \tag{18.25}$$

By working backward, the parameter is estimated as

$$\dot{\hat{\sigma}}_2 = e_{\sigma_2} - w_4^2 \tag{18.26}$$

Substitute (18.25) and (18.26) into (18.24). Then we are led to

$$\dot{V}_4 = -x_1^2 - e_{\sigma_1}^2 - e_{\gamma}^2 - w_2^2 - e_{\varepsilon_1}^2 - w_3^2 - e_{\varepsilon_2}^2 - e_{\alpha}^2 - w_4^2 - e_{\sigma_2}^2$$

which is a negative definite function.

By the stability theory due to Lyapunov, it is seen that the Colpitts oscillator provided by Eq. (18.1) is asymptotically stable.

18.3.2 Numerical Simulation

For the numerical simulation, the initial conditions of the parameters are taken as

$$\begin{aligned} \hat{\sigma}_1(0) &= 10.9546, & \hat{\sigma}_2(0) &= 5.9353, \\ \hat{\alpha}(0) &= 3.8765, & \hat{\gamma}(0) &= 2.1654, \\ \hat{\varepsilon}_1(0) &= 7.8762, & \hat{\varepsilon}_2(0) &= 9.9876 \end{aligned}$$

with the initial conditions for the modified Colpitts oscillator *x*_1 (0) = 1.9124, *x*_2 (0) = 1.3942, *x*_3 (0) = 1.3125 and *x*_4 (0) = 1.9873.

Figure 18.4 depicts the parameter estimation of the modified Colpitts oscillator.

Figure 18.5 depicts the stability of the modified Colpitts oscillator.

18.4 Synchronization of Modified Chaotic Colpitts Oscillator

The synchronization of a chaotic system is another way of explaining the sensitivity based on the initial conditions. One has to design *master-slave* or *drive-response* coupling between the two chaotic systems such that the time evolution becomes ideal.

In general, the two dynamic systems involved in the synchronization are called the master and slave systems, respectively. A well-designed controller will make the trajectory of the slave system track and trajectory of the master system, that is, the two systems will be synchronous.

The following subsection contains the detailed explanation of the synchronization process for the modified Colpitts oscillator using non-linear control.

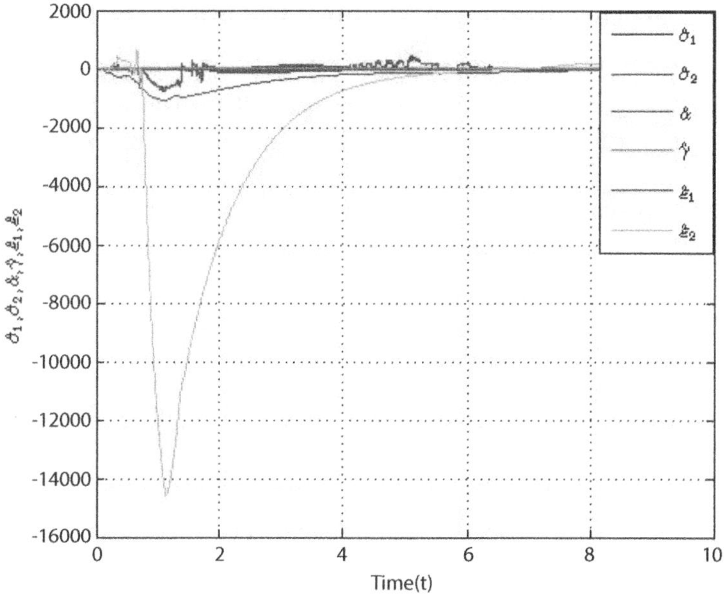

Figure 18.4 The parameter estimation of the modified Colpitts oscillator.

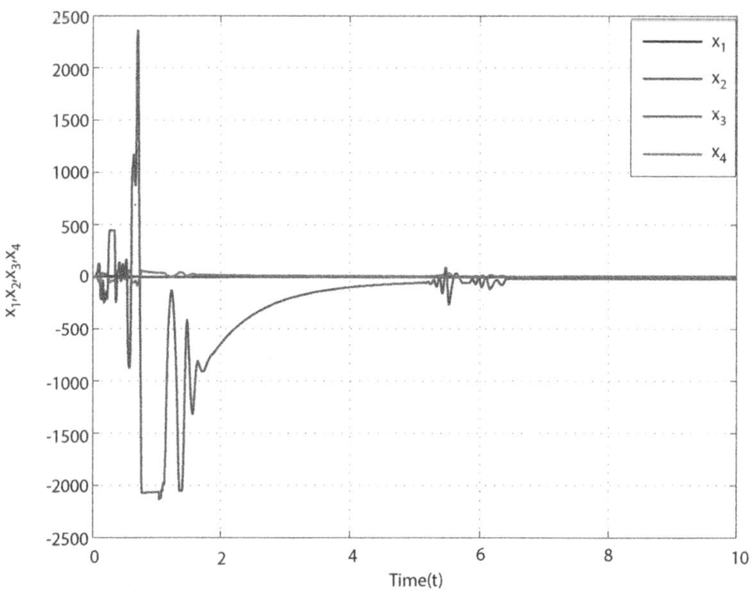

Figure 18.5 The stability of the modified Colpitts oscillator.

18.4.1 Synchronization of Modified Chaotic Colpitts Oscillator using Non-Linear Feedback Method

The synchronization of modified Colpitts oscillator is now taken up. The drive-response formalism is utilized. The identical synchronization is elaborated between the modified Colpitts oscillators.

The chaos synchronization basically requires the global asymptotic stability of the error dynamics

$$\text{i.e., } \lim_{t\to\infty} \| e(t) \| = 0.$$

The modified Colpitts oscillator is taken as drive system, which is described by

$$\begin{aligned} \dot{x}_1 &= \sigma_1(-x_1 - x_2) + x_4 - \gamma\phi_1(x_3) \\ \dot{x}_2 &= -\varepsilon_1\sigma_1 x_1 - \varepsilon_1\sigma_1 x_2 + \varepsilon_1 x_4 \\ \dot{x}_3 &= \varepsilon_2 x_4 - \varepsilon_2(1-\alpha)\gamma\phi_2(x_1) \\ \dot{x}_4 &= -x_1 - x_2 - x_3 - \sigma_2 x_4 \end{aligned} \tag{18.27}$$

where x_1, x_1, x_3 and x_4 are state variables, σ_1, σ_2, ε_1, ε_2, γ, α are positive parameters, $\phi_1(x_3) = \frac{2a}{\neq}\sin^{-1}\left(\sin\left(\frac{2\neq}{p}(x_3)\right)\right)$ and $\phi_2(x_1) = \frac{2a}{\neq}\sin^{-1}\left(\sin\left(\frac{2\neq}{p}(x_1)\right)\right)$.

The modified Colpitts oscillator is also taken as the response system which is described by

$$\begin{aligned} \dot{y}_1 &= \sigma_1(-y_1 - y_2) + y_4 - \gamma\phi_1(y_3) + u_1 \\ \dot{y}_2 &= -\varepsilon_1\sigma_1 y_1 - \varepsilon_1\sigma_1 y_2 + \varepsilon_1 y_4 + u_2 \\ \dot{y}_3 &= \varepsilon_2 y_4 - \varepsilon_2(1-\alpha)\gamma\phi_2(y_1) + u_3 \\ \dot{y}_4 &= -y_1 - y_2 - y_3 - \sigma_2 y_4 + u_4 \end{aligned} \tag{18.28}$$

where $\phi_1(y_3) = \frac{2a}{\pi}\sin^{-1}\left(\sin\left(\frac{2\pi}{p}(y_3)\right)\right)$, $\phi_2(y_1) = \frac{2a}{\pi}\sin^{-1}\left(\sin\left(\frac{2\pi}{p}(y_1)\right)\right)$.

The synchronization error occurring in the system is defined by

$$e_i = y_i - x_i,\ i = 1,2,3,4 \tag{18.29}$$

The resulting error dynamics of the system is governed by the set of equations

$$\begin{aligned}
\dot{e}_1 &= -\sigma_1 e_1 - \sigma_1 e_2 + e_4 - \gamma\phi_1(y_3) + \gamma\phi_1(x_3) + u_1 \\
\dot{e}_2 &= -\varepsilon_1\sigma_1 e_1 - \varepsilon_1\sigma_1 e_2 + \varepsilon_1 e_4 + u_2 \\
\dot{e}_3 &= \varepsilon_2 e_4 - \varepsilon_2(1-\alpha)\gamma(\phi_2(\mathrm{y}_1) - \phi_2(x_1)) + u_3 \\
\dot{e}_4 &= -e_1 - e_2 - e_3 - \sigma_2 e_4 + u_4
\end{aligned} \tag{18.30}$$

where $u = (u_1, u_2, u_3, \mathrm{u}_4)^T$ is the non-linear controller to be designed so as to synchronize the states of identically modified Colpitts oscillator.

Now the objective is to find the control law u_i, $i = 1,2,3,4$ for stabilizing the error variable of the system (30) at the origin.

Let the energy source function Lyapunov be chosen as

$$V = \frac{1}{2}\sum_{i=1}^{4} e_i^2 \tag{18.31}$$

The derivative of (18.31) with respect to is provided by

$$\dot{V} = \sum_{i=1}^{4} e_i \dot{e}_i \tag{18.32}$$

Substituting (18.29) and (18.30) into (18.32) we are led to the relation

$$\begin{aligned}
\dot{V} = {} & e_1(-\sigma_1 e_1 - \sigma_1 e_2 + e_4 - \gamma\phi_1(y_3) + \gamma\phi_1(x_3) + u_1) \\
& + e_2(-\varepsilon_1\sigma_1 e_1 - \varepsilon_1\sigma_1 e_2 + \varepsilon_1 e_4 + u_2) \\
& + e_3(\varepsilon_2 e_4 - \varepsilon_2(1-\alpha)\gamma(\phi_2(y_1) - \phi_2(x_1)) + u_3) \\
& + e_4(-\mathrm{e}_1 - \mathrm{e}_2 - \mathrm{e}_3 - \sigma_2 e_4 + u_4)
\end{aligned}$$

The controllers are defined by

$$\begin{aligned}
u_1 &= \sigma_1 e_2 - e_4 + \gamma\big(\phi_1(y_3) - \phi_1(x_1, x_3)\big) \\
u_2 &= \varepsilon_1\sigma_1 e_1 - \varepsilon_1 e_4 \\
u_3 &= \varepsilon_2(1-\alpha)\gamma(\phi_2(y_1) - \phi_2(x_1)) - \varepsilon_2 e_4 - e_3 \\
u_4 &= e_1 + e_2 + e_3
\end{aligned}$$

Therefore the relation (18.32) becomes

$$\dot{V} = -\sigma_1 e_1^2 - \varepsilon_1 \sigma_1 e_2^2 - e_3^2 - \sigma_2 e_4^2$$

which is a negative definite function.

Thus, by Lyapunov stability theory, the error dynamics provided by (18.30) is found to be globally asymptotically stable for all initial conditions $e(0) \in R^4$.

Thus, the states of the drive and response system synchronize globally and asymptotically.

18.4.2 Numerical Simulation

For numerical simulation, the initial conditions of the drive system are chosen as 0.09124, 0.3942, 0.0125, 0.9823 and the initial conditions for the response system are taken as 0.9546, 0.9353, 0.8765, 0.1654.

The following Figure 18.6 depicts the chaos synchronization of the modified Colpitts oscillator and Figure 18.7 depicts the error dynamics of chaotic Colspitts oscillator.

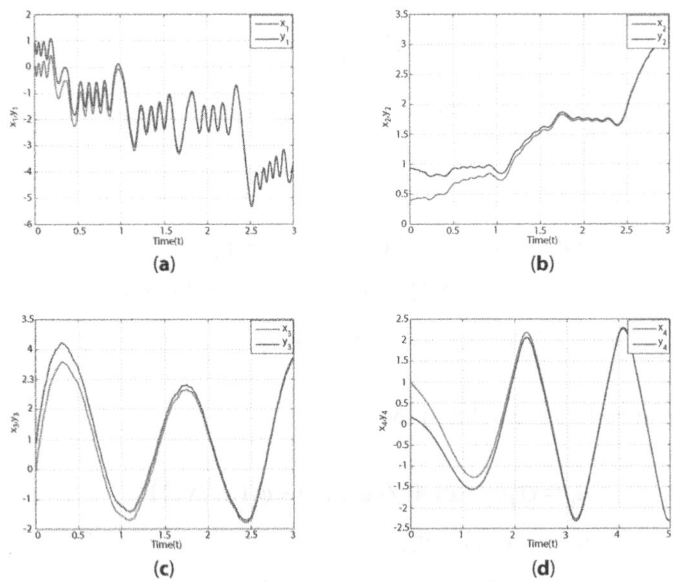

Figure 18.6 Synchronization of the modified colpitts oscillator. (a) Synchronization between x_1 and y_1. (b) Synchronization between x_2 and y_2. (c) Synchronization between x_3 and y_3. (d) Synchronization between x_4 and y_4.

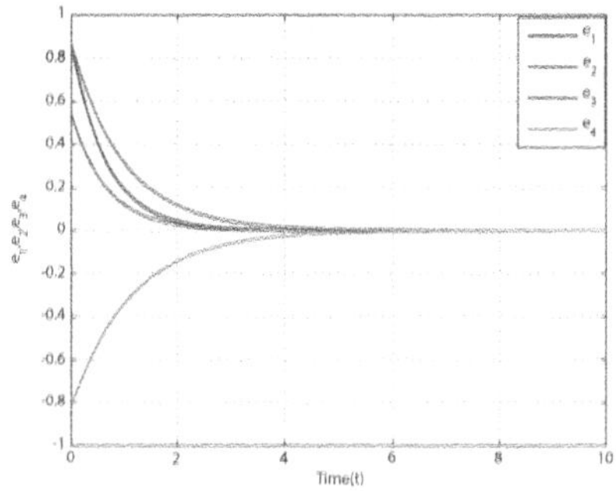

Figure 18.7 Error dynamics of chaotic colspitts oscillator.

18.5 The Synchronization of Colpitts Oscillator via Backstepping Control

The backstepping technique is a cyclic procedure through a suitable Lyapunov function along with a feedback controller. It leads to the global stability synchronization of the strict feedback chaotic systems. In this section, the backward backstepping method is employed for the proposed system.

18.5.1 Analysis of the Error Dynamics

The error dynamics system is taken as

$$\begin{aligned}
\dot{e}_4 &= -e_1 - e_2 - e_3 - \sigma_2 e_4 + u_1 \\
\dot{e}_3 &= \varepsilon_2 e_4 - \varepsilon_2(1-\alpha)\gamma\left(\phi_2(y_1) - \phi_2(x_1)\right) + u_2 \\
\dot{e}_2 &= -\varepsilon_1\sigma_1 e_1 - \varepsilon_1\sigma_1 e_2 + \varepsilon_1 e_4 + u_3 \\
\dot{e}_1 &= -\sigma_1 e_1 - \sigma_1 e_2 + e_4 - \gamma\left(\phi_1(y_3) - \phi_1(x_3)\right) + u_4
\end{aligned} \tag{18.33}$$

Now the objective is to find the control laws u_i ($i = 1,2,3,4$) for stabilizing the error variables of the system (33) at the origin.

First consider the stability of the system

$$\dot{e}_4 = -e_1 - e_2 - e_3 - s_2 e_4 + u_1 \tag{18.34}$$

where is e_3 considered as virtual controller provided by

$$e_3 = \beta_1(e_4) \text{ and } \beta_1(e_4) = 0$$

The Lyapunov function is defined as

$$V_1 = \frac{1}{2}e_4^2 \tag{18.35}$$

The derivative of V_1 with respect to is obtained as

$$\dot{V}_1 = e_4\dot{e}_4 \tag{18.36}$$

If $\beta_1 = 0$ and $u_1 = e_1 + e_2$, then we obtain

$$\dot{V}_1 = -\sigma_2 e_4^2 \tag{18.37}$$

which is a negative definite function.

Hence the system (34) is globally asymptotically stable.

The function $\beta_1(e_4)$ is an estimator when e_3 is considered as virtual controller.

The relation between e_3 and β_1 is defined by

$$\omega_2 = e_3\,\beta_1 = e_3$$

Consider the subsystem (e_4, ω_2) given by

$$\begin{aligned} \dot{e}_4 &= -\omega_2 - \sigma_2 e_4 \\ \dot{\omega}_2 &= \varepsilon_2 e_4 - \varepsilon_2(1-\alpha)\gamma\big(\phi_2(y_1) - \phi_2(x_1)\big) + u_2 \end{aligned} \tag{18.38}$$

Let e_2 be a virtual controller in system (38).

Assume that when $e_2 = \beta_2(e_4, \omega_2)$, the system (38) is rendered globally asymptotically stable.

Consider the Lyapunov function defined by

$$V_2 = V_1 + \frac{1}{2}\omega_2^2$$

The derivative of V_2 with respect to t is

$$\dot{V}_2 = e_4\dot{e}_4 + \omega_2\dot{\omega}_2$$

If $\beta_2 = 0$ and $u_2 = -(\varepsilon_2 - 1)e_4 + \varepsilon_2(1 - \alpha)\gamma(\phi_2(y_1) - \phi_2(x_1)) + e_2 - \omega_2$, then we obtain

$$\dot{V}_2 = -\sigma_2 e_4^2 - \omega_2^2$$

which is a negative definite function.

Hence by Lyapunov stability theory, the system is stable.

Let us consider the relation between e_2 and β_2 defined by

$$\omega_3 = e_2 - \beta_2 = e_2$$

Now the subsystem $(e_4, \omega_2, \omega_3)$ is considered as

$$\begin{aligned} \dot{e}_4 &= -\omega_2 - \sigma_2 e_4 \\ \dot{\omega}_2 &= e_4 + \omega_3 - \omega_2 \\ \dot{\omega}_3 &= -\varepsilon_1\sigma_1 e_1 - \varepsilon_1\sigma_1\omega_3 + \varepsilon_1 e_4 + u_3 \end{aligned} \tag{18.39}$$

Consider the function V_3 due to Lyapunov function defined by

$$V_3 = V_2 + \frac{1}{2}\omega\frac{2}{3}$$

On differentiating V_3 with respect to t, we get

$$\dot{V}_3 = e_4\dot{e}_4 + \omega_2\dot{\omega}_2 + \omega_3\dot{\omega}_3$$

If $\beta_3 = 0$ and $u_3 = -\omega_2 - \varepsilon_1 e_4$, then we obtain

$$\dot{V}_3 = -\sigma_2 e_4^2 - \omega_2^2 - \varepsilon_1\sigma_1\omega_3^2$$

which is a negative definite function.

Now the relation between e_1 and β_3 is defined as

$$\omega_4 = e_1 - \beta_3 = e_1$$

Let us consider the subsystem (e_4, ω_2, ω_3, ω_4) provided by

$$\begin{aligned}
\dot{e}_4 &= -\omega_2 - \sigma_2 e_4 \\
\dot{\omega}_2 &= e_4 + \omega_3 - \omega_2 \\
\dot{\omega}_3 &= -\varepsilon_1\sigma_1\omega_4 - \varepsilon_1\sigma_1\omega_3 - \omega_2 \\
\dot{\omega}_4 &= -\sigma_1\omega_4 - \sigma_1\omega_3 + e_4 - \gamma\big(\phi_1(y_3) - \phi_1(x_3)\big) + u_4
\end{aligned} \tag{18.40}$$

Consider the Lyapunov function

$$V_4 = V_3 + \frac{1}{2}\omega_4^2$$

The derivative of V_4 with respect to is

$$\dot{V}_4 = e_4\dot{e}_4 + \omega_2\dot{\omega}_2 + \omega_3\dot{\omega}_3 + \omega_4\dot{\omega}_4$$

If $\beta_4 = 0$ and $u_4 = \varepsilon_1\sigma_1\omega_3 + \sigma_1\omega_3 - e_4 + \gamma(\phi_1(y_3) - \phi_1(x_3))$, then we obtain

$$\dot{V}_4 = -\sigma_2 e_4^2 - \omega_2^2 - \varepsilon_1\sigma_1\omega_3^2 - \sigma_1\omega_4^2$$

which is a negative definite function.

Hence by Lyapunov stability theory, the system is stable.

18.5.2 Numerical Simulation

For solving the system of differential equations (18.33) with the backstepping controls u_1, u_2, u_3 and u_4, the fourth-order Runge–Kutta method is used and numerical simulation is carried out. We have

$$\begin{aligned}
u_1 &= e_1 + e_2 \\
u_2 &= -(\varepsilon_2 - 1)e_4 + \varepsilon_2(1-\alpha)\gamma(\phi_2(y_1) - \phi_2(x_1)) + e_2 - \omega_2 \\
u_3 &= -\omega_2 - \varepsilon_1 e_4 \\
\text{and } u_4 &= \varepsilon_1\sigma_1\omega_3 + \sigma_1\omega_3 - e_4 + \gamma\big(\phi_1(y_3) - \phi_1(x_3)\big)
\end{aligned}$$

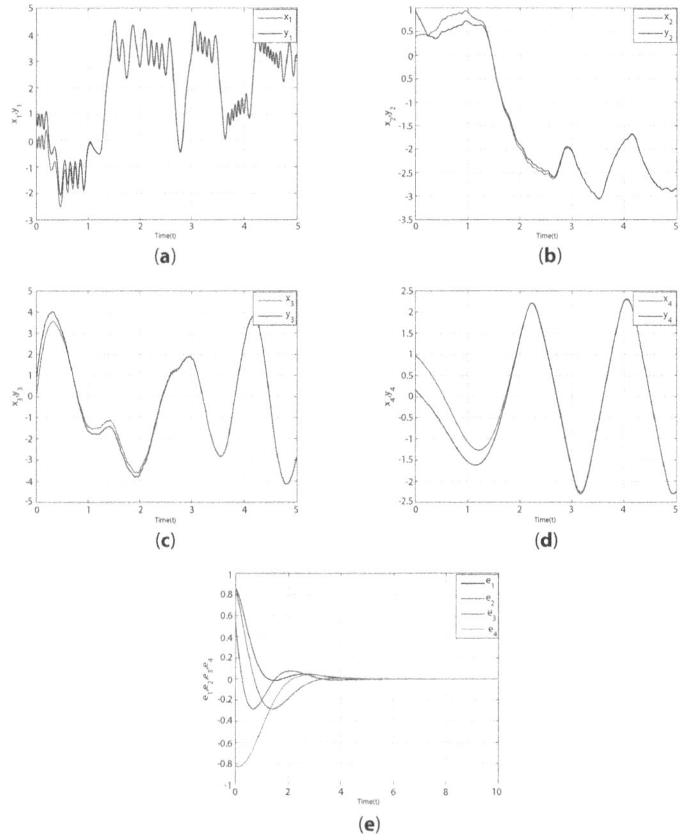

Figure 18.8 Synchronization of identical modified Colpitts oscillator, error plot for identical modified Colpitts oscillator. (a) Synchronization between x_1 and y_1. (b) Synchronization between x_2 and y_2. (c) Synchronization between x_3 and y_3. (d) Synchronization between x_4 and y_4. (e) Error Dynamics of modified Colpitts oscillator.

The initial values of the drive system (27) are chosen as x_1 (0) = 0.09124, x_2 (0) = 0.3942, x_3 (0) = 0.0125, x_4 (0) = 0.9873. The initial values of the response system (28) are taken as y_1 (0) = 0.9546, y_2 (0) = 0.9353, y_3 (0) = 0.8765, y_4 (0) = 0.1654.

Figure 18.8 portrays the chaos synchronization of identical drive and response systems provided by Equations (18.27) and (18.28), respectively.

18.6 Circuit Implementation

In order to verify the dynamical properties of the modified Colpitts oscillator, an operational amplifier circuit is designed in accordance with the equation (18.1). The circuit is designed by linear resistance and linear capacitors. The allowable voltage range of operational amplifiers leads to the appropriate variables proportional compression transformation to the state variables of the system. According to the circuit diagrams, the corresponding oscillation circuit equation is described as follows:

$$\begin{aligned}
\dot{x}_1 &= \sigma_1(-x_1 - x_2) + x_4 - \gamma\phi_1(x_3) \\
\dot{x}_2 &= -\varepsilon_1\sigma_1(-x_1 - x_2) + \varepsilon_1 x_4 \\
\dot{x}_3 &= \varepsilon_2(x_4 - (1-\alpha)\gamma\phi_2(x_1)) \\
\dot{x}_4 &= -x_1 - x_2 - x_3 - \sigma_2 x_4
\end{aligned}$$

where $\phi_1(x_3) = \frac{2a}{\pi}\sin^{-1}\left(\sin\left(\frac{2\pi}{p}(x_3)\right)\right)$, $\phi_2(x_1) = \frac{2a}{\pi}\sin^{-1}\left(\sin\left(\frac{2\pi}{p}(x_1)\right)\right)$

and the parameter values are

$$\sigma_1 = \frac{R_2(R_5+R_8)}{R_5R_1C_1R_3(R_6+R_7)} = \frac{R_{36}(R_{31}+R_{32})}{R_{37}R_{31}(R_{34}+R_{35})}, \quad \sigma_1 = \frac{R_{64}R_{76}R_{78}}{R_{63}C_4R_{65}R_{75}R_{77}},$$

$$\varepsilon_1 = \frac{R_{28}R_{37}}{R_{27}C_2R_{29}R_{36}}, \quad \varepsilon_1 = \frac{R_{42}R_{46}}{R_{41}C_3R_{43}R_{45}},$$

$$\gamma = \frac{R_2R_{20}}{R_1C_1R_3R_{17}} = \frac{R_{58}}{R_{55}}, \quad \alpha = \frac{R_{46}-R_{45}}{R_{46}}$$

Op amp circuit diagram of chaotic variables x_1, x_2, x_3 and x_4 are presented in diagrams Figure 18.9, Figure 18.10, Figure 18.11 and Figure 18.12 respectively.

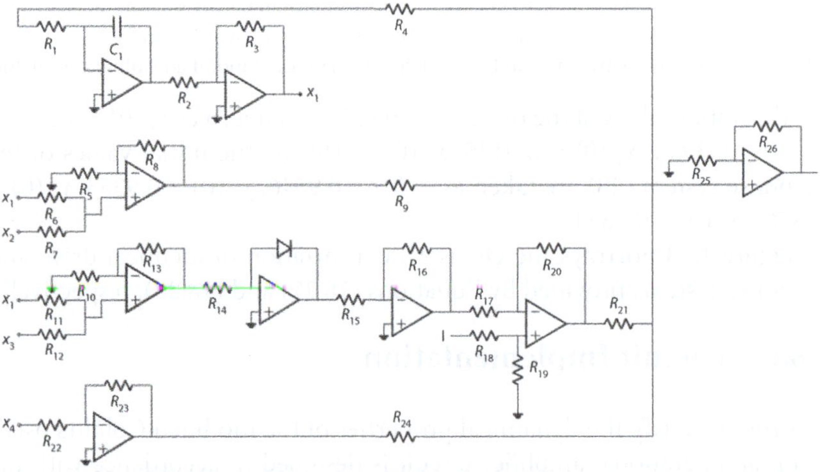

Figure 18.9 Op amp circuit diagram of chaotic variable x_1.

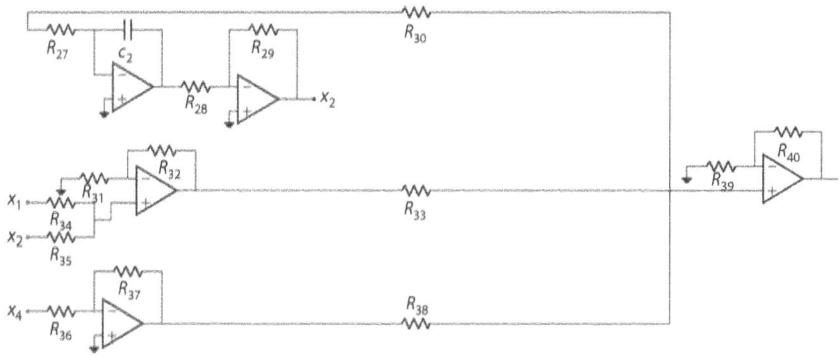

Figure 18.10 Op amp circuit diagram of chaotic variable x_2.

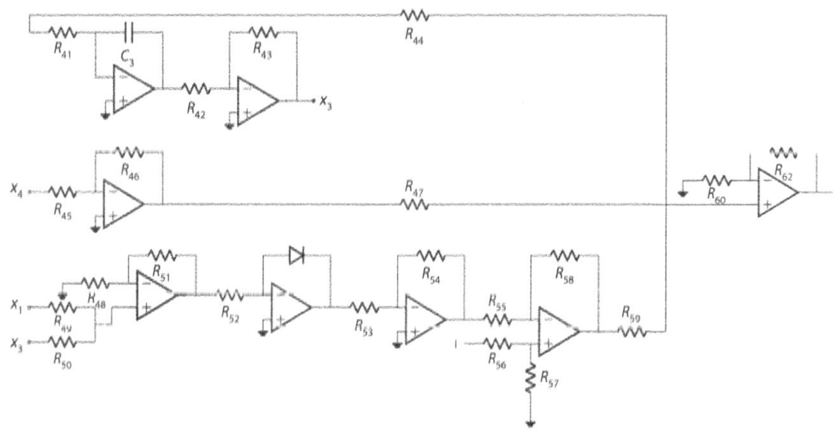

Figure 18.11 Op amp circuit diagram of chaotic variable x_3.

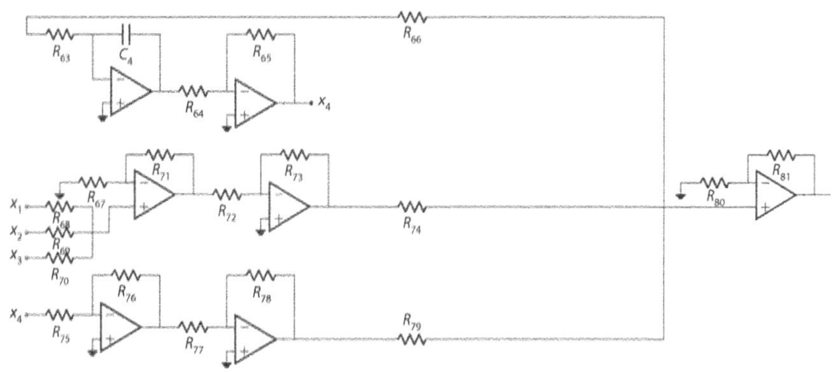

Figure 18.12 Op amp circuit diagram of chaotic variable x_4.

18.7 Conclusion

In this paper, the Colpitts oscillator with triangular wave non-linearity is analyzed. The qualitative properties of the modified Colpitts oscillator is analyzed in this study. It exhibits the chaotic and hyperchaotic nature for some specified initial conditions and parameters. By Wolf method, the Lyapunov exponent is calculated. For some initial conditions, it exhibits the dissipative nature. The adaptive backstepping control technique is used to control the system. Synchronization, the non-linear and backstepping control are utilized. Numerical simulations support the results. MATLAB is used for numerical simulation.

References

1. Kennedy, M.P., 1994. Chaos in the Colpitts Oscillator. *IEEE Transactions on Circuits and Systems I: Fundamental Theory and Applications*, 41(11), pp. 771-774.
2. Vaidyanathan, S., Rajagopal, K., Volos, C.K., Kyprianidis, I.M. and Stouboulos, I.N., 2015. Analysis, Adaptive Control and Synchronization of a Seven-Term Novel 3-D Chaotic System with Three Quadratic Nonlinearities and its Digital Implementation in LabVIEW. *J Eng Sci Technol Rev*, 8(2), pp. 130-141.
3. Kvarda, P., 2001. Identifying the Deterministic Chaos by using the Lyapunov Exponents. *Radioengineering* -Prague, 10(2), pp. 38-38.
4. Lai, Y.C. and Grebogi, C., 1999. Modeling of Coupled Chaotic Oscillators. *Physical Review Letters*, 82(24), p. 4803.
5. Deng, H. and Wang, D., 2017. Circuit Simulation and Physical Implementation for a Memristor-Based Colpitts Oscillator. *AIP Advances*, 7(3), p. 035118.
6. Čenys, A., Tamaševičius, A., Baziliauskas, A., Krivickas, R. and Lindberg, E., 2003. Hyperchaos in Coupled Colpitts Oscillators. *Chaos, Solitons & Fractals*, 17(2-3), pp. 349-353.
7. Kim, C.M., Rim, S., Kye, W.H., Ryu, J.W. and Park, Y.J., 2003. Anti-Synchronization of Chaotic Oscillators. *Physics Letters A*, 320(1), pp. 39-46.
8. Elwakil, A.S. and Kennedy, M.P., 1999. A Family of Colpitts-Like Chaotic Oscillators. *Journal of the Franklin Institute*, 336(4), pp. 687-700.
9. Vaidyanathan, S., Sambas, A. and Zhang, S., 2019. A New 4-D Dynamical System Exhibiting Chaos with a line of Rest Points, its Synchronization and Circuit Model. *Archives of Control Sciences*, 29.
10. Volos, C.K., Pham, V.T., Vaidyanathan, S., Kyprianidis, I.M. and Stouboulos, I.N., 2015. Synchronization Phenomena in Coupled Colpitts Circuits. *Journal of Engineering Science Technology Review*, 8(2).

11. Fujisaka, H. and Yamada, T., 1983. Stability Theory of Synchronized Motion in Coupled-Oscillator Systems. *Progress of Theoretical Physics*, 69(1), pp. 32-47.
12. Corron, N.J., Pethel, S.D. and Hopper, B.A., 2000. Controlling Chaos with Simple Limiters. *Physical Review Letters*, 84(17), p. 3835.
13. Effa, J.Y., Essimbi, B.Z. and Ngundam, J.M., 2009. Synchronization of Improved Chaotic Colpitts Oscillators using Nonlinear Feedback Control. *Nonlinear Dynamics*, 58(1-2), p. 39.
14. Mishra, S., Singh, A.K. and Yadava, R.D.S., 2020. Effects of Nonlinear Capacitance in Feedback LC-Tank on Chaotic Colpitts Oscillator. *Physica Scripta*, 95(5), p. 055203.
15. Vaidyanathan, S. and Rasappan, S., 2014. Global Chaos Synchronization of n-Scroll Chua Circuit and Lur'e System using Backstepping Control Design with Recursive Feedback. *Arabian Journal for Science and Engineering*, 39(4), pp. 3351-3364.
16. Fotsin, H.B. and Daafouz, J., 2005. Adaptive Synchronization of Uncertain Chaotic Colpitts Oscillators Based on Parameter Identification. *Physics Letters A*, 339(3-5), pp. 304-315.
17. Sarkar, S., Sarkar, S. and Sarkar, B.C., 2014. On the Dynamics of a Periodic Colpitts Oscillator Forced by Periodic and Chaotic Signals. *Communications in Nonlinear Science and Numerical Simulation*, 19(8), pp. 2883-2896.
18. Kammogne, S.T. and Fotsin, H.B., 2011. Synchronization of Modified Colpitts Oscillators with Structural Perturbations. *Physica scripta*, 83(6), p. 065011.
19. Kammogne, S.T. and Fotsin, H.B., 2014. Adaptive Control for Modified Projective Synchronization-Based Approach for Estimating all Parameters of a Class of Uncertain Systems: Case of Modified Colpitts Oscillators. *Journal of Chaos*, 2014.
20. Pecora, L.M. and Carroll, T.L., 1990. Synchronization in Chaotic Systems. *Physical Review Letters*, 64(8), p. 821.
21. Ahmad, I. and Srisuchinwong, B., 2017, June. A simple Two-Transistor 4D Chaotic Oscillator and its Synchronization via Active Control. In *2017 IEEE 26th International Symposium on Industrial Electronics (ISIE)* (pp. 1249-1254). IEEE.
22. Bumelienė, S., Tamaševičius, A., Mykolaitis, G., Baziliauskas, A. and Lindberg, E., 2006. Numerical Investigation and Experimental Demonstration of Chaos from Two-Stage Colpitts Oscillator in the Ultrahigh Frequency Range. *Nonlinear Dynamics*, 44(1-4), pp.167-172.
23. Wu, F.Q., Ma, J. and Ren, G.D., 2018. Synchronization Stability between Initial-Dependent Oscillators with Periodical and Chaotic Oscillation. *Journal of Zhejiang University-Science A*, 19(12), pp. 889-903.
24. Li, G.H., Zhou, S.P. and Yang, K., 2007. Controlling Chaos in Colpitts Oscillator. *Chaos, Solitons & Fractals*, 33(2), pp. 582-587.

25. Park, J.H., 2008. Adaptive Control for Modified Projective Synchronization of a Four-Dimensional Chaotic System with Uncertain Parameters. *Journal of Computational and Applied Mathematics*, 213(1), pp. 288-293.
26. Rehan, M., 2013. Synchronization and Anti-Synchronization of Chaotic Oscillators under input Saturation. *Applied Mathematical Modelling*, 37(10-11), pp. 6829-6837.
27. Liao, M.C., Chen, G., Sze, J.Y. and Sung, C.C., 2008. Adaptive Control for Promoting Synchronization Design of Chaotic Colpitts Oscillators. *Journal of the Chinese Institute of Engineers*, 31(4), pp. 703-707.
28. Rasappan, S. and Vaidyanathan, S., 2013. Hybrid Synchronization of n-Scroll Chaotic Chua Circuits using Adaptive Backstepping Control Design with Recursive Feedback. *Malaysian Journal of Mathematical Sciences*, 7(2), pp. 219-246.
29. Hahn, W., 1967. *Stability of Motion* (Vol. 138). Berlin: Springer.
30. Wolf, A., Swift, J.B., Swinney, H.L. and Vastano, J.A., 1985. Determining Lyapunov Exponents from a Time Series. *Physica D: Nonlinear Phenomena*, 16(3), pp. 285-317.

Index

Also of Interest

Check out these other titles from Scrivener Publishing

SIMULATION AND ANALYSIS OF MATHEMATICAL METHODS IN REAL TIME ENGINEERING APPLICATIONS, Edited by T. Ananth Kumar, E. Golden Julie, Y. Harold Robinson, and S. M. Jaisakthi, ISBN: 9781119785378. Written and edited by a group of international experts in the field, this exciting new volume covers the state of the art of real time application of computer science using mathematics. *NOW AVAILABLE!*

INTRODUCTION TO DIFFERENTIAL GEOMETRY WITH TENSOR APPLICATIONS, By Dipankar De, ISBN: 9781119795629. This is the only volume of its kind to explain, in precise and easy-to-understand language, the fundamentals of tensors and their applications in differential geometry and analytical mechanics with examples for practical applications and questions for use in a course setting. *DUE OUT IN SPRING 2022!*

i-Smooth Analysis: Theory and Applications, by A.V. Kim, ISBN 9781118998366. A totally new direction in mathematics, this revolutionary new study introduces a new class of invariant derivatives of functions and establishes relations with other derivatives, such as the Sobolev generalized derivative and the generalized derivative of the distribution theory. *NOW AVAILABLE!*

Open Ended Problems: A Future Chemical Engineering Approach, by J. Patrick Abulencia and Louis Theodore, ISBN 9781118946046. Although the primary market is chemical engineers, the book covers all engineering areas so those from all disciplines will find this book useful. *NOW AVAILABLE!*

SYSTEMS WITH DELAYS: Analysis, Control, and Computations, By A.V. Kim and A. V. Ivanov, ISBN: 9781119117582. This mathematical treatise presents a constructive theory of systems with delays, also known as delay differential equations (DDE), which includes standard theory and new results on numerical treatment of time-delay systems and LQR control design theory. *NOW AVAILABLE!*

Printed and bound by CPI Group (UK) Ltd, Croydon, CR0 4YY
17/09/2026
14964246-0005

Printed and bound by CPI Group (UK) Ltd, Croydon, CR0 4YY

17/09/2026

14964246-0005